Spectroscopy
and the Dynamics
of Molecular Biological Systems

ACADEMIC PRESS RAPID MANUSCRIPT REPRODUCTION

Spectroscopy and the Dynamics of Molecular Biological Systems

Edited by

Peter M. Bayley

National Institute for Medical Research
The Ridgeway
London, England

Robert E. Dale

Paterson Laboratories
Christie Hospital and Holt Radium Institute
Manchester, England

1985

ACADEMIC PRESS
(Harcourt Brace Jovanovich, Publishers)
London Orlando San Diego New York
Toronto Montreal Sydney Tokyo

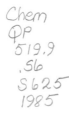

ACADEMIC PRESS INC. (LONDON) LTD.
24–28 Oval Road
LONDON NW1 7DX

United States Edition published by
ACADEMIC PRESS, INC.
Orlando, Florida 32887

BRITISH LIBRARY CATALOGUING IN PUBLICATION DATA
Spectroscopy and the dynamics of molecular
    biological systems.
    1. Molecular biology   2. Spectrum analysis
    I. Bayley, Peter M.   II. Dale, R.E.
    574.8′8′028   QH506

    ISBN 0-12-083240-2
LIBRARY OF CONGRESS CATALOGING-IN-PUBLICATION DATA
Main entry under title:
Spectroscopy and the dynamics of molecular biological
    systems.
    "Composed of a series of reviews based initially on
lectures given in April 1983 at a meeting in London
of the British Biophysical Society"—Pref.
    Includes index.
    1. Macromolecules—Analysis—Congresses.
2. Biopolymers—Analysis—Congresses.   3. Molecular
spectra—Congresses.   4. Molecular dynamics—Congresses.
5. Membrane proteins—Congresses.   I. Bayley, Peter M.
II. Dale, R. E.   III. Title.

QP519.9.S6S625   1985   574.19′24   85-47779
ISBN 0-12-083240-2 (alk. paper)

PRINTED IN THE UNITED STATES OF AMERICA

85 86 87 88   9 8 7 6 5 4 3 2 1

CONTENTS

CONTRIBUTORS

Numbers in parentheses indicate the pages on which the authors' contributions begin.

D. AXELROD (163), *Department of Physics and Biophysics Research Division, University of Michigan, Ann Arbor, Michigan 48109, USA*

V. A. BARNETT (239), *Department of Biochemistry, University of Minnesota Medical School, Minneapolis, Minnesota 55455, USA*

K. BECK (177), *Max-Planck-Institut für Biophysik, 6000 Frankfurt am Main 70, Federal Republic of Germany*

J. M. BEECHEM (259), *Department of Biology and the McCollum-Pratt Institute, The Johns Hopkins University, Baltimore, Maryland 21218, USA*

V. A. BLOOMFIELD (1), *Department of Biochemistry, University of Minnesota, St. Paul, Minnesota 55108, USA*

J. BORDAS (321), *European Molecular Biology Laboratory, Hamburg Outstation, D-2000 Hamburg 52, Federal Republic of Germany*

L. BRAND (259), *Department of Biology and the McCollum-Pratt Institute, The Johns Hopkins University, Baltimore, Maryland 21218, USA*

D. CHAPMAN (119), *Department of Biochemistry and Chemistry, Royal Free Hospital School of Medicine, London NW3 2PS, England*

R. J. CHERRY (79), *Department of Chemistry, University of Essex, Colchester CO4 3SQ, England*

A. F. CORIN (53), *Abteilung Molekulare Biologie, Max-Planck-Institut für Biophysikalische Chemie, D-3400 Göttingen, Federal Republic of Germany*

R. E. DALE (259), *Paterson Laboratories, Christie Hospital and Holt Radium Institute, Manchester M20 9BX, England*

L. DAVENPORT (259), *Department of Biology and the McCollum-Pratt Institute, The Johns Hopkins University, Baltimore, Maryland 21218, USA*

T. M. EADS (239), *Department of Biochemistry, University of Minnesota Medical School, Minneapolis, Minnesota 55455, USA*

V. ECK (351), *Fritz-Haber-Institut der Max-Planck-Gesellschaft, D-1000 Berlin 33, Federal Republic of Germany*

P. B. GARLAND (95), *Department of Biochemistry, University of Dundee, Dundee DD1 4HN, Scotland*

A. GENZ (351), *Fritz-Haber-Institut der Max-Planck-Gesellschaft, D-1000 Berlin 33, Federal Republic of Germany*

J. F. HOLZWARTH (351), *Fritz-Haber-Institut der Max-Planck-Gesellschaft, D-1000 Berlin 33, Federal Republic of Germany*

J. B. HOPKINS (379), *Department of Chemistry, Louisiana State University, Baton Rouge, Louisiana 70803, USA*

B. R. JENNINGS (21), *Electro-Optics Group, Physics Department, Brunel University, Uxbridge UB8 3PH, England*

P. JOHNSON (95), *Department of Biochemistry, University of Dundee, Dundee DD1 4HN, Scotland*

T. M. JOVIN (53), *Abteilung Molekulare Biologie, Max-Planck-Institut für Biophysikalische Chemie, D-3400 Göttingen, Federal Republic of Germany*

J. R. KNUTSON (259), *Department of Biology and the McCollum-Pratt Institute, The Johns Hopkins University, Baltimore, Maryland 21218, USA*

A. A. KOWALCZYK (259), *Institute of Physics, Nicolas Copernicus University, 87-100 Toruń, Poland*

K. M. LINDAHL (239), *Department of Biochemistry, University of Minnesota Medical School, Minneapolis, Minnesota 55455, USA*

E. MANDELKOW (321), *Max-Planck-Institute for Medical Research, D-6900 Heidelberg, Federal Republic of Germany*

E.-M. MANDELKOW (321), *Max-Planck-Institute for Medical Research, D-6900 Heidelberg, Federal Republic of Germany*

D. MARSH (209), *Abteilung Spektroskopie, Max-Planck-Institut für Biophysikalische Chemie, D-3400 Göttingen, Federal Republic of Germany*

E. D. MATAYOSHI (53), *Abteilung Molekulare Biologie, Max-Planck-Institut für Biophysikalische Chemie, D-3400 Göttingen, Federal Republic of Germany*

H. M. McCONNELL (197), *Stauffer Laboratory of Physical Chemistry, Stanford University, Stanford, California 94305, USA*

D. A. MOMONT (239), *Department of Biochemistry, University of Minnesota Medical School, Minneapolis, Minnesota 55455, USA*

I. H. MUNRO (307), *SERC, Daresbury Laboratory, Warrington WA4 4AD, England*

E. K. MURRAY (119), *Department of Biochemistry and Chemistry, Royal Free Hospital School of Medicine, London NW3 2PS, England*

R. PETERS (177), *Max-Planck-Institut für Biophysik, 6000 Frankfurt am Main 70, Federal Republic of Germany*

W. RENNER (321), *Max-Planck-Institute for Medical Research, D-6900 Heidelberg, Federal Republic of Germany*

P. M. RENTZEPIS (379), *Bell Laboratories, Murray Hill, New Jersey 07974, USA*

C. J. RESTALL (119), *Department of Biochemistry and Chemistry, Royal Free Hospital School of Medicine, London NW3 2PS, England*

R. RIGLER (35), *Department of Medical Biophysics, Karolinska Institutet, S-104 01 Stockholm, Sweden*

T. C. SQUIER (239), *Department of Biochemistry, University of Minnesota Medical School, Minneapolis, Minnesota 55455, USA*

D. D. THOMAS (239), *Department of Biochemistry, University of Minnesota Medical School, Minneapolis, Minnesota 55455, USA*

D.G. WALBRIDGE (259), *Department of Biology and the McCollum-Pratt Institute, The Johns Hopkins University, Baltimore, Maryland 21218, USA*

B. R. WARE (133), *Department of Chemistry, Syracuse University, Syracuse, New York 13210, USA*

PREFACE

This book reviews a range of modern methods for the study of the dynamic processes of individual biological molecules and of molecules in integrated macromolecular and membrane systems. Interest in the dynamic properties of biological structures has grown substantially in the past decade and is well exemplified by the fundamental progress recently made in understanding the relationship between protein structure and protein function. X-ray crystallography has established the details of the three-dimensional structure of proteins to unparalleled resolution, resolving the components of secondary structure, the helical and sheet conformations recognised from previous studies of synthetic polypeptides, and the specific but irregular conformation of the tertiary structural components. The detailed description of molecular interactions achieved in the resulting molecular structure is directly related to the large information content of the diffraction pattern and the precision in measuring it over a considerable period of time. Necessarily, the final models correspond to the time-average of the structure, and the possible time-dependent component of the information is largely suppressed. Parallel solution studies have for many years been interpreted on the basis of the potentiality for conformational change being an essential property of the protein structure, both as regards mechanism of action and also the potentiality for control of this action by the binding of additional "effector" molecules. The resolution of individual steps in the catalytic pathways of enzymes by analysis of steady-state kinetics, and more directly by the study of individual reaction steps using the methods of fast reaction techniques, has emphasized the time-dependent transformations of the typical enzyme during the time-span of a few milliseconds duration of the catalytic process.

There is therefore general agreement that proteins should be considered dynamic structures. Similar arguments can be advanced for all the other groups of biological macromolecules, for model biological lipid bilayer systems and for molecules and macromolecules in their attachment to, or being embedded in, the cell surface. Some motions, for example the rotations of individual groups or the vibrations of certain side chains, may be without direct biological significance, but may indeed be analogues of functionally mobile structures. In order to assess the relevance of the dynamic properties, highly sophisticated experimental methods are clearly required for investigations over a wide range of the time domain. Ideally, measurements should be performed on molecules which are in their truly native state, even devoid of "labels," using monitoring signals of high specificity, and reflecting properties which are amenable to exact theoretical formulation and detailed molecular interpretation. At the molecular level, such observations should be able to produce experimental information on the shortest time-scale which will inspire attempts at simulation by the powerful techniques of molecular dynamics calculations. At the supramolecular level, the formulation of the experimental approach and the interpretation of the observations depends upon the availability of theoretical treatments of the hydrodynamics of real molecules. This is an area where, despite major conceptual advances made in recent years, much remains to be done in extending the treatment from classical models.

In addition to a proper theoretical background, we also require the associated experimental capability for making time-resolved measurements of dynamic properties over a wide time range where the system under study is contained within an environment as close as possible to the natural biological context. Given the diversity of biological systems, this implies the existence of a wide range of problems for investigation. Recent developments include lateral and rotational diffusional motion of molecules in isotropic or anisotropic solution, as well as two-dimensional diffusional motion, lateral and rotational, for molecules constrained by a biological membrane. A long-term goal is the description of the dynamic properties of large assemblies of molecules, organised into the supramolecular structures which are involved in the time-dependent structural and functional evolution of that most complicated molecular machine—the biological cell. Our lack of knowledge of the physical nature of the environment encountered in a true biological system is, at present, a major deficiency in the interpretation of *in vivo* systems.

A wide range of physical methods has become available for the study of the dynamic properties of molecules. Many of these have the common property of being dependent upon the interaction of electromagnetic radiation with matter, and the potential area of concern is extremely large. Some selectivity in subject matter has therefore been necessary in composing this volume; in particular, two important areas at the extremes of the electromagnetic spectrum—X-ray diffraction and nuclear magnetic resonance—are not treated here. A recent volume "Mobility and Function in Proteins and Nucleic Acids" (CIBA Symposium 93; CIBA Foundation, London, 1983) deals with recent advances in applications of both these techniques to dynamic studies of biological molecules.

This volume describes recent studies of the dynamic properties of biological macromolecules, macromolecular assemblies and supramolecular structures. In particular, emphasis is given to the application of modern fast time-resolved optical spectroscopic techniques involving lasers and sychrotron sources, since their unusual power and time-structure provide major advantages for biophysical investigations. The volume is composed of a series of reviews, based initially on lectures given in April 1983 at a meeting in London of the British Biophysical Society. The techniques and topics presented here include the hydrodynamics of complex macromolecules, laser correlation spectroscopy, triplet-state probes of molecular motions and chemical kinetics, fluorescence and phosphorescence depolarization, fluorescence photobleaching techniques, electron paramagnetic resonance techniques, time-resolved fluorescence spectroscopy, synchrotron-based studies of time-resolved optical spectroscopy and time-resolved X-ray scattering in macromolecular self-assembly, and laser techniques for nanosecond temperature jump relaxation and picosecond optical spectroscopy.

Even with such diversity of interests and techniques, it is not possible to present a fully comprehensive account of this rapidly developing area of biophysical research. However, the reader will detect a common purpose in these contributions, and a common methodological approach based ultimately upon our theoretical under-standing of the behaviour of ideal molecules in well-defined systems. In real experimental investigations, the important areas of biophysical research involve observations of nonideal molecules in imperfectly defined systems. Moreover the observations, by their very nature, can provide only an incomplete picture of actual behaviour. This fact enforces the need for the careful refinement of methods and techniques, and the development of the associated theory, plus the application of several experimental techniques to the same problem. In this regard, it is perhaps important to point out that apparently contradictory evidence as to the nature and degree of motion within an assembly may not in fact be incompatible, when the time-ranges of the different methods are taken into account. The methods described here allow investigations over the range of picoseconds to hours—some sixteen orders of magnitude. This suggest that dynamics, particularly in the fastest ranges now opened to investigation with the new technology, will continue to provide an enormously fruitful area of research for a considerable time to come. The editors hope that this volume will assist in introducing those unfamiliar with this area to recent developments in the techniques and their biophysical applications. For those already involved with some of these techniques, we hope that there will be additional stimulus in this fuller illustration of the potentiality of the present methods and inspiration for the development of even more powerful methods for future research.

We should like to record our thanks primarily to the contributors to this volume whose work this represents. Also our thanks go to the staff of Academic Press, London, for their considerable patience and expert assistance in the preparation of this publication and to David Clark and Edward Manser for their work in compiling the index.

P. M. Bayley*
R. E. Dale

*Dr. Bayley's mailing address is as follows:
National Institute for Medical Research
The Ridgeway
Mill Hill
London MW7 1AA, England

Hydrodynamic Properties of Complex Macromolecules

V.A. BLOOMFIELD

INTRODUCTION

For many years, at least since the days of Svedberg and Tiselius, hydrodynamic techniques have been among the most important means used to characterize biological macromolecules. Experimental measurements of diffusion and sedimentation coefficients, electrophoretic mobility, rotational diffusion, intrinsic viscosity and non-steady-state visco-elastic properties have yielded information about molecular weight, size, shape, charge, flexibility and interactions.

To interpret transport coefficients in terms of molecular para-meters, a sound theoretical base is required. Until fairly recently, most biological macromolecules have been modelled for hydrodynamic purposes as spheres or ellipsoids, and the interpretation of their hydrodynamic behaviour made in terms of theories for such model struc-tures developed by Stokes, Einstein, Perrin, and Simha.

As we have learned more about the actual structures of biological macromolecules and supramolecular complexes, through x-ray diffrac-tion and especially electron microscopy, the inadequacy of such simple models has become manifest. How can one fit a complex bac-terial virus, such as T4 with its large prolate icosahedral head, its long tail connecting the head to a hexagonal baseplate, and its six long, thin, kinked tail fibres attached to the baseplate, into the Procrustean bed of an ellipsoid of revolution?

SPECTROSCOPY AND THE DYNAMICS
OF MOLECULAR BIOLOGICAL SYSTEMS

1

At the same time, tools for measuring hydrodynamic properties have become more powerful, sensitive, versatile, and discriminating. The techniques discussed in this book, dynamic laser light scattering and other correlation methods, fluorescence recovery after photobleaching, singlet and triplet anisotropy decay, electric field orientation, and magnetic resonance lineshape and relaxation analysis, and their application not just to dilute solutions but to semiordered arrays, membranes and cell surfaces, have given hydrodynamicists new motivation to extend the scope and realism of their models.

In this article I shall review some of the major developments in hydrodynamic theory and its applications to complex biological macromolecules. My aim is to show how, given the basic theoretical framework along with modern powerful, inexpensive computational capabilities, realistic calculations can be made for rigid molecules of almost arbitrary complexity. For macromolecules with restricted flexibility or embedded in a membrane, matters are less satisfactory. Even here, however, important progress has been made, which I shall briefly review.

HYDRODYNAMIC THEORY - THE POINT SOURCE APPROACH

Conventional hydrodynamic theory solves the Navier-Stokes equation in the low Reynolds number limit subject to the boundary conditions of zero relative velocity between the solvent, treated as a continuum, and the surface of the particle of interest. With the "stick" boundary condition, appropriate for particles large relative to solvent molecules, both the normal and tangential relative velocity components are zero. With the "slip" boundary condition, which may be more suitable for very small particles, the tangential component is nonzero. This approach is useful, and the equations can be solved exactly, for spheres and ellipsoids with smooth, simple surface contours. But it becomes impossibly complex for particles with less regular shapes.

In this circumstance, an alternative formulation of low Reynolds number hydrodynamics developed by Oseen (1927) becomes useful. This is essentially a Green's function approach, in which the frictional resistance of a particle is treated as a sum of point source resistances. The forces associated with these point sources of friction

cause velocity perturbations in the solvent. They are distributed so
that the perturbations sum to a net zero relative velocity at the
particle surface, thus satisfying the boundary conditions. Oseen's
approach was translated, simplified, brought to the attention of a
wider audience, and applied to some important cases of biological
interest, notably circular cylinders with flat ends, by Burgers
(1938).

The Oseen-Burgers formulation is particularly appropriate for polymers: to a first approximation each monomer or subunit can be idealized as a single point source of frictional resistance. This approach
was taken by Kirkwood and Riseman (1948) to create a general theory
of polymer hydrodynamics. The theory was notably applied to rods
composed of a linear array of subunits, and to random coils composed
of subunits connected by bonds idealized as universal joints. Zimm
(1956) combined the Kirkwood-Riseman hydrodynamic approach with the
normal coordinate analysis of chain motions (Rouse, 1953) to create
the current standard theory of flexible polymer chain hydrodynamics.

As they are most commonly formulated, the Oseen-Burgers-Kirkwood-
Riseman equations relate the force $\underline{F}_i$ on subunit i to its frictional
coefficient ζ_i and its velocity relative to solvent, $\underline{u}_i - \underline{v}'_i$:

$$\underline{F}_i = \zeta_i \, (\underline{u}_i - \underline{v}'_i) \qquad (1)$$

The unperturbed solvent velocity is $\underline{u}_i$, the perturbation arising
from all the other subunits is $\underline{v}'_i$. The perturbation produced by
subunit j at the location of subunit i is the product of the force
$\underline{F}_j$ and the hydrodynamic interaction tensor $\underline{\underline{T}}_{ij}$:

$$\underline{v}'_i = \Sigma \, \underline{\underline{T}}_{ij} \underline{F}_j \qquad (2)$$

Equations 1 and 2 constitute a set of simultaneous equations for
the forces $\underline{F}_i$, and thus for the total force on the N particles in the
complex:

$$\underline{F} = \Sigma \, \underline{F}_i = \zeta \underline{u} \qquad (3)$$

which defines the frictional coefficient ζ of the complex assembly.

If the frictional elements are treated as points, the hydrodynamic interaction tensor is:

$$\underset{=}{T}_{ij} = (1/8\pi\eta R_{ij})[\underset{=}{I}+\underset{-}{R}_{ij}\underset{-}{R}_{ij}/R_{ij}^2] \qquad (4)$$

where $\underset{-}{R}_{ij}$ is the vector from i to j, η is the solvent viscosity, and $\underset{=}{I}$ is the unit tensor. To simplify the equations, it has been common to preaverage $\underset{=}{T}_{ij}$ over all orientations before computing the forces:

$$<\underset{}{T}_{ij}> = 1/6\pi\eta R_{ij} \qquad (5)$$

Results obtained with these equations have been tested in a variety of situations (Bloomfield et al., 1967a,b). While they work well with relatively compact or symmetrical objects, they have some obvious inadequacies when applied to highly asymmetric structures with sub-units of quite unequal sizes, such as bacteriophages T2 and T4. Therefore, some improvements were called for if the approach was to be applied with confidence to a wide variety of biomolecular structures.

IMPROVEMENTS IN ACCURACY AND CALCULATIONAL EFFICIENCY

Finite size of the frictional elements

Equation (4) lacks any reference to subunit size. This implicitly means that the frictional force is concentrated at the centre of the subunit, rather than at its surface. While this is satisfactory if the subunits are far from one another, it is less so if they are close. An improved version of the hydrodynamic interaction tensor, essentially carrying to higher order the series expansion of the full hydrodynamic equations, was developed by Rotne and Prager (1969) and independently by Yamakawa (1970). As slightly generalized by Garcia de la Torre and Bloomfield (1977a) to encompass subunits of different radii σ_i, the improved hydrodynamic interaction tensor is:

$$\underset{=}{T}_{ij} = (1/8\pi\eta R_{ij}) \times \{\underset{=}{I}+[\underset{-}{R}_{ij}\underset{-}{R}_{ij}/R_{ij}^2]+[(\sigma_i^2+\sigma_j^2)/R_{ij}^2][(\underset{=}{I}/3)-(\underset{-}{R}_{ij}\underset{-}{R}_{ij}/R_{ij}^2)]\}$$

$$(6)$$

Avoidance of preaveraging

The approximation of preaveraging the hydrodynamic interaction tensor before solution of the simultaneous equations for the forces greatly reduces computational demands, by decoupling the x y z coordinates, but at the cost of potentially significant inaccuracies. An interesting examination of the consequences of preaveraging for random chains has been presented by Zimm (1980). With the large digital computers commonly available in university computer centres, it is now possible to solve directly the 3Nx3N simultaneous equations for structures containing up to about 100 subunits. Memory becomes limiting for larger structures. For relatively small N, standard matrix inversion techniques may be used. For larger complexes, the Gauss-Seidel iteration procedure has been most useful. However, this method may be unstable if the subunits differ widely in size.

Many biomolecular complexes have some symmetry, which can be used to great advantage in reducing the size of the required calculations. For example, the T-even bacterial viruses appear to have six-fold symmetry, so the number of elements of the matrix to be stored and manipulated is reduced by a factor of 36.

Diagonal approximation

If there is insufficient computer time or memory to solve the equations exactly, a useful approximation is to neglect the off-diagonal elements of the hydrodynamic interaction tensor (Garcia de la Torre and Bloomfield, 1977a). The rationale is that diagonal elements are of order $1/R$, while off-diagonal elements are of order $1/R^3$. This reduces the problem to three independent NxN matrices, a reduction of nine in required memory and a corresponding factor in time. When tested against solutions to the full equations, the diagonal approximation is generally quite accurate. It is exact for a linear array of elements of arbitrary size.

Rotational and shearing motion

A peculiarity of the subunit-modelling approach is that when a spherical element is on, or very near, the centre of rotation, its contribution to the rotational frictional coefficient and intrinsic viscosity is improperly computed as zero or very small. This is true even

with the improved hydrodynamic interaction tensor, in which the subunit size is explicitly represented, since the frictional resistance
is still taken to be exerted at the centre of the subunit. If the
subunit is a dominant one, such as a single large sphere representing
the head of a bacterial virus, the rotational diffusion coefficient
or intrinsic viscosity of the entire complex may be grossly miscalcu-
lated.

A simple computational stratagem my be employed to overcome this
problem (Wilson and Bloomfield (1979a). The single large subunit is
replaced with a small number of smaller spheres, none of whose centres
is on the rotation axis. The nature of the substitute array is not
critical: cubes, octahedrons, and trigonal bipyramids all give similar
results. The subunits are chosen either to have collectively the
same volume or the same rotational diffusion coefficient as the
parent sphere. This approach gives markedly better agreement with
the known theoretical values for short prolate ellipsoids of revolution and with experimental results for bacteriophages.

Extrapolation from course to fine grain

To model a structure with a smooth surface requires a large number of
small subunits. Likewise, to represent accurately a long, thin
structure such as the thin tail fibres of T4 phage or the rodlike
structure of myosin requires many frictional elements. To minimize
calculational requirements, it is efficient to model such structures
with several choices of a relatively small number of large subunits,
then extrapolate from these coarse representations to the desired
smooth or fine-grained one. This is the basis of the shell model
approach to hydrodynamics (Bloomfield *et al.*, 1967a,b) and has been
shown by Swanson *et al.* (1978) to yield extremely accurate results.
For rods, one chooses the number and size of the subunits so that the
length of the linear array remains constant while its radius decreases,
then extrapolates to the desired axial ratio. This approach has been
applied with good success to myosin (Garcia de la Torre and Bloomfield, 1980).

Closed-form equations

By expanding the simultaneous equations for the forces in powers of
the hydrodynamic interaction and truncating the series after a few

terms, it is possible to derive approximate equations for the transport coefficients that depend simply on the subunit sizes and positions. Such equations have been developed to various levels of approximation for translation, rotation, and intrinsic viscosity; they are reviewed by Garcia de la Torre and Bloomfield (1981). In general, they are not as accurate as the computational methods described above, and they may be quite inaccurate for certain types of structures. However, they are generally considerably more satisfactory than arbitrary ellipsoid of revolution models for complex structures, and require very little calculational effort.

RESULTS FOR RIGID BIOMOLECULAR COMPLEXES

Simple models

Oligomeric subunit arrays   Oligomeric proteins frequently have subunits of roughly equal size and approximately spherical shape, and occur in geometrically regular arrays. This type of structure is naturally suited to the hydrodynamic treatment outlined above. Table 3 of the review by Garcia de la Torre and Bloomfield (1981) contains values of the translational and rotational diffusion coefficients and intrinsic viscosities for such arrays, containing from two to eight subunits. Most regular geometries are included. The values are expressed relative to the hydrodynamic coefficient for a single sphere with the same total volume as the subunits in the array. This Table may eliminate the need for independent calculations in studies of small oligomers.

Cylindrical rods   Many biomolecular structures, such as tobacco mosaic virus (TMV), filamentous bacteriophages, myosin, microtubules, and microfilaments, are long rods of constant diameter. They can be represented as right circular cylinders. It is not possible to solve the hydrodynamic properties of such cylinders in closed form. Consequently, they have often been replaced by prolate ellipsoids of revolution. Such replacement works reasonably well for translational motion, if the equivalent ellipsoid is chosen to have the same length and volume as the cylinder, so its semi-minor axis is $\sqrt{(3/2)}$ that of the cylinder radius. However, for rotational or shearing motion the

narrowing of the ends of the ellipsoid tends greatly to underestimate
the frictional resistance.

The most commonly used hydrodynamic treatment for cylinders is
that developed by Broersma (1960a,b) who expanded the end effects
associated with blunt ends in powers of the logarithm of the axial
ratio. The coefficients in the expansion have recently been revised
slightly (Newman et al., 1977). For long cylinders, essentially
exact hydrodynamic properties have been calculated by Yamakawa (1975),
Yamakawa and Fujii (1973), and Norisuye et al. (1979). The asymptotic
behaviours of the Broersma and Yamakawa treatments do not agree.

For this reason, Tirado and Garcia de la Torre (1979,1980) have
developed a shell model treatment of cylinders of arbitrary axial
ratio, using the improved hydrodynamic interaction tensor and avoid-
ance of preaveraging discussed earlier. Their results fit Broersma's
low axial ratio results as well as Broersma's own equations do, and
agree asymptotically with the Yamakawa results. For isotropic translational diffusion, averaged over the three orthogonal orientations
of the cylinder with respect to the direction of motion, the diffusion
coefficient D_T for a cylinder of length L and axial ratio p may be
written:

$$3\pi\eta L D_T/KT = \ln p + \gamma \tag{7}$$

where:

$$\gamma = 0.312 + 0.567/p + 0.100/p^2 \tag{8}$$

while for rotation of the long axis about the short axis, the rotational diffusion coefficient D_R is:

$$\pi\eta L^3 D_R/3KT = \ln p + \delta \tag{9}$$

with:

$$\delta = -0.662 + 0.917/p - 0.050/p^2 \tag{10}$$

These equations should now become the standards in work on rodlike
biopolymers.

Dumb-bells and lollipops Some biopolymers, such as amphipathic membrane proteins with globular lipophilic and hydrophilic domains separated by a thin linker, may be modelled as dumb-bells. Such complexes

as tailed bacteriophages, with a spherical head attached to a long
tail, may be modelled as lollipops. These structures are difficult
to model convincingly as ellipsoids of revolution, but fit naturally
into the spherical subunit scheme. The large elements at the ends of
the dumb-bells or lollipops are treated as single large spheres,
while the linker or tail is a straight rod of smaller spheres. Polynomial equations have been obtained (Garcia de la Torre and Bloomfield, 1977a,b 1981; Bloomfield *et al.*, 1979) for the diffusion
coefficients of such structures in terms of the ratios of large to
small sphere radii and of rod length to large sphere radius. Such
equations make it possible to compute hydrodynamic properties accurately and simply with a pocket calculator.

More complex structures

Multisubunit structures: proteins and viruses: small
globular proteins: The ultimate test of hydrodynamic theories for
complex macromolecules is whether they agree with experiments. The
most direct comparison is with rigid molecules, since the additional
complexities of averaging over internal coordinates do not arise.
With the availability of atomic-resolution structures of globular
proteins, derived from x-ray crystallography, it has become possible
to put shell model or point source hydrodynamics to a demanding test.
Each atom on the surface may be represented by a frictional element,
and the translational frictional coefficient compared with experiment.
Rotational diffusion coefficients of requisite accuracy are generally
not available, and intrinsic viscosities are insensitive to size and
shape so long as the protein is roughly spherical.

When calculations are performed with the simple Kirkwood equations,
the calculated hydrodynamic radius is too small (Teller and de Haen,
1975), and a uniform 2.8 $\overset{o}{A}$-thick layer of water molecules is required
to achieve agreement with experiment (Teller *et al.*, 1979). With the
more rigorous theoretical methods developed recently, a uniform
hydration layer yields too large a hydrodynamic size; agreement is
obtained when only the charged surface residues are hydrated (Teller
et al., 1979).

These results, while encouraging, point up the difficulty in using continuum hydrodynamics when the macromolecule is not much bigger than the solvent molecule. An alternative way of considering the flow of solvent around a biomolecular surface, using classical hydrodynamics, has been presented by Kuntz and Kauzmann (1974).

Bacterial viruses: At the other extreme we have the bacterial viruses, with dimensions so large, several thousand ångströms, that hydration layer effects are quite negligible. The T-even bacteriophage are probably the most complex biomolecular structures to have been modelled hydrodynamically. Attempts to reconcile their hydrodynamic properties with the structures visible in electron micrographs have led to many of the advances in hydrodynamic theory in the past 15 years.

In the first place, the various parts of a phage are of very different size: from the 1000 Å head to the 40 Å-thick fibres. To model this structure with spherical subunits of the same diameter would have required more computer memory and time than practical. This was one of the major motives for developing the generalized theory with different subunit sizes. Even the fibres themselves, over 1400 Å long, require a large number of subunits. This led to the procedure of modelling with a small number of larger spheres, then extrapolating to the proper axial ratio. The six-fold symmetry around the long axis of the phage encouraged the formulation of methods that took advantage of symmetry to minimize computational requirements.

The tail fibre transition from an extended to a retracted configuration emphasizes the concept of the centre of rotational resistance. The centre of mass of a T-even phage is very near the centre of the head; but the centre of resistance moves from about 200 Å distal to the centre of mass in the configuration with fibres retracted, to about 600 Å distal when the fibres are extended (Wilson and Bloomfield, 1979b). When the fibres are retracted, the head is sufficiently near the centre of rotation that the rotational diffusion coefficient is grossly overestimated by subunit hydrodynamic theory if the head is modelled as a single large sphere. This led to the strategy, described earlier, of replacing such a large sphere with an equivalent arry of a few smaller spheres (Wilson and Bloomfield, 1979a).

Perhaps the most striking result of the hydrodynamic study of phage, however, is the realization of the importance of hydrodynamic interaction in the volume outlined by the extended tail fibres. Even though this volume is very sketchily outlined indeed, the included solvent is effectively immobilized. This non-free-draining effect can be convincingly demonstrated by macroscopic model experiments scaled to equivalent Reynolds number conditions (Douthart and Bloomfield, 1968). The early versions of subunit hydrodynamic theory were quite incapable of explaining this effect (Bloomfield et al., 1967b). Only with the theoretical and computational advances described above can the effects of the tail fibres be rationalized.

One of the most complex and interesting phage structures is the baseplate. Its hexagonal symmetry and some of its structural detail are visible in electron microscopy, but its size and, particularly, the state of extension of the short tail fibres and spikes, is often unclear. Hydrodynamic calculations on various baseplate models, and comparison with experimental S and D values, have enabled determination of fibre and spike disposition under normal solution conditions (Aksiyote-Benbasat and Bloomfield, 1982).

Dynein: Another situation in which hydrodynamic techniques and modelling can be quite profitably used has been found in recent studies on the microtubule-associated protein, dynein, in collaboration with Professor Kenneth Johnson at Pennsylvania State University. Electron microscopy shows a three-headed structure, with each head connected to an ill-defined base by a long, thin rod. When we compared the sedimentation and diffusion coefficients with those predicted for a molecule of the measured molecular weight, structure, and dimensions, it became clear that dynein is bigger than it appears in micrographs. Since the heads and connecting rods are fairly well-defined, it is likely that the base, whose shape and boundaries are fuzzy, is considerably larger than initially thought.

These examples illustrate a general point. Hydrodynamics can provide only one or a few parameters, quite insufficient to define a complex biomolecular structure. Electron microscopy is very powerful in illuminating structure and topology, but is subject to various artifacts that may render dimensions doubtful. Concurrent use of microscopy and hydrodynamics, however, allows refinement of dimensions in the context of a basic structural model.

Bent, rodlike molecules: Rigid bending angle: A number of rodlike biopolymers, such as myosin and the tail fibres of T-even bacteriophage, are bent. Bends may also be introduced into synthetic helical polymers through the introduction of a disruptive monomer. A fair amount of work has been devoted in the past several years to comparing the hydrodynamic properties of such bent rods to straight rods of the same total length L. The differences will be functions of the lengths of the two parts, $L_1 + L_2 = L$, and of the bending angle θ. While most calculations have been carried out for general L_1 and L_2, the differences are greatest for $L_1 = L_2 = L/2$.

Since differences or ratios of hydrodynamic coefficients for bent and straight rods are rather insensitive to the axial ratio or details of modelling, calculations have commonly constructed rods from arrays of touching spheres. Wilemski (1977) used the simple Kirkwood-Riseman theory to compute D_T and $[\eta]$, obtaining results that are strictly valid only in the limit of very long rods. However, Garcia de la Torre and Bloomfield (1978), using the more rigorous methods discussed above, found that Wilemski's results were adequate even for finite axial ratios in the range encountered experimentally. The most important result is that D_T is quite insensitive to bending, except for very small θ, when the rod is bent almost double; while $[\eta]$ decreases rapidly with θ. This makes intrinsic viscosity a much more sensitive probe of bending.

To compute rotational diffusion coefficients, careful attention must be paid to definition of the centre of rotation. This does not in general coincide with the position of the bend or with the centre of mass. It must be computed by finding the point at which the rotational frictional coefficient is minimized, or by some equivalent scheme (Happel and Brenner, 1973; Harvey and Garcia de la Torre, 1980). When this is done, reliable results for the components of the rotational diffusion coefficient tensor are obtained. Another complication is translation-rotation coupling, whose neglect may introduce inaccuracies of a few per cent in calculated translational and

rotational diffusion coefficients (Garcia Bernal and Garcia de la
Torre, 1980; Wegener, 1981).

It should be recognized that there are three orthogonal components
to D_R, whose relative magnitudes will vary greatly depending on θ.
The diffusion coefficient for rotation of the long axis about a
diameter will increase by roughly a factor of eight (proportional to
L^3), as θ goes from π to 0. However, the diffusion coefficient for
rotation of an axis perpendicular to the long axis will decrease
precipitously as θ goes from π to $\pi/2$. The number of rotational
relaxation times that may actually be distinguished in an electro-
optical or emission anisotropy decay experiment will depend on the
relative magnitudes of the Ds. Generally, no more than two separable
relaxations should be expected (Belford et al., 1972; Wegener et al.,
1979; Mellado and Garcia de la Torre, 1982).

Variable bending angle: If the bending angle is free to vary over
some range, one may speak of a hinged rod. In principle, the orientation and velocity of a hinged rod in a flow field should influence
its bending angle and vice versa. However, to a good approximation,
it seems adequate to compute the hydrodynamic coefficients by calculating them for fixed angles, then averaging them over the angular
distribution. The propriety of replacing a flexible chain by an
ensemble of rigid bodies has been examined by Zimm (1982).

Garcia de la Torre and Bloomfield (1980) used this approach to
investigate flexibility in the connection between the two rodlike
segments, S2 and LMM, of myosin. They calculated the hydrodynamic
coefficients for $\theta = \pi$ and $\pi/2$. The experimental values were bracketed
by the calculated ones, and were definitely different from those of
$\theta = \pi$. This buttresses the electron microscopic observation of
variable angles to indicate the flexibility of the myosin rod.

Star-branched molecules with rigid arms: The hydrodynamic properties of a bent or hinged rod can be viewed as the sum of two types of
contributions: those from each half-rod, and those from interaction
between the half-rods. The validity of such a decomposition arises
from the pairwise additivity of hydrodynamic interaction. This
enables a useful extension to star-branched molecules with rigid
arms. If there are f equivalent branches, each of length L, then

some hydrodynamic property X may be written:

$$X = fX_{11}(L) + f(f-1)X_{12}(L\theta) \tag{11}$$

where X_{11} and X_{12} are the self and cross terms, which may be obtained
from the theories for bent rods.

This approach was developed by Bloomfield (1983) to compute the
properties of crosslinked mucus glycoproteins, or mucins, which
appear on average to have f=4. It complements the normal coordinate
theory of Zimm and Kilb (1959) for random coiled branched polymers.
Comparison of the sedimentation coefficients and intrinsic viscosities
of monomeric and tetrameric mucins suggests that the monomers are
more rigid than flexible, a conclusion at variance with the properties
of the monomers alone.

Macromolecules with porous regions Many biological macromolecules
and complexes have regions which are only sparsely occupied by poly-
meric subunits. This is commonplace, of course, with random coil
polymers; but more regular structures may also have significant
porosity.

Globular proteins: Even the assumption of an impermeable protein
surface may be open to some question. The analysis by Kauzmann *et*
al. (1974) of protein packing densities suggests that the outer few
ångströms of typical globular proteins are less dense than the
interior core, an observation consistent with a variety of indications
that small molecules can penetrate well into the protein interior
without unfolding or denaturation. This two-layer model has been
combined with a porous sphere hydrodynamic treatment to provide a
possible explanation for observed values of the Scheraga-Mandelkern β
parameter that are below the theoretical minimum for impermeable
spheres (McCammon *et al.*, 1975a).

Mucins: Mucus glycoproteins are important examples of biopolymers
with a porous surface layer. The structure of the mucin monomer has
been likened to a bottle brush, with a long polypeptide backbone
densely substituted with oligosaccharides O-linked to serine and
threonine (Allen, 1978; Roberts, 1978). The chain lengths of these
oligosaccharide "bristles" typically range from 2-20, averaging about

eight residues (Roussel *et al.*, 1975). In treating the hydrodynamic properties of the mucins, which must be viewed locally as cylinders whether their overall geometries are random coil or rodlike, it is important to have an accurate estimate of their effective radii.

Since solvent can penetrate some distance into the permeable sugar side-chain region, the effective radius a must be somewhat less than the extended length L of the oligosaccharide plus the polypeptide radius R. A quantitative estimate may be made by assuming that the two-layer theory of McCammon *et al.* (1975a), derived for spheres, also holds for circular cylinders. The important quantity is κ, the inverse hydrodynamic shielding depth in the porous region. This is related to the number density d and hydrodynamic radius b of the monomeric frictional elements:

$$\kappa = (6\pi bd)^{1/2} \tag{12}$$

The porous sphere hydrodynamic theory then allows calculation of a, relative to R'=R+L, for various values of κR' and R/R'. When suitable values for mucins are used, a/R' is found to be about 0.8 (Bloomfield, 1983).

Protein-vesicle complexes: One of the most important biochemical applications of porous sphere hydrodynamics is to complexes between small phospholipid vesicles and extrinsic binding proteins. These complexes are often used as models for protein-membrane interactions. With modern techniques for vesicle preparation and fractionation, monodisperse samples can be made with radii less than 200 Å. Since the protein length is typically 50-100 Å, binding of a few protein molecules to a vesicle results in a marked increased in Stokes radius. This increase in hydrodynamic size, combined with a measurement of the number of proteins bound per vesicle, can be interpreted to yield information on the geometry of binding of the protein to the membrane.

For such interpretation to be valid, two subtleties must be taken into account. These are related to the porous sphere hydrodynamics discussed above, but have been developed in somewhat different form. The first is that the effective radius of the complex does not extend to the outer tip of the bound protein, but is somewhat less. The relation is a remarkably simple one (Bloomfield *et al.*, 1967a; McCammon *et al.*, 1975b). If the radius of the vesicle is R, and a is the Stokes radius of the complex fully coated with protein, then:

$$a = R(1 + 0.25\beta) \tag{13}$$

where β is the ratio of protein to vesicle radius. The second is that a/R depends on the degree of coverage of the vesicle surface by protein. The functional dependence of a/R has been obtained by numerical simulation by Bloomfield *et al.* (1967a); it can also be deduced from the two-layer porous sphere theory. The function increases sharply at low extents of coverage, but levels off above about 50%.

This approach has been applied to two important protein-membrane systems. With the blood-clotting proteins prothrombin and factor X, it was possible to show that the protein projects radially outward from the vesicle surface, with no penetration into the bilayer (Lim *et al.*, 1977). With cholera toxin, a multisubunit protein with a central A subunit that projects from one side of a plane defined by five B subunits, we found that the A subunit is inserted into the membrane when the B subunits bind to their ganglioside G_{M1} receptors (Dwyer and Bloomfield, 1982).

DIFFUSION IN TWO-DIMENSIONAL MEMBRANES

With the development of the spectroscopic tools discussed in this symposium, it has become possible to measure translational and rotational diffusion coefficients of proteins and other molecules inserted in membrane bilayers. Interpretation of the results, however, is not as simple as might have been expected, because there is no solution to the two-dimensional equations for steady translational motion in the slow viscous flow regime. This is known as the Stokes paradox (Lamb, 1932).

For a sphere of radius a immersed in a three-dimensional continuum solvent of viscosity η, the translational and rotational diffusion coefficients are well known to be:

$$D_T = KT/6\pi\eta a, \quad D_R = KT/8\pi\eta a^3 \tag{14}$$

and the ratio is

$$D_T/D_R = (4/3)\,a^2.\tag{15}$$

For two dimensions, with a cylinder of radius a moving with its axis normal to the surface of a membrane of thickness h, the rotational diffusion coefficient (Saffman and Delbrück, 1975) is:

$$D_R = KT/4\pi\eta a^2 h.\tag{16\mu}$$

Saffman and Delbrück considered several ways to obtain a finite, velocity-independent translational diffusion coefficient: give the membrane a finite size, consider the viscosity η' of the fluid on either side of the membrane, or use the Langevin equation together with the equations for unsteady flow.

Under conditions generally appropriate for biological membranes, the appropriate choice is the second, to give the outer fluid a finite viscosity. When that is done, with the further assumption that $\eta' \ll \eta$, the result is:

$$D_T = (KT/4\pi\eta h)\,[\ln(\eta h/\eta' a) - \gamma]\tag{17}$$

where γ here is Euler's constant. The ratio of translational to rotational coefficients is then:

$$D_T/D_R = [\ln(\eta h/\eta' a) - \gamma]a^2.\tag{18}$$

This ratio is approximately four times that for three-dimensional motion given by Equation (15).

Several experimental tests of the Saffman-Delbrück theory have been made. Equation (18) allows estimates of the protein diameter to be made from simultaneous measurements of D_T and D_R. The value obtained for bacteriohodopsin is in reasonable agreement with its known structure (Peters and Cherry, 1982). These workers also found that a 12-fold increase in viscosity of the aqueous phase reduced D_T by only 50%, in general accord with the logarithmic dependence on η' predicted by Equation (17). They pointed out, however, that lipid

protein (l/p) ratios need to be considered in interpreting experimental results, since boundary layer, steric crowding, and protein aggregation effects can all be significant at lower l/p.

Another prediction of Equation (17) - that translational diffusion
should depend only in a weak, logarithmic fashion on protein radius -
was qualitatively confirmed by Vaz et al. (1982). However, measurements of D_T for a fluorescent lipid derivative were consistent with
continuum hydrodynamic theory only if the non-slip boundary condition
was replaced with a slip boundary condition at the bilayer midplane.

In a more detailed calculation, Saffman (1976) allowed the bathing
fluids on the two sides of the membrane to have different values η'
and η''. Hughes et al. (1982) have recently improved on Saffman's
solution. They solved the equations for all values of the dimension-
less parameter $\varepsilon = (\eta' + \eta'')a/\eta h$. The results indicate that mobility
experiments are not determined simply by the membrane microviscosity
- the bathing fluid viscosities are also important.

ACKNOWLEDGEMENT

Preparation of this review was supported in part by NSF grant PCM 81-
18107.

REFERENCES

Aksiyote-Benbasat, J. and Bloomfield, V.A. (1982). *Biopolymers* 21,
    797-804.
Allen, A. (1978). *Brit. Med. Bull.* 345, 28-33.
Belford, G.G., Belford, R.L. and Weber, G. (1972). *Proc. Natl. Acad.
    Sci. USA* 69, 1392-1393.
Bloomfield, V.A. (1983). *Biopolymers* 22, 2141-2154.
Bloomfield, V.A., Dalton, W.O. and Van Holde, K.E. (1967a). *Biopoly-
    mers* 5, 135-148.
Bloomfield, V.A., Garcia de la Torre, J. and Wilson, R.W. (1979). *In*
    "Electrooptics and Dieletrics of Macromolecules and Colloids" (Ed.
    B.R. Jennings), pp. 183-195, Plenum, New York.
Bloomfield, V.A., Van Holde, K.E. and Dalton, W.O. (1967b). *Biopoly-
    mers* 5, 149-159.
Broersma, S. (1960a). *J. Chem. Phys.* 32, 1626-1631.
Broersma, S. (1960b). *J. Chem. Phys.* 32, 1632-1635.
Burgers, J.M. (1938). *In* "Second Report on Viscosity and Plasticity"
    Chapter 3. Nordemann, Amsterdam.
Douthart, R.J. and Bloomfield, V.A. (1968). *Biochemistry* 7, 3912-
    3917.
Dwyer, J.D. and Bloomfield, V.A. (1982). *Biochemistry* 21, 3227-3231.
Garcia Bernal, J.M. and Garcia de la Torre, J. (1980). *Biopolymers*
    19, 751-766.

Garcia de la Torre, J. and Bloomfield, V.A. (1977a). *Biopolymers* 16, 1747-1763.
Garcia de la Torre, J. and Bloomfield, V.A. (1977b). *Biopolymers* 16, 1765-1778.
Garcia de la Torre, J. and Bloomfield, V.A. (1978). *Biopolymers* 17, 1605-1627.
Garcia de la Torre, J. and Bloomfield, V.A. (1980). *Biochemistry* 19, 5118-5123.
Garcia de la Torre, J. and Bloomfield, V.A. (1981). *Quart. Rev. Biophys.* 14, 81-139.
Happel, J. and Brenner, H. (1973). *In* "Low Reynolds Number Hydro-dynamics" Chapter 5, Prentice-Hall, New York.
Harvey, S.C. and Garcia de la Torre, J. (1980). *Macromolecules* 13, 960-964.
Hughes, B.D., Pailthorpe, B.A., White, L.R. and Sawyer, W.H. (1982). *Biophys. J.* 37, 673-676.
Kauzmann, W., Moore, K. and Schultz, D. (1974). *Nature* 248, 447-449.
Kirkwood, J.G. and Riseman, J.R. (1948). *J. Chem. Phys.* 16, 565-573.
Kuntz, I.D. Jr. and Kauzmann, W. (1974). *Adv. Protein Chem.* 28, 239-345.
Lamb, H. (1932). *In* "Hydrodynamics" Sect. 343, Cambridge Univ. Press, New York.
Lim, T.K., Bloomfield, V.A. and Nelsestuen, G.L. (1977). *Bio-chemistry* 16, 4177-4181.
McCammon, J.A., Deutch, J.M. and Bloomfield, V.A. (1975a). *Biopoly-mers* 14, 2479-2487.
McCammon, J.A., Deutch, J.M. and Felderhoff, B.U. (1975b). *Biopoly-mers* 14, 2613-2623.
Mellado, P. and Garcia de la Torre, J. (1982). *Biopolymers* 21, 1857-1872.
Newman, J., Swinney, H.L. and Day, L.A. (1977). *J. Mol. Biol.* 116, 593-606.
Norisuye, T., Motokawa, M. and Fujita, H. (1979). *Macromolecules* 12, 320-323.
Oseen, C.W. (1927). *In* "Hydrodynamik" Akademisches Verlag, Leipzig.
Peters, R. and Cherry, R.J. (1982). *Proc. Natl. Acad. Sci. USA* 79, 4317-4321.
Roberts, G.P. (1978). *Brit. Med. Bull.* 345, 39-41.
Rotne, J. and Prager, S. (1969). *J. Chem. Phys.* 50, 4831-4837.
Rouse, P.E. (1953). *J. Chem. Phys.* 21, 1272-1280.
Roussel, P., Lamblin, G., Degand, P., Walker-Nasir, E. and Jeanloz, R.W. (1975). *J. Biol. Chem.* 250, 2114-2122.
Saffman, P.G. (1976). *J. Fluid Mech.* 73, 593-602.
Saffman, P.G. and Delbrück, M. (1975). *Proc. Natl. Acad. Sci. USA* 72, 3111-3113.
Swanson, E., Teller, D.C. and De Haen, C. (1978). *J. Chem. Phys.* 68, 5097-5102.
Teller, D.C. and De Haen, C. (1975). *Fed. Proc.* 34, 598.
Teller, D.C., Swanson, E. and De Haen, C. (1979). *Adv. Enzymol.* 61, 103-124.
Tirado, M.M. and Garcia de la Torre, J. (1979). *J. Chem. Phys.* 71, 2581-2588.

Tirado, M.M. and Garcia de la Torre, J. (1980). *J. Chem. Phys.* <u>73</u>, 1986-1993.
Vaz, W.L.C., Criado, M., Madeira, V.M.C., Schoellman, G. and Jovin, T.M T.M. (1982). *Biochemistry* <u>21</u>, 5608-5612.
Wegener, W.A. (1981). *Biopolymers* <u>20</u>, 303-326.
Wegener, W.A., Dowben, R.M. and Koester, V.J. (1979). *Biophys. J.* <u>21</u>, 114a.
Wilemski, G. (1977). *Macromolecules* , 28-34.
Wilson, R.W. and Bloomfield, V.A. (1979a). *Biopolymers* <u>18</u>, 1205-1211.
Wilson, R.W. and Bloomfield, V.A. (1979b). *Biopolymers* <u>18</u>, 1543-1549.
Yamakawa, H. (1970). *J. Chem. Phys.* <u>53</u>, 436-443.
Yamakawa, H. (1975). *Macromolecules* <u>8</u>, 339-342.
Yamakawa, H. and Fujii, M. (1973). *Macromolecules* <u>6</u>, 407-414.
Zimm, B.H. (1956). *J. Chem. Phys.* <u>24</u>, 269-278.
Zimm, B.H. (1980). *Macromolecules* <u>13</u>, 592-602.
Zimm, B.H. (1982). *Macromolecules* <u>15</u>, 520-525.
Zimm, B.H. and Kilb, R.W. (1959). *J. Polym. Sci.* <u>37</u>, 19-42.

Electric Field Orientation and Macromolecular Relaxation

B.R. JENNINGS

INTRODUCTION

The majority of macromolecules are non-spherical and hence are geometrically anisotropic. Also, they are either polar or have an
asymmetric electric polarizability, resulting in molecular electrical
anisotropy. In addition, their optical characteristics such as
refractive index, the ability to absorb light, and fluorescence are
all related to directional characteristics within constituent groups
of the molecules. Hence, in the main, macromolecules are also optically anisotropic. It is this combination of optical, electrical and
geometrical anisotropy that is the foundation of electro-optical
phenomena. Briefly stated, the principle of electro-optical characterization of macromolecules is as follows. A short pulsed electric
field is applied to a dilute molecular solution and one or other of
the optical properties of the medium is monitored. The applied field
couples with the electrical dipole moments of the molecules and causes
orientational order. Owing to the anisotropic optical properties,
this results in a change in these properties for the medium as a
whole. By measuring such changes, one can infer the molecular origins
of the characteristics responsible. Typically, pulses of either ac
or dc electric fields are used and the rapid transient changes in
birefringence, scattered light intensity, fluorescence, dichroism or
optical activity are recorded.

SPECTROSCOPY AND THE DYNAMICS
OF MOLECULAR BIOLOGICAL SYSTEMS

21

In this chapter some of the simpler aspects of electro-optical
methods and the type of molecular data they can yield are introduced.
Representative illustrative experimental results from the literature
are reviewed for three very different biopolymer systems when studied
via three different electro-optical phenomena: firstly the cartilage
polysaccharide, proteoglycan, and its interaction with hyaluronic
acid using the electric birefringence method; secondly, the effect of
antibiotics on bacteria observed via the measurements of electrically
induced light scattering; and thirdly, the binding of the active
mutagenic diol-epoxide derivatives of the ubiquitous hydrocarbon
benzo[a]pyrene with the genetic material deoxyribonucleic acid (DNA)
was calculated from electrically induced measurements of the polarized components of the fluorescence.

ELECTRIC BIREFRINGENCE

The variability of refractive index along the main geometric axes of
a molecule results in the material being birefringent. When such
molecules are dissolved in a dilute solution, they adopt a random
array and the medium as a whole is not birefringent. The orientational order which accompanies application of an applied field
results in the inherent birefringence of the molecules being partially
imposed upon the overall medium (Fig. 1). Hence application of an

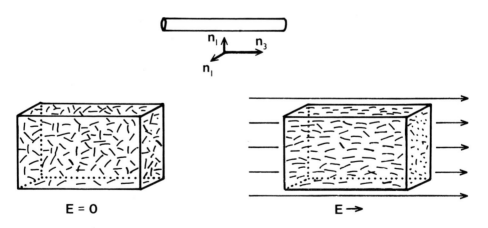

FIG. 1 The electro-optic principle. An electric field E aligns molecules and imposes the inherent molecular anisotropy in an optical
property on the medium. In this case the property is the refractive
index which gives rise to birefringence.

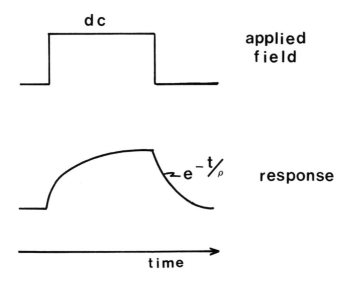

FIG. 2 The principle of the fast transient method. The optical response to a rectangular pulsed field is transient in nature as the molecules align in their frictional medium with a rotational relaxation time ρ.

electrical pulse is accompanied by an increasing birefringence of the solution until orientation saturation is reached and an equilibrium alignment in the field is obtained. Termination of the pulse is accompanied by disorientation of the molecules back to their chaotic random array. The birefringence of the medium is thus transient in nature. The principle is illustrated in Fig. 2.

Birefringence is readily recorded by shining laser light through a polarizer prior to its incidence on a cell containing the test solution (Fredericq and Houssier, 1973). Light transmitted through the cell then passes through an analyzing polarizer before falling on a photodetector. The polarizer and analyzer are almost "crossed" relative to each other, and an electric field is applied transverse to the light beam in an azimuth of 45° to the polarization of the incident beam. Typically the sample cell is a few centimetres long and the field is applied by means of a pair of steel electrodes placed either side of the light beam. A few millilitres of sample are all that is required. In principle, this optical system transits no light if the sample is non-birefringent. When the pulsed field is

applied to the cell electrodes, the sample birefringence is detected
by the increasing amplitude of transmitted light which follows the
orientation rate of the molecules to a steady-state value. Termination of the pulse is accompanied by the decay of the birefringence
back to zero. The complete birefringence transient is recorded via
the photodetector and is either displayed on an oscilloscope or fed
to a transient digitizer.

Two parameters are of value. The first is the amplitude of the
birefringence which, by suitable theory, can be related to the inherent optical anisotropy of the molecules (Peterlin and Stuart, 1939).
It also leads to information on the magnitude of the permanent dipole
moment (μ) and the anisotropy ($\Delta\alpha$) of the polarizability (α) of the
molecules. The decay of the birefringence after field cessation provides a direct measure of the molecular size as the decay is of the
form $\exp[-(t/\rho)]$ where ρ is the molecular rotational relaxation time
(Benoit, 1951), which is itself a function of a^3 where a is the major
molecular dimension (Perrin, 1934). It is this cubic dependence which
accounts for the extreme sensitivity of all transient electro-optical
methods for molecular size determination. One recalls from the Debye
(1929) concept of dipolar rotation, that the permanent dipole moment
is far more frequency dependent than the induced polarizing mechanisms. Hence the birefringence response to a dc voltage contains
contributions from both μ and $\Delta\alpha$, while the response to a pulse of
high frequency ac field contains only the induced dipole moment contribution. In this manner μ and $\Delta\alpha$ can be isolated and evaluated
discretely.

An illustration of the versatility of the method is given by data
from the author's laboratory on the complex cartilage polysaccharide
proteoglycan (PG). Cartilage is composed of a number of macromolecules which interact together to form the complicated matrix responsible for its elastic characteristics. This matrix is composed of
collagen fibrils which are interspaced and interbound with proteoglycan molecules. Recently it was found that small but definite
amounts of the polysaccharide hyaluronic acid are generally associated
with the cartilage matrix. A proposal has been made that the proteoglycan molecules (Fig. 3a), which are themselves extended, highly

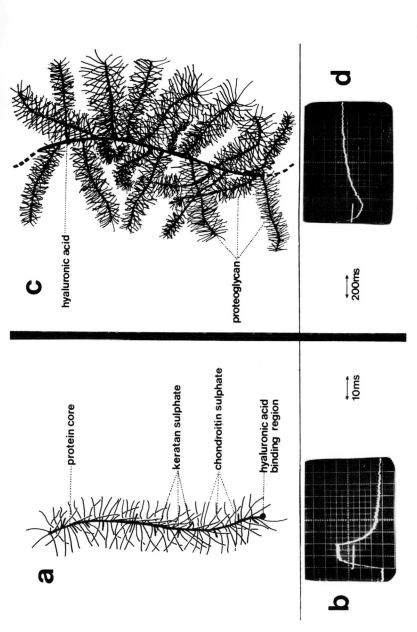

FIG. 3 Electric birefringence of proteoglycan (PG) and the proteoglycan/hyaluronic acid complex. (a) the 350 nm long PG molecule and (b) the positive transient birefringence response for a concentration of 210 μg uronic acid/ml., with E = 360 Vcm^{-1} giving ρ = 3.6 ms, (c) PG molecules associated with a 3 μm hyaluronic acid chain and (d) the resulting birefringence for E = 360 Vcm^{-1} giving ρ = 650 ms.

charged hydrophilic polysaccharides, bind laterally to the hyaluronic acid to form complex extended bottle brush-type molecules as represented in Fig. 3c. It is this aggregate which is thought to infill the collagen superstructure.

By conventional biochemical means, analysis of such a structure is difficult. From electric birefringence data (Foweraker *et al.*, 1977), one is able to confirm the proposed model with speed and ease. In Fig. 3b, transient responses are shown for dilute solutions of aqueous proteoglycan. Two things should be noted. Firstly, the birefringence is positive, and secondly, the time scale for the relaxation process is of the order of 4 ms. Upon the addition of a small amount of hyaluronic acid (HA) of molecular mass 7×10^5, the response shown in Fig. 3d was obtained. Here it can be seen that the birefringence was essentially negative, and that the rotational relaxation time has now been extended some 20-fold. The explanation is as follows. From the experimental result displayed in Fig. 3b, the positive birefringence of the individual PG molecules was established and their relaxation time was consistent with a flexible molecule some 300 nm in length. The negative birefringence seen in Fig. 3d was compatible with the fact that the PG molecules associated laterally with the extended HA molecules and that their major axis then formed the minor axis of the complex. In addition it was the length of the HA molecule that then controlled the major dimension of the new complex, and this greatly extended length resulted in a much longer relaxation time.

It has been suggested that the absence of the associate in the cartilage accompanies osteoarthrotic disorder as the system is readily attacked proteolytically (McDevitt and Muir, 1975). Current studies are aimed at determining whether HA is (i) reduced or absent, (ii) does not bind to PG, or (iii) lacks the link protein necessary for the association under conditions relating to medical malady.

The obvious use of birefringence as a means of indicating the nature of PG association with HA is apparent. The analysis of the effects of various chemotherapeutic agents on the PG/HA association is an obvious extension to such studies. It is hoped that for such complicated interacting systems, electric birefringence may develop into a powerful laboratory tool.

ELECTRIC FIELD-INDUCED LIGHT SCATTERING

The imposed reorientation of asymmetric molecules results in a change
in the time-averaged intensity of the scattered light from a dilute
molecular solution (Wippler, 1956). Hence, by recording the intensity scattered at an angle θ to the forward direction of the incident
light beam, one can monitor the orientation and alignment directly.
With light scattering, however, the scattered intensity prior to the
application of the field is used as a standard means of obtaining the
molecular mass, radius of gyration and hence the size of the molecules
(Kerker, 1969). In the presence of the electric field, relaxation
data lead to ρ and so to further information on size, whilst the
amplitude of the changes (ΔI) yield μ and Δα. In this case, the
apparatus simply consists of a small cylindrical cell which contains
a few ml of the solution. This is irradiated by a well-defined collimated beam of light. A detecting limb records the light scattered at
angle θ via a photodetector. Pulsed electric fields are again used
and the transient changes in ΔI are analyzed to give the electrical
characteristics and ρ for the molecules. In this case the theory is
more complicated and fuller details are given elsewhere (Jennings,
1981).

Bacteria scatter light strongly. They are complicated, multicomponent micro-organisms for which an exact description of the
scattering characteristics is difficult. In addition, they always
exist as polydisperse suspensions. *E. coli* are renowned bacteria of
roughly ellipsoidal shape with an axial ratio of 1.5 and with major
dimensions of the order of a few μm. Their large size means that
only very small electric fields are needed to order them orientationally. Figure 4 presents the transient scattered intensity change
in a dilute *E. coli* suspension using an ac field. From the transient,
both the average relaxation time and the anisotropy of the electrical
polarizability could be evaluated (Jennings and Morris, 1975).

In conventional physical chemistry, two origins are well known for
polarizability. These are the electronic polarization, in which the
electrons are distorted about nuclear sites, and the atomic polarization, in which constituent atoms of molecules are displaced relative
to each other. In aqueous and conducting media, a third contribution

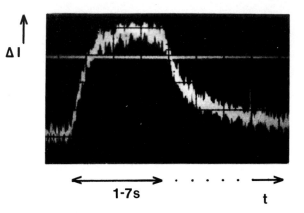

FIG. 4 Transient change in the scattered light intensity for a dilute *E. coli* suspension. Data for a 1.7s pulse of an ac field at a frequency of 600 Hz and an amplitude of 60 V cm^{-1}, using a scattering angle of 30° with $\lambda = 1.06$ μm in the near infrared.

which has been known to colloid scientists for many years is also possible. This is a distortion of the electric double layer and the counter-ion cloud which surrounds charged macromolecules (Dukhin and Shilov, 1974). During recent years it has become increasingly evident that it is this third polarization mechanism which is predominant in the dielectric properties of aqueous macromolecular solutions. One would then suspect that any charged or polar additive might influence the particle/solvent interface and manifest itself via an apparent change in $\Delta\alpha$. In Fig. 5, the effect of increasing concentration of antibiotic in a dilute suspension of *E. coli* is shown (Morris *et al.*, 1976). An increasing amount of antibiotic elicits a continual decrease in the observed anisotropy of the polarizability. Furthermore, the shape of the curve is reminiscent of Langmuir absorption isotherms in which the change is initially drastic but the curve then slowly falls towards a constant asymptotic value.

The addition of various antibiotics at concentration c results in a series of such graphs. In the light of these, it has been of value to do two things. Firstly each graph has been characterized using an equation of the type:

$$\Delta\alpha/\Delta\alpha_0 = 1 - \left(\frac{\kappa\ \beta\ c}{1 + \beta c} \right) \tag{1}$$

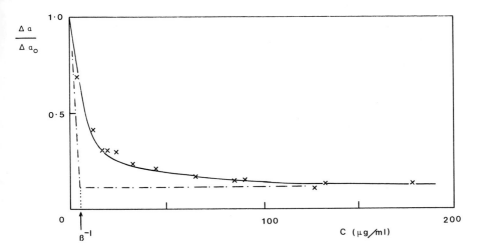

FIG. 5 Change in the anisotropy of electrical polarizability (Δα) of
E. coli with increasing concentration of antibiotic - in this case
streptomycin sulphate. The concentration factor β^{-1} is indicated.

in which β is the inverse of the concentration at which the approximate knee of the curve occurs, whilst κ designates the high concen-
tration limiting value of $^{\Delta\alpha}/\Delta\alpha_0$. In Table 1, the parameter β^{-1} has
been recorded for each *E. coli*/antibiotic system. The Table also
gives values of the minimum inhibitory concentration (MIC) for comparison. This concentration represents the minimum amount of antibiotic which must be lawned on to specific cultures to prevent
bacterial replication. It is a conventional parameter for quantifying
the efficiency of antibiotic action. The remarkable similarity in
the sequences of β^{-1} and MIC values for each antibiotic is surprising

TABLE 1 Comparison of the concentration parameter β^{-1} with the minimum inhibitory concentrations (MIC) for antibiotic-*E. coli* interactions.

Antibiotic	MIC (μg/ml)	β^{-1} (μg/ml)
Streptomycin	9.0	6.25
Paramomycin	1.0	5.0
Sisomycin	0.25	0.2
Gentamycin	0.2	0.1

but gratifying, and seems to indicate that whatever the site of action
of the antibiotic, it must initially proceed via the bacterial surface. It is this property which is apparently evidenced through the
electrical characteristic $\Delta\alpha$. It would thus appear that fast transient measurements of ΔI for bacteria in the presence of antibiotics
offer a convenient means of estimating the suitability of an antibiotic activity against bacteria. The correlation of β^{-1} with the
minimum inhibitory concentration also suggests that these electro-
optic experiments might have potential as active assays for antibiotics.

ELECTROFLUORESCENCE

Fluorescence is a two-fold optical property which involves absorption
of incident light and its re-emission at a different wavelength. For
any active chemical group (fluorophore), the absorption and emission
processes are optimized along specific directions in space. Maximal
fluorescence is thus achieved when the incident light is polarized
parallel to the absorption moment. If the absorption and emission
moments do not coincide, the fluorescent light is depolarized (Pesce
et al., 1971). If the polarized components of the fluorescence are
recorded when an electric field imposes orientational order on an
array of macromolecules whose characteristics in an electric field
are known, the specific binding geometry of any fluorophore to the
macromolecule can be evaluated (Czekalla, 1960). In this way, the
binding characteristics of drugs and carcinogens to DNA have been
determined.

The apparatus requires a fairly high-powered laser which operates
in the blue or ultra-violet region of the spectrum (Ridler and
Jennings, 1977). An argon-ion laser is suitable. Light from such a
laser, attenuated and strictly polarized in a selected direction, is
allowed to fall on a test sample held between a pair of electrodes
placed such that the electric fields acts vertically. For vertically
polarized incident light, the applied field direction and the light
polarization states are then coincident. Two detection limbs monitor
the fluorescence and record those components of fluorescence which
are polarized in the horizontal and vertical quadrature azimuths.

The transient changes in each of these components when a pulsed field
is applied between the electrodes is recorded and used to evaluate
the average orientation of the active fluorophores in space. Fore-
knowledge of the orientation of the carrier macromolecules in the
field enables the binding geometry of the fluorophore to the substrate
to be determined.

Illustrative data are given for benzo[a]pyrene, a ubiquitous carcinogen, to DNA. A dilute solution of 10^6 molecular mass DNA was
introduced to a dilute solution of each of two forms of benzo[a]-
pyrene (BP). The first was the native hydrocarbon whilst the second
was an enantiomeric diol-epoxide derivative of BP. It is known that
if carcinogenesis is to result, the host mammal must metabolize the
native BP into the diol-epoxide (BPDE) derivative (Simms *et al.*,
1974), which then combines covalently with the DNA. Of the four possible steroisomers of BPDE, only one (the + anti) form is thought to
be the ultimate carcinogen (Buening *et al.*, 1978). In these studies
a racemic mixture, +/- anti BPDE, was used.

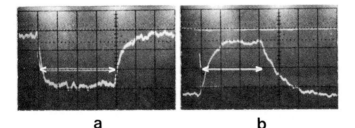

a   **b**

FIG. 6 Transient changes in one polarized component of fluorescence
for DNA tagged with (a) native benzo[a]pyrene and (b) +/- anti benzo-
[a]pyrene diol-epoxide. Data for the vertically polarized component
incident and emitted, with E = 13kV.cm^{-1}, λ = 351 and 364 nm for
incident light, λ > 400 nm for the emission.

In Fig. 6, two transients are shown (Ridler and Jennings, 1982)
for the vertically polarized fluorescence components with vertically
polarized incident light. Frames (a) and (b) are for DNA tagged with
adsorbed BP and bound BPDE respectively. The field was in the vertical direction. The fluorescence response was different in the two
systems. For the BP/DNA system, the vertically and horizontally
polarized components of the emission had relative values of -0.22 and

-0.14. The equivalent parameters for the BPDE/DNA system were +0.10 and +0.04. With the BP complex, both components of the fluorescence decreased, indicating a diminished overall fluorescence. The absorption transmission moments were thus oriented predominantly horizontally. Furthermore, the horizontally polarized output was less reduced than the vertical, indicating that the emitted light was polarized predominantly in the horizontal direction. Thus BP binds to DNA with both its absorption and emission transmission moments predominantly in the horizontal plane when DNA is aligned vertically. This is compatible with the intercalation binding model for BP in which the planar dye molecules bind perpendicular to the DNA long axis.

The BPDE molecules bind in a very different manner. Inspection of the data indicates that the changes in the polarized fluorescence components were different from those for BP in both amplitude and sign. In fact, the BPDE molecules bind to DNA with an inclination very close to that of the projected angle of the minor helical groove.

It is thus apparent that the carcinogenic form of BPDE and native BP bind to DNA in very different manners, and that the electrofluorescence method has the sensitivity to indicate this. The biochemical analysis of this behaviour is awaited. Nevertheless, it is hoped that this electro-optical method will be extended to form a valuable laboratory tool to aid our understanding of the binding characteristics of drugs and carcinogens to DNA.

CONCLUSIONS

Various other electro-optical methods exist. Changes have been recorded in the optical activity (Tinoco, 1959), dichroism (Fredericq and Houssier, 1973), and the photon correlation spectra of macromolecules and colloids (Ware and Flygare, 1971). Each of these adds specific information. Studies have been made on proteins, polysaccharides, vesicles, membranes, cells and a host of other polymer and biopolymer systems. Electro-optical methods have the advantages of being extremely fast and sensitive to small changes in molecular geometry. They have the disadvantage that the use of electric fields limits the methods to media which are of low electrical conductivity. Extremes of pH and ionic strength are thus excluded.

From the examples presented herein one sees that the methods are
useful for characterizing complicated materials and interacting systems. It is hoped that more biophysicists will apply them to a
greater number of systems.

ACKNOWLEDGEMENTS

The original research covered in this chapter was financed via grants
from the Arthritis and Rheumatism Council, the Ministry of Defence,
the SERC and the MRC, to whom thanks are expressed.

The assistance of colleagues in the Electro-Optics group is grate-
fully acknowledged. Permission to use in part or full Figs. 3 and 6
has been graciously given by Messrs. J. Wiley and Sons and Elsevier
Biomedical Press respectively.

REFERENCES

Benoit, H. (1951). *Ann. Phys.* (Paris), 6, 561-607.
Buening, M.K., Wislocki, P.G., Levin, W., Yagi, H., Thakkar, D.R.,
   Akaqi, H., Koreeda, M., Jerina, D.M. and Conney, A.H. (1978).
   Proc. Natl. Acad. Sci. USA 75, 5358-5361.
Czekalla, J. (1960). *Z. Electrochem.* 64, 1221-1228.
Debye, P. (1929). *In* "Polar Molecules", Chemical Catalogue Co.,
   reprinted by Dover Publications, New York.
Dukhin, S.J. and Shilov, V.N. (1974). *In* "Dielectric Phenomena and
   the Double Layer in Disperse Systems and Polyelectrolytes", J.
   Wiley and Sons, New York.
Foweraker, A.R., Isles, M., Jennings, B.R., Hardingham, T.E. and
   Muir, H. (1977). *Biopolymers* 16, 1367-1369.
Fredericq, E. and Houssier, C. (1973). *In* "Electric Dichroism and
   Electric Birefringence", Clarendon Press, Oxford.
Jennings, B.R. (1981). *In* "Molecular Electro-optics" (Ed. S. Krause)
   pp. 181-212, Plenum Press, New York.
Jennings, B.R. and Morris, V.J. (1975). *J. Colloid Interface Sci.*
   50, 352-358.
Kerker, M. (1969). *In* "The Scattering of Light", Academic Press, New
   York.
McDevitt, C.A. and Muir, H. (1975). *Ann. Rheum. Disease* 34, suppl. 2,
   137-138.
Morris, V.J., Jennings, B.R., Pearson, N.J. and O'Grady, F. (1976).
   Microbios 17, 133-139.
Pesce, A.J., Rosen, C.G. and Pasby, T.L. (1971). *In* "Fluorescence
   Spectroscopy", Marcel Dekker, New York.
Perrin, F. (1934). *J. Phys. Radium* 5, 497-511.
Peterlin, A. and Stuart, H.A. (1939). *Z. Phys.* 112, 129-147.

Ridler, P.J. and Jennings, B.R. (1977). *J. Phys. E.* (Sci. Instrum.)
     <u>10</u>, 558-563.
Ridler, P.J. and Jennings, B.R. (1982). *FEBS Lett.* <u>139</u>, 101-104.
Simms, P., Grover, P.L., Swaisland, A., Pal, K. and Hewer, A. (1974).
     Nature <u>252</u>, 326-327.

Tinoco, I. (1959). *J. Amer. Chem. Soc.* <u>81</u>, 1540-1544.
Ware, B.R. and Flygare, W.H. (1971). *Chem. Phys. Lett.* <u>12</u>, 81-85.
Wippler, C. (1956). *J. Chim. Phys.* <u>53</u>, 316-327.

Progress in Intensity Correlation Spectroscopy and Analysis of Structure and Dynamics of Biopolymers

R. RIGLER

INTRODUCTION

Correlation spectroscopy has become a very powerful tool in analyzing dynamic properties of molecular systems in solution. The technique depends on the properties of laser light: monochromaticity, coherence and directionality, which enable one to observe molecular motion from the measurement of intensity fluctuations due to the interference of Doppler-shifted scattered light or due to fluctuations in particle numbers in small volume elements.

The phenomenon of interference fluctuations caused by coherently scattered light is often called quasielastic light scattering (QELS) or dynamic light scattering. This approach has found many important applications in physics, chemistry and biology (Cummins and Pike, 1974, 1977; Chu, 1979). The speed and ease with which hydrodynamic properties such as diffusion coefficients can be determined allows one to document the time dependence of conformational changes in macromolecules occurring during their biological function. Developments in our laboratory are proceeding in this direction. The time resolution in monitoring structural transitions is comparable to or even better than time-resolved low angle X-ray scattering studies with synchrotron light, as will be exemplified in the following.

The observation of fluctuations in particle number has been used less frequently for measuring molecular motions (Weissman, 1981). Its

application, however, is of interest, since separation between the
translational and rotational parts of diffusion is easily attained.
In a modified form fluorescence emission instead of light scattering,
can be utilized to follow the diffusion of molecular species specifically labelled with fluorescent groups in a complex mixture. This
technique is known as fluorescence correlation spectroscopy (FCS).
A closely related technique involves the observation of particle
motion using fluorescence photobleaching recovery (FPR), which is
treated in detail elsewhere in this volume.

THEORETICAL BACKGROUND

The time dependence of intensity fluctuations is obtained by calculating the autocorrelation function $G(t)$ of the intensity $I(t)$:

$$G(t) = <I(t)I(0)> = \lim_{T\to\infty} \frac{1}{2T} \int_{-T}^{T} <i(t+\tau)i(\tau)> d\tau \tag{1}$$

$G(t)$ can be decomposed in a constant part $<I>^2$ and a time-dependent
part $(<\delta I(t).\delta I(0)>$:

$$G(t) = <I>^2 + <\delta I(t)\delta I(0)> \tag{2}$$

where $<I>$ is the average photon flux and $\delta I(t)$ the fluctuating part
of scattered or emitted light.

Since I is proportional to $<N>$, the average number of particles in
the scattering region, Equation 2 can be expanded to:

$$G(t) \propto <N>^2 + <N>^2 C[g(t)]^2 + <\delta N(t)\delta N(0)> \tag{3}$$

or, after normalization by $<N>^2$:

$$G^*(t) = 1 + C[g_c(t)]^2 + (<\delta N(t)\delta N(0)>)/<N>^2 \tag{4}$$

Here $g_c(t)$ is the (field) autocorrelation function for the interference fluctuation term and $C \backsim 1/N_{coh}$ is a proportionality factor dependent on the observed number of coherence areas (N_{coh}).
$<\delta N(t)\delta N(0)> = g_N(t)$ describes the autocorrelation function of the
number fluctuation part.

Interference fluctuations: coherent scattering

The original observation of Doppler-shifted scattered light due to
Brownian motion stems from measurements of the broadening of the
power spectrum of scattered light using frequency mixing between
incident and scattered light in a heterodyne arrangement (Cummins *et
al.*, 1964). It could be shown that interference between scattered
light leads to self-beating and intensity fluctuations with frequency
components related to diffusion (Ford and Benedek, 1965). Instead of
the power spectrum, its Fourier transform, the autocorrelation function, can be measured using real-time correlators. These time- and
space-dependent intensity fluctuations can be envisaged as concentration fluctuations over an interference grid with wavelength Λ
determined by the wavelength of light and the scattering angle θ:

$$\Lambda = \lambda/(2n\sin\theta/2)$$

A general treatment of $g_c(t)$, assuming translational and rotational
motion to be uncoupled, has been given by Pecora (1968) with the
result:

$$g_c(t) = \sum_{\ell \text{ even}} B_\ell \exp(-(DK^2+\ell(\ell+1)D_R)t) \qquad (5)$$

where D_T and D_R are the translational and rotational diffusion coefficients and $K=2\pi\Lambda^{-1}$. ℓ takes the values $0,2,4\ldots$ and B_ℓ depends
upon the product of the wave vector K and particle length L, KL. For
the simplest case, $\ell=0,2$:

$$g_c(t) = B_0\exp[-D_TK^2t]+B_1\exp[-(D_TK^2+6D_R)t] \qquad (6)$$

which means that g(t) contains an angle (wavevector)-dependent translational term and an angle-independent rotational diffusion term.

From the angular dependence of g(t) and that of B_1 (Pecora, 1968),
D_T and D_R can be obtained. Alternatively information on D_R can be
obtained from the fluctuation of depolarized light, and it can be
shown (Pecora, 1968; Caroli and Parodi, 1969) that:

$$g_c(t) = \exp[-(DK^2 + 6D_R)t]$$

which means that at zero angle only D_R is determined. Several experiments using this approach have been performed notably by Wada *et al.* (1969) and Schmitz and Schurr (1973).

Number fluctuations: non-coherent scattering

Analysis of $<\delta N(t)\delta N(0)>$ requires specification of the volume element in which particle fluctuation occurs. This is usually attained via focussed laser beam propagating in the z-direction with a Gaussian spatial intensity profile in the x,y direction of radius ω_1 (cross section). If the third dimension (z) is observed with a small slit of half width ω_2 the scattered intensity $I(r)$ depends upon the position (r) in the volume element as:

$$I(r) = I_0\exp\{-[2(x^2+y^2)/\omega^2{}_1]\}\exp[-(2z^2/\omega^2{}_2)] \qquad (7)$$

Translational motion across the volume element leads to intensity (number) fluctuation. The normalized autocorrelation function $g_N^T(t)$ can be calculated (Schaefer, 1973; Berne and Pecora, 1976):

$$g_N^T(t) = (<\delta N(t)\delta N(0)>)/<N>^2 = (1/<N>)[1+(4Dt/\omega_1^2)]^{-1}[1+(4D_T/\omega^2{}_2)]^{-1/2}$$

$$(8)$$

When measuring number fluctuations by variations of the scattered intensity or of laser-excited fluorescence in a FCS experiment the same formalism holds (cf. Magde *et al.*, 1972, 1974, 1978b).

A special and very interesting solution for the FCS case has been derived by Asai and Ando (1976) using standing light waves propagating in the z direction with a Gassian intensity profile in the x,y plane for excitation:

$$I = I_0\exp[-2(x^2+y^2)/\omega^2{}_1][\sin^2(2\pi z/\lambda)] \qquad (9)$$

which yields a normalized autocorrelation function:

$$g_N^T(t) = (1/<N>)[1+(4D_T t/\omega^2)]^{-1}[1+(1/2)\exp\{-(16\pi^2 D_T/\lambda^2)t\}]$$

$$(10)$$

This solution provides an important improvement since the two terms are well separated in time and the diffusion coefficient can be determined accurately from the exponential term without knowledge of the beam diameter. Exact determination of the beam diameter has been the subject of various publications (Sorscher and Klein, 1980; Schneider and Webb, 1981).

Rotational motion

The autocorrelation function $g_N^R(t)$ for rotational motions of particles with reference to the vector of exciting light was first solved for the case of fluorescence detection (Ehrenberg and Rigler, 1974; Aragon and Pecora, 1975). In this case, the orientational fluctuation of molecules are observed via their anisotropic emission from the excited state. The autocorrelation function thus contains a contribution from rotational motion (D_R) as well as from the excited-state to ground-state transition (lifetime τ).

For 4π or 2π observation of the emitted radiation, rotational motion of bodies of arbitrary shape can be described in terms of the eigenfunction (a_j) and eigenvalues (E_j) of a symmetric rotor (van Winter, 1954; Favro, 1960).

For fluorescence lifetimes τ short compared to rotational reorientation ($1/\tau >> D_R$):

$$g_N^R(t) = (1/<N>)\{\sum_{j=1}^{5} a_j \exp[-E_j t] - (9/5)\exp[-t/\tau]\} \qquad (11)$$

values for a_j and E_j as function of the rotational diffusion coefficients around the main axes of revolution (D_x, D_y, D_z) are given elsewhere (Ehrenberg and Rigler, 1972).

For the symmetric rotor with $D_x = D_y$, D_z

$$E_1 = 5D_x + D_z, \quad E_2 = 2D_x + 4D_z \quad \text{and} \quad E_3 = 6D_x \quad \text{and} \quad \sum_{j=1}^{3} a_i = 4/5 \qquad (12)$$

The first term of Equation (11) is identical with that derived for measurements of rotational motion from fluorescence depolarization experiments (Ehrenberg and Rigler, 1972; Rigler and Ehrenberg, 1973). As an additional term, the fluorescence lifetime is observed which appears as a multiplicative factor in the pulsed experiment (where the rotation of the excited state is observed) and usually confines orientational measurements to short time scales (nanosecond range). Translational diffusion can easily be included, and with emission and diffusion processes well separated in time $(1/\tau \gg 6D_R \gg 4D_T/\omega^2)$ we obtain for the simple case of a spherical rotor:

$$g_N^{RT}(t) = 1/<N>\{(4/5)\exp[-6D_Rt]+[1+(4D_T/\omega^2)]^{-1}-(9/5)\exp[-t/\tau]\}$$

$$(13)$$

Like translational motion, rotational motion can also be observed from light scattering fluctuations provided the scattering is anisotropic. Usually the contribution of these fluctuations is so small as compared to interference scattering that they can only be observed when the interference terms can be effectively suppressed. Methods of cancelling interference fluctuations have been devised by cross correlating the intensities scattered at different angles. Experimental solutions using two detectors at 180° opposite to each other (Griffith and Pusey, 1979) or closely spaced (Kam and Rigler, 1982) have been demonstrated and the cross correlation function $g^R(t)$ has been derived:

$$\tilde{g}_N^R(t) \propto \sum_{\ell=even} A_\ell \exp[-\ell(\ell+1)]D_Rt \qquad (14)$$

with $A_\ell = S_{\ell 0}(K)S_{\ell 0}(K')[2\ell+1/2]P_1[\cos\phi]$.

$S_{\ell 0}(K)$ are the spherical harmonics coefficients (Kam, 1977) of the scattering intensity function observed at wave vectors K and K', P is a Legendre Polynomial and ϕ the angle between K and K'.

For $\ell=0,2$ one obtains:

$$\tilde{g}_T^R(t) = A_0+A_2\exp[-6D_Rt] \qquad (15)$$

In comparison with the case of fluorescence emission, the situation here is more complicated due to the existence of the spherical harmonic terms of D_R and detailed evaluation of anisotropic rotational diffusion is more difficult.

When cross correlating intensity fluctuations observed at different scattering angles, only number fluctuation terms due to rotational and translational motion are observed. Combination of these two processes leads to the product of the cross correlation functions for translational and rotational diffusion (Equations 8 and 14):

$$g_N^{RT} = (1/<N>)[1+4Dt/\omega^2]^{-3/2}\{A_0+A_2\exp[-6D_R t]\} \tag{16}$$

with $\omega = \omega_1 = \omega_2$.

EXPERIMENTAL RESULTS

Fluorescence correlations

The advantage of analysing time-dependent number fluctuations using fluorescence emission rests in its large sensitivity (detection of a few hundred particles) and in the possibility of studying a single molecular species amongst a variety of others. Thus, diffusion properties near to ideal conditions (zero concentration) as well as self diffusion can be studied.

A typical example concerns the translational motion of tRNA marked with a fluorescent label at a known site (Plumridge *et al.*, 1980). From analysis of the rotational motion it became evident that the phenylalanyl-specific tRNA from yeast must occur in at least two Mg^{2+} dependent configurations with different hydrodynamic properties (Ehrenberg *et al.*, 1979; Rigler and Wintermeyer, 1983). A set-up for fluorescence correlation spectroscopy constructed in this laboratory is shown in Fig. 1a, and its modification for fluorescence photobleaching recovery in Fig. 1b. Measurement of the autocorrelation function of fluorescence fluctuation (Fig. 2a) demonstrates that the translational diffusion coefficient is decreasing to an asymptotic value at 20 mM Mg^{2+}. For comparison, the same experiment was performed measuring the fluorescence photobleaching recovery (Fig. 2b). Good agreement within experimental uncertainties is observed between the two approaches (Fig. 2c), despite the fact that the concentration of tRNA differs by about three orders of magnitude. These results confirm the notion of a concentration-independent conformational equilibrium with hydrodynamic implications rather than a Mg^{2+}-induced aggregation of tRNA.

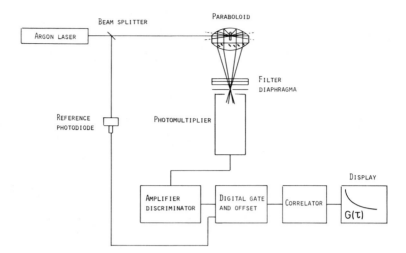

FIG. 1A Set-up for fluorescence correlation measurements. Modified after Rigler *et al.*, 1979.

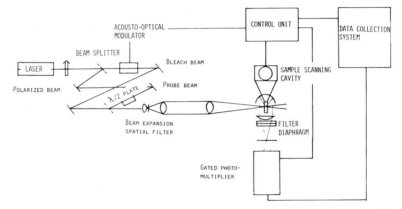

FIG. 1B Set-up for photobleaching recovery measurements. Double beam arrangement for separation of translational and rotational motion (Wegener and Rigler, 1984).

Various examples of the measurement of translational diffusion by FCS have been given in the literature ranging from free fluorescent dyes (Elson and Magde, 1974; Rigler *et al.*, 1979; Kandler *et al.*, 1982) to nucleic acids (Elson and Magde, 1974; Icenogle and Elson, 1983a,b), proteins (Borejdo, 1979) and membranes (Elson *et al.*, 1976). An interesting comparison of FCS and scattering fluctuation experiments for evaluating self-diffusion has been given recently (Andries *et al.*, 1983).

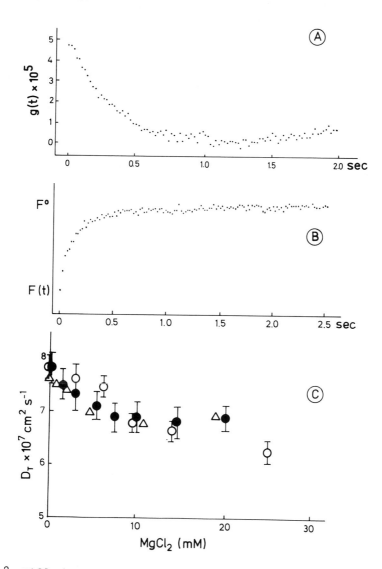

FIG. 2 Diffusion measurements of phenylalanyl-tRNA. (A) FCS-auto-
correlation function of tRNA^Phe labelled with tetramethylrhodamine
[2x10^-9 M] (data from Rigler and Grasselli, 1980). (B) Fluorescence
photobleaching recovery of the same tRNA [1x10^-6 M]. (C) Diffusion
constant of tRNA^Phe measured by FCS (O), FPR (●) and by fluorescence depolarization (△) (data from Ehrenberg et al., 1979).

FCS can be used for studying rotational motion in a virtually
unlimited time range since it is independent of the lifetime of the
excited state (Ehrenberg and Rigler, 1974, 1976; Rigler and Ehrenberg,
1976). An example for the possibility of measuring low frequency
rotations has been given by Borejdo in the case of muscle fibres
(Borejdo, 1979; Borejdo et al., 1979). The instability of fluorescent
dyes to high laser intensities has, in general, prevented measurements
of rotational motions on shorter time scales. However the use of dye
compounds with high stability (Kändler et al., 1982), as well as the
availability of efficient hardware correlators available on the market
or as laboratory constructions (Sirk et al., 1979; Kask et al., 1979;
Thomas et al., 1983) have opened up the μs to ms time range which was
previously confined to triplet probes (see other contributions in
this volume).

Just as translational diffusion can be investigated by both FCS
and FPR, rotational motion can also be analysed by FPR. The theoretical background and protocols for separating translational and
rotational motion have recently become available (Wegener and Rigler,
1984).

Finally, the usefulness of FCS in the analysis of fluctuations
around chemical equilibria has been demonstrated by the work of Elson
and collaborators (Elson and Magde, 1974; Icenogle and Elson, 1983a,
b).

Cross correlation scattering

Number fluctuations can also be observed from the fluctuation intensity of scattered light, provided the coherent part usually prevailing
can be reduced. As evident from Equation 3, this can be achieved by
increasing the number of coherence areas with appropriate optical
arrangements as well as by cross correlating the intensity fluctuation
of the same volume element scattered in different directions (Griffith
and Pusey, 1979; Kam and Rigler, 1982).

Given appropriate dimensions of the volume element illuminated by
the laser beam translational and rotational diffusion can usually be
separated. A set-up for cross correlation measurements developed
in our laborabory is shown in Fig. 3 and allows measurement of.
rotational motion and angular dependence. As an example, the cross

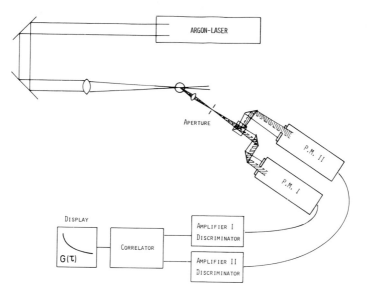

FIG. 3   Set-up for cross correlation laser light scattering.

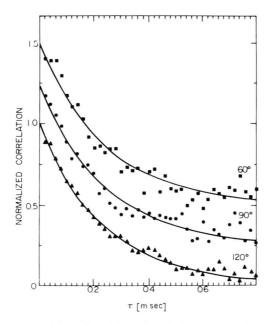

FIG. 4   Cross correlation function for tobacco mosaic virus-DNA at various scattering angles (Kam and Rigler, 1982).

correlation function due to rotational motion of tobacco mosaic virus around its short axis of symmetry is given (Fig. 4) which is independent of the angle of observation in contrast to flexible polymers like DNA (Kam and Rigler, 1982).

The advantage of being able to discriminate rotational components from translational ones is evident. A case investigated by us with a close linkage between rotational and translational motion is the swimming motion of human spermatozoa (Rigler and Thyberg, 1984).

Time-resolved intensity correlations

The ease and sensitivity with which intensity correlations can be measured in short time intervals has prompted us to use correlation spectroscopy as a tool for investigating conformational dynamics in biopolymers. The idea of this approach is to collect correlation functions in successive time intervals after initiation of a reaction leading to conformational transitions (Fig. 5).

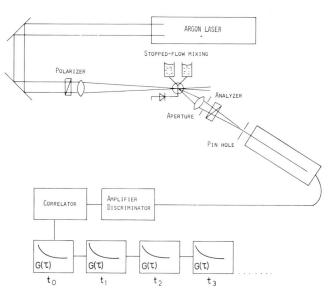

FIG. 5 Set-up for DLS experiments with stopped-flow mixing and time-resolved storage of autocorrelation functions.

FIG. 6A Stopped-flow cell with thermal isolation for low temperature work.

FIG. 6B Functional scheme of stopped-flow cell with pressurized sample compartments and electromagnetic exit valve.

FIG. 6A

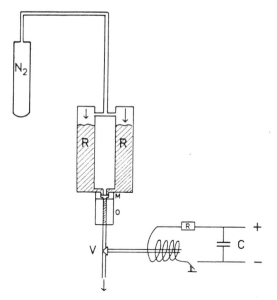

FIG. 6B

In the set-up developed by us, the reaction is initiated in a stopped-flow cell which can be operated by remote control. For short time intervals, with insufficient photon statistics of the auto-correlation curve, multiple shots are superimposed. Stopped-flow mixing in the cell (Figs. 6a,b) is initiated and repeated under the control of a microprocessor which also stores and averages the time sequence of autocorrelation functions.

For tRNA the translational diffusion coefficient, which changes due to conformational transitions, can be measured with a precision of a few per cent in a time interval of 100 ms and at a concentration of 1 mg/ml (Fig. 7). At present we are investigating the time dependence of conformational transitions in tRNA induced by the interaction of the specific codon with the anticodon sequence (Nilsson *et al.*, 1982).

Another case where time-resolved correlation spectroscopy has proven to be very informative is in the polymerization of fibrinogen into fibrinogen polymers catalysed by thrombin (Larsson *et al.*, 1983, 1984). After initiation of the polymerization by thrombin,

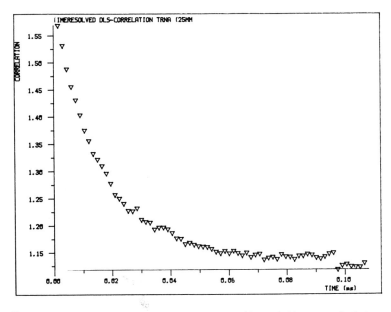

FIG. 7 Autocorrelation function of tRNA (1 mg/ml) recorded in a time interval of 100 ms with 20 mM $MgCl_2$ with the stopped-flow cell. 13 shots superimposed and averaged.

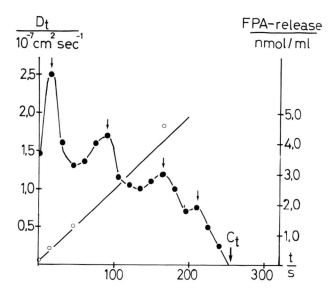

FIG. 8 Time dependence of diffusion constant D_t of fibrinogen during polymerization of fibrinogen. The polymerization is monitored by the release of fibrinopeptide A (Larsson *et al.*, 1983).

fibrinogen undergoes significant conformational transitions as judged by the fluctuations observed in the diffusion constant in the time interval before gelation occurs (Fig. 8).

The possibilities provided by time-resolved intensity correlation measurements are closely comparable to those offered by time-resolved X-ray scattering using synchrotron radiation (Stuhrmann, 1982). In our experience, intensity correlation spectroscopy appears to be at least as sensitive and informative.

ACKNOWLEDGEMENTS

The instrumental development was supported by grants from the K. and A. Wallenberg Foundation and Swedish Natural Science Research Council.

REFERENCES

Andries, C., Guedens, W. and Clauwaert, J. (1983). *Biophys. J.* <u>43</u>, 345-354.
Aragon, S.R. and Pecora, R. (1975). *Biopolymers* <u>14</u>, 119-138.
Asai, H. and Ando, T. (1976). *J. Phys. Soc. Japan* <u>40</u>, 1527-1528.
Berne, B.J. and Pecora, R. (1976). *In* "Dynamic Light Scattering", Wiley, New York.
Borejdo, J. (1979). *Biopolymers* <u>18</u>, 2807-2820.

Borejdo, J., Putnam, S. and Morales, M.F. (1979). *Proc. Natl. Acad. Sci. USA* 76, 6346-6350.
Caroli, C. and Parodi, O. (1969). *J. Phys. B.* 2, 1229-1234.
Chu, B. (1979). *Physica Scripta* 19, 458-470.
Cummins, H.Z., Knable, N. and Yeh, Y. (1964). *Phys. Rev. Lett.* 12, 150-153.
Cummins, H.Z. and Pike, E.R. (1974). *In* "Photon Correlation and Light Beating Spectroscopy". Nato Advanced Study Institute, Series B, Vol. 3, Plenum Press, New York.
Cummins, H.Z. and Pike, E.R. (1977). *In* "Photon Correlation and Velocimetry". Nato Advanced Study Institute, Series B, Vol. 23, Plenum Press, New York.
Ehrenberg, M. and Rigler, R. (1972). *Chem. Phys. Lett.* 14, 539-544.
Ehrenberg, M. and Rigler, R. (1974). *Chem. Phys.* 4, 390-401.
Ehrenberg, M. and Rigler, R. (1976). *Quart. Rev. Biophys.* 9, 69-81.
Ehrenberg, M., Rigler, R. and Wintermeyer, W. (1979). *Biochemistry* 18, 4588-4599.
Elson, E.L. and Magde, D. (1974). *Biopolymers* 13, 1-27.
Elson, E.L., Schlessinger, J., Koppel, D.E., Axelrod, D. and Webb, W. W. (1976). *In* "Membranes and Neoplasia: New Approaches and Strategies" (Ed V.T. Marchesi), P. 137. Alan R. Liss, Inc., New York
Favro, L.D. (1960). *Phys. Rev.* 119, 53-62.
Ford, N.C.,Jr. and Benedek, G.B. (1965). *Phys. Rev. Lett.* 15, 649-653.
Griffith, W.G. and Pusey, P.N. (1979). *Phys. Rev. Lett.* 43, 1100-1104.
Icenogle, R.D. and Elson, E.L. (1983a). *Biopolymers* 22, 1919-1948.
Icenogle, R.D. and Elson, E.L. (1983b). *Biopolymers* 22, 1949-1966.
Kam, Z. (1977). *Macromolecules* 10, 927-934.
Kam, Z. and Rigler, R. (1982). *Biophys. J.* 39, 7-13.
Kask, P., Kändler, T., Sirk, A., Karu, T. and Lippmaa, E. (1979). *Füüs. Matem.* 28, 221-226.
Kändler, T., Kask, P., Piksarv, Sirk, A. and Lippmaa, E. (1982). *Füüs. Matem.* 31, 314-320.
Larsson, U., Blomback, B., Rigler, R. and Nilsson, L. (1983). *In* Abstracts, IX International Congress on Thrombosis and Haemostasis, p. 355, F.K. Schattauer Verlag, Stuttgart.
Larsson, U., Blomback, B., Rigler, R. and Nilsson, L. (1984). Submitted for publication.
Magde, D., Elson, E.L. and Webb, W.W. (1972). *Phys. Rev. Lett.* 29, 705-708.
Magde, D., Elson, E.L. and Webb, W.W. (1974). *Biopolymers* 13, 29-61.
Magde, D., Webb, W.W. and Elson, E.L. (1978). *Biopolymers* 17, 361-376.
Nilsson, L., Rigler, R. and Laggner, P. (1982). *Proc. Natl. Acad. Sci. USA* 79, 5891-5895.
Pecora, R. (1968a). *J. Chem. Phys.* 48, 4126-4128.
Pecora, R. (1968b). *J. Chem. Phys.* 49, 1036-1043.
Plumbridge, J.A., Baumert, H.G., Ehrenberg, M. and Rigler, R. (1980). *Nucleic Acids Res.* 8, 827-843.
Rigler, R. and Ehrenberg, M. (1973). *Quart. Rev. Biophys.* 6, 139-199.
Rigler, R. and Ehrenberg, M. (1976). *Quart. Rev. Biophys.* 9, 1-19.
Rigler, R. and Grasselli, P. (1980). *In* "Lasers in Biology and Medicine" (Eds F. Hillenkamp, R. Pratesi and C.A. Sacchi) pp. 151-164. Plenum Press, New York.

Rigler, R., Grasselli, P. and Ehrenberg, M. (1979). *Physica Scripta* 19, 486-490.

Rigler, R. and Thyberg, P. (1984). *Cytometry* 5, 327-332

Rigler, R. and Wintermeyer, W. (1983). *Ann. Rev. Biophys. Bioeng.* 13, 475-505.

Schaefer, D.W. (1973). *Science* 180, 1293-1295.

Schmitz, K.S. and Schurr, J.M. (1973). *Biopolymers* 12, 1543-1564.

Schneider, M.B. and Webb, W.W. (1981). *Appl. Optics* 20, 1382-1388.

Sirk, A., Kask, P., Kändler, T., Karu, T., Puskar, J. and Lippmaa, E. (1979). *Füüs Matem.* 28, 227-232.

Sorscher, S.M. and Klein, M.P. (1980). *Rev. Sci. Instrum.* 51, 98-102.

Stuhrmann, H.B. (1982). *In* "Uses of Synchrotron Radiation in Biology" Academic Press, London.

Thomas, J.C., Lum, Y.T. and Kennett, D. (1983). *Rev. Sci. Instrum.* 54, 1346-1355.

van Winter, C. (1954). *Physica* 20, 274-292.

Wada, A., Ford, N.C. and Karascz, F.E. (1971). *J. Chem. Phys.* 55, 1798-1802.

Wegener, W.A. and Rigler, R. (1984). *Biophys. J.* In press.

Weissman, M.B. (1981). *Ann. Rev. Phys. Chem.* 32, 205-232.

Triplet-state Spectroscopy for Investigating Diffusion and Chemical Kinetics

A.F. CORIN, E.D. MATAYOSHI and T.M. JOVIN

INTRODUCTION

Present concepts in cytology depict the biological cell as a complex system of chemical fluctuations which include a vast number of conformational, rotational, and translational motions in biomacromolecular structures. An understanding of these dynamic processes is crucial for forming a complete picture of the normal and abnormal functional status exhibited by this basic unit of living tissue. Nucleic acid, protein and lipid components interact to form the complex supra-molecular assemblies constituting chromatin and membranes. These structures undergo motions on time-scales ranging from picoseconds to hours. Numerous spectroscopic techniques have been employed to probe the nature and time course of these processes. One such tool is luminescence. Fluorescence techniques cover a range of 1-100 nanoseconds while methods based on delayed luminescence, which exploit the triplet state, make the microsecond and millisecond time ranges accessible. Examples illustrating the latter are drawn from current research in our laboratory concerning the dynamics of nucleic acid structures and membrane associated proteins.

As the cell is not an isolated system but must interact extensively with its environment, it has developed mechanisms for regulating the flow of information and nutrients to the inside and the release of products and information to the outside. Such functions are performed

SPECTROSCOPY AND THE DYNAMICS
OF MOLECULAR BIOLOGICAL SYSTEMS

by numerous membrane-associated protein systems. The use of lumines-
cent probes covalently or noncovalently attached to the macromolecular
structures in question provides information on the size, shape,
translational and rotational behaviour as well as the state of aggregation. Most rotational motions observed for such systems are found
to occur in the microsecond to second domain as a result of the high
viscosity of the lipid bilayer medium (see chapters in this volume
by Cherry and by Garland and Johnson; also Matayoshi *et al.*, 1983),
associations with various cytoskeletal elements which anchor membrane
structures, and other protein-protein and protein-lipid interactions.
Intrinsic probes such as tryptophan and tyrosine, and extrinsic labels
conjugated to proteins or to specific ligands (e.g. hormones, toxins)
can be used to detect local or domain flexibility as well as global
motions. Similarly, the interactions of various planar organic
molecules exhibiting both fluorescence and delayed luminescence with
polynucleotide polymers can be employed to probe the structure and
solution dynamics of nucleic acids. In addition, as demonstrated
here with preliminary data, the chemical kinetics of drug-DNA interactions can be explored as a result of temporal coupling with photophysical decay processes.

PHOTOPHYSICAL PRINCIPLES AND INSTRUMENTATION

The luminescence of organic molecules is the result of photophysical
relaxation processes from the excited singlet S_1 and triplet T_1
states which are formed rapidly by internal conversion and intersystem crossing after the absorption of light. A simple representation
of these events is given by the scheme shown in Fig. 1. Following
the absorption of a photon a chromophore is excited from the ground
singlet state S_0 to a more energetic electronic state, from which it
decays by nonradiative and radiative processes. The latter include
direct deactivation (prompt fluorescence) and delayed luminescence
(phosphorescence and delayed fluorescence). The transition from the
excited singlet to the lower energy triplet state (intersystem crossing) is formally forbidden as it requires a change in spin multipli-
city. However, spin orbital coupling, which can be greatly enhanced
by the introduction of internal or external heavy atoms, increases
the probability of intersystem crossing which in turn increases the

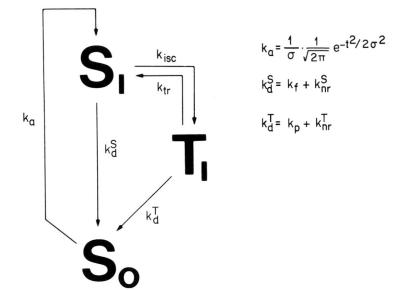

$$k_a = \frac{1}{\sigma} \cdot \frac{1}{\sqrt{2\pi}} e^{-t^2/2\sigma^2}$$

$$k_d^S = k_f + k_{nr}^S$$

$$k_d^T = k_p + k_{nr}^T$$

FIG. 1 Electronic transitions between the singlet ground state S_0 and the first excited singlet S_1 and triplet T_1 states. The rate constant for absorption k_a is expressed as a Gaussian profile of an exciting light pulse. The decay from S_1 (k_d^S) and T_1 (k_d^T) is represented as the sum of radiative (f - fluorescence, p - phosphorescence) and nonradiative (generally multicomponent) pathways (k_{nr}^S, k_{nr}^T). Intersystem crossing from S_1 to T_1 is denoted by the composite rate constant k_{isc} and the reverse reaction (thermal reactivation of T_1 and intersystem crossing) by k_{tr}. (Reproduced with permission from Jovin et al., 1981).

triplet population. Thermal reactivation from T_1 to S_1 followed by subsequent radiative decay results in delayed fluorescence having an energy spectrum identical to that of the prompt fluorescence, but a lifetime characteristic of the triplet state. Deactivation of the excited state by prompt fluorescence is usually efficient and thus fast (typically ns) whereas emission from T_1 is slower (µs to s) and, therefore, must compete with numerous nonradiative processes. The latter events, which are generally dominant in fluid media (Turro et al., 1978), include thermal deactivation upon collision with solvent molecules and quenching upon collision with paramagnetic molecules such as ground state O_2 (hence, oxygen purging is usually a prerequisite to observation of delayed emission). The longer lifetimes found for delayed luminescence renders this form of radiative decay useful for studying systems via traditional emission techniques (decay kinetics, depolarization) in the time domain >1 μs (Austin *et al.*, 1979; Moore *et al.*, 1979).

Decay laws

As previously demonstrated (Jovin *et al.*, 1981) the various deactivation processes shown in Fig. 1 lead to the following expressions for the lifetimes τ_1 and τ_2 of prompt and delayed emission:

$$\tau_1^{-1} = \lambda_1 = k_d^{\ s} + k_{isc}$$

$$\tau_2^{-1} = \lambda_2 = k_{nr}^{\ T} + k_p + k_{tr} (1 - \phi_{isc})$$

(1)

where the kinetic rate constants are defined in Fig. 1 and expressions for the quantum yields are given below in Equation (5). At all times t,

$$\text{fluorescence intensity} \propto k_f \, S_1(t)$$

$$\text{phosphorescence intensity} \propto k_p T_1(t)$$

(2)

where the proportionality factors account for the chemical concentrations of ground state species, extinction coefficients, and instrumental parameters (photomultiplier spectral sensitivity, optical collection efficiency, polarization bias, electronic bandwidth and amplification) that determine the absolute and relative signal intensities. Luminescence intensities are proportional to the excitation light intensity provided that the system is pumped well below saturation.

The ratio T_1/S_1 increases rapidly with the product $\tau_1^{-1}\sigma$ (Jovin *et al.*, 1981), where σ reflects the width of the Gaussian excitation pulse (Fig. 1). It follows that the ratio of the triplet to the singlet population can be increased by a "long" excitation pulse. The resulting reduction in the prompt fluorescence relieves problems caused by insufficiently effective gating as well as enhancing the signal by pumping the triplet state, an option currently being explored in our laboratory using a gated CW argon ion laser. If the pulse duration is chosen to be of the order of the triplet lifetime then it is necessary to deconvolute the exciting pulse from the delayed emission response function. However, for t >> σ the delayed emission is monoexponential and can be described by the simple scheme:

$$S_1(t) \cong S_d \exp[-\lambda_2 t] \qquad T_1(t) \cong \phi_{isc} \exp[-\lambda_2 t] \qquad (3)$$

Thus, for the delayed emission:

$$\text{fluorescence/phosphorescence} \propto [\phi_f k_{tr}/k_p] \qquad (4)$$

$$\text{time-integrated phosphorescence} \propto [\phi_p' \phi_{isc}/(1 - \phi_{tr}' \phi_{isc})]$$

where the following definitions apply for the previous expressions:

$$
\begin{aligned}
S_d &= k_{tr}/k_{isc} \; \phi_{isc} \\
\phi_f &= k_f/(k_f + k_{nr}^S + k_{isc}) = k_f \tau_1 \\
\phi_{isc} &= k_{isc} \tau_1 \\
\phi_{tr}' &= k_{tr}/(k_d^T + k_{tr})
\end{aligned}
\qquad (5)
$$

The primed quantum yields refer only to those segments of the deactivation pathways indicated in Fig. 1, not to the entire process beginning with the absorption event.

The triplet state can also be exploited by monitoring the absorption properties of the singlet or triplet state (see chapter in this volume by Cherry; Kawato and Kinosita, 1981), generally with greater time resolution than can be achieved with delayed luminescence (Hogan et al., 1982; Austin et al., 1983). The photoselected depletion of the ground state can also be measured by the polarized prompt fluorescence: "fluorescence depletion" (see chapter in this volume by Garland and Johnson).

Anisotropic Emission

Excitation of a random distribution of absorbing chromophores with linearly polarized light generates a transient population of spatially anisotropic excited state molecules. The two polarized components (parallel, $I_{||}$ and perpendicular, $I_{\perp}$) of the resulting luminescence

define the emission anisotropy r(t), which for a time-resolved
measurement is given by:

$$r(t) = [I_{||}(t) - I_{\perp}(t)]/[I_{||}(t) + 2I_{\perp}(t)] \qquad (6)$$

The numerator represents the difference function D(t), and the denominator represents the total emission intensity S(t). For a system
containing several emitting species each characterized by an excited-
state lifetime τ_i

$$S_i(t) = S_{oi}\exp[-t/\tau_i] \qquad (7)$$

$$r_i(t) = \sum_j \alpha_{ij}\exp[-t/\phi_{ij}] \qquad (8)$$

$$r(t) \equiv <r(t)> = \sum_i S_i(t) \cdot r_i(t)/\sum_i S_i(t) \qquad (9)$$

As is indicated in Equation (8), the anisotropy function for a single
chemical component can be expressed as a sum of exponential decays.
The relaxation of a rigid rotating spherical molecule in an isotropic
medium is monoexponential, but an ellipsoid of revolution requires
three and an irregular body five exponential terms (Rigler and Ehrenberg, 1973). More complex expressions for $r_i(t)$ are obtained if
internal flexibility must be considered, or if the surrounding medium
is anisotropic. For nonspherical rigid bodies, the rotational correlation times ϕ_{ij} consist of linear combinations of the rotational
diffusion constants for motions about different molecular axes.
Equations (7) - (9) make it clear than even for systems consisting
only of spheres, in which n>1 chemical species with n lifetimes and
n diffusion constants are present, the decay of r(t) can be complex
and non-exponential.

Although expressions have been derived for various models (see
references cited in Jovin *et al.*, 1981; Austin *et al.*, 1983), real
experimental data are limited in resolution and the anisotropy decay
is generally analyzed simply by:

$$r(t) = \sum_i r_{oi}\exp[-t/\phi_i] + r_\infty \qquad (10)$$

where i≤2. Depending on what is known about the complexity of the
biochemical system under study, the obtained ϕ_i can sometimes be
interpreted as arising either from individual chemical species, or
from the exponential terms of the expression describing diffusion of
a single species. In the former case the coefficients r_{oi} will
reflect relative population sizes as well as the fast (submicrosecond)
relaxations due to local motion of the probe, whereas in the latter
the r_{oi} can further be interpreted in terms of the orientations of
the absorption and emission transition dipoles relative to the rotation axis, according to the particular model used. The constant term
r_∞ is nonzero if the rotational motions are restricted in space. It
can, for example, be expressed in terms of the familiar order parameter S used in magnetic resonance spectroscopy, or in the model of
a rigid body undergoing restricted diffusion (wobbling) within a
cone, it can be related to the cone angle (Kinosita et al., 1977;
see discussion in Jovin et al., 1981).

Triplet state spectrometer

Triplet-state spectroscopic measurements in our laboratory are performed with a versatile spectrometer the essential features of which
have been described in previous publications (Austin et al., 1979;
Jovin et al., 1981; Matayoshi et al., 1983). A schematic of the most
recent version of the instrument is presented in Fig. 2. A number
of experimental arrangements for the instrument are shown which permit easy conversion among several modes of operation: linear dichroism, fluorescence-detected ground-state depletion, observation of the
parallel and perpendicular polarized components of emission decay,
and collection of time-dependent polarized emission and excitation
spectra. A nitrogen-pumped tunable dye laser providing vertically
polarized light pulses of 3-5 ns FWHM is used to excite luminescence
which is collected at right angles to the excitation source. The two
polarized components of the emission are recorded by rotating a sheet
polarizer $90°$ after a preselected number of laser flashes. The EMI
9817 QCB photomultiplier is electronically gated off for up to 2 µs
after the laser pulse in order to suppress any intense prompt fluorescence. Optical filters or a monochromator can be inserted on the
emission side. Data records consisting of up to 2048 channels of

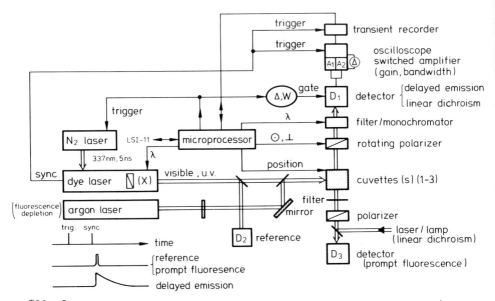

FIG. 2 Schematic of the triplet-state spectrometer. Symbols: ⇑,
light path; ↑, electronic path; Δ, adjustable delay; W, variable
width; polarizer (⊙, ⊥: parallel and perpendicular orientations);
(X), frequency doubling crystal (optional); λ, wavelength. Notes
on components: amplifier/filter-real time switch (see text for more
detail); transient recorder, three models with maximal sampling frequencies of 2–100 MHz (Biomation 802, 1010, 8100); computer (LSI-11,
Digital Equipment Corp.); microprocessor (Intel 8085); D_1-D_3, detectors (photomultipliers or photodiodes).

parallel and perpendicular polarized components of the emitted light
are time averaged by strobing the contents of a Biomation model 8100
or model 1010 transient recorder into a microprocessor after each
laser flash. Successive transients are summed in the microprocessor
and the total is transferred to a DEC LSI-11/23 computer for subsequent data editing and analysis. The data editor software package
allows one to perform numerous arithmetic operations on the digitized
data records such as background subtraction using a pretrigger portion
of the data record or a separate blank record, and construction of the
time dependent functions: S(t), D(t) and r(t). Nonlinear regression
analysis of all three functions (allowing weighted fits of up to
four exponentials plus a constant on a maximum of 2048 data points)
is according to procedures and computer programs developed by G.
Striker (1982) and E. Matayoshi. The goodness of fit is determined
by inspection of plots of (1) the experimental data points overlayed

with the fit, (2) the residuals, and, if necessary, (3) the autocorrelation of the residuals.

For collection of transient decay curves, the output current of the photoamplifier is directed into one of two amplifier plug-ins (model 7A22 or 7A15A) of a Tektronix model 7603 oscilloscope. In addition, a simple modification to the oscilloscope allows the automatic transfer of signal processing from one vertical amplifier to the other at any designated time following the excitation pulse. This feature permits an increase in gain and a decrease in bandwidth, thereby effecting enhancement of the signal quality during the later stages of the transient decay process. It is particularly useful for rescuing the late time course of an anisotropy decay function as illustrated in Fig. 3. Figure 3D depicts the anisotropy decay of eosin in a D,L-arabinose glass: after ∿3 ms the signal-to-noise ratio is so small as to obscure the time course of the anisotropy decay. Figures 3A and 3C show the same experiment using the dual gain option; after ∿1 ms the signal was transferred from an amplification setting of 1 millivolt/division and a bandwidth of 10 MHz to a channel with ten-fold greater gain (0.1 millivolt/division) and an electronic filter of 3 KHz. With the increased amplification and filtering, the later stages of the anisotropy decay of the immobilized eosin (i.e. after 1 ms) can plainly be seen to continue on its downward excursion.

Time-dependent polarized luminescence spectra can also be obtained. The emission is scanned with a monochromator and signal averaging at each step is conducted by summing successive integrated transients acquired after each laser pulse. The integration can be performed between any specified channels so that the transient decay can be divided into any desired number of such consecutive time windows and stored in the microprocessor.

DYNAMICS OF MEMBRANE-ASSOCIATED PROTEINS

The rotational motions of intrinsic membrane proteins have been studied using both intrinsic and extrinsic probes. The former include cofactors and visual pigments, and the latter consist of chromophores conjugated either directly to the macromolecule in question or

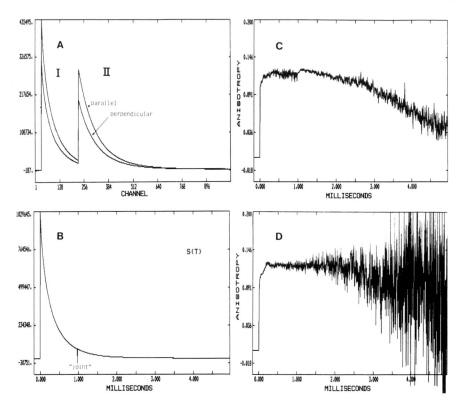

FIG. 3 Dual gated amplifier detection mode. Eosin immobilized in
D,L-arabinose glass. (A) Parallel and Perpendicular components of
the emitted light, 5 μs/channel decay: part I, amplification 1 mv/cm,
bandwidth 10 Mz; and part II, amplification 0.1 mv/cm, bandwidth
3 KHz; (B), (C) Total intensity and anisotropy decay respectively -
calculated from the raw data in (A). The gains in the S(t) curve
have been normalized. (D) Anisotropy decay calculated from raw taken
in the single amplifier mode, amplification 1 mv/cm, bandwidth 10 MHz.

indirectly via coupling to a specific ligand, such as a toxin or an antibody. Numerous studies illustrating the use of extrinsic probes are listed in Table 1. In general, one observes:

  (a) rapid depolarizing motions below the time range of the triplet measurements (local probe and segmental protein motions);

  (b) a monoexponential or multiexponential decay of anisotropy in the μs-ms domain;

  (c) a finite limiting anisotropy;

  (d) changes in these parameters depending upon time-dependent chemical reactions such as binding or clustering.

These characteristics have been interpreted in terms of rotational motions hindered by the anisotropic structure of the lipid bilayer, which favours uniaxial over librational (wobbling) movements (see chapter in this volume by Cherry). Formalisms for the quantitative analysis of these two limiting cases are available (Saffmann and Delbruck, 1975; and Kawato and Kinosita, 1981, respectively).

DNA-LIGAND INTERACTIONS: PROTEIN-DNA COMPLEXES

Anisotropy decay of molecules which twist and bend (DNA)

Biological macromolecules can be regarded as rigid structures only as a first approximation, an unfortunate circumstance in view of the fact that most theoretical treatments of rotational diffusion (Garcia de al Torre and Bloomfield, 1981) do not incorporate provisions for segmental flexibility. Double-helical DNA constitutes a particularly good example of a molecule which, depending upon molecular weight (length), exhibits at one extreme the hydrodynamic properties of a rigid rod and at the other extreme that of a flexible coil (i.e. short segments versus long viral or genomic DNA). In addition to the end-over-end and spinning rotational motions, twisting and bending of the DNA helix also contribute to rotational depolarization. These

TABLE 1 Membrane-associated systems studied in our laboratory by time-resolved phosphorescence anisotropy

Membrane protein	Probe	System	Correlation times	Temp. (°C)	References
Concanavalin A receptors	eosin	Friend-virus-transformed cells	immobile	4-37	Austin et al.(1979)
Band 3, anion transport protein	eosin	erythrocyte ghosts	26; 203 μs; 1.0, >10	4-37	Matayoshi et al.(1983; 1984; in preparation
Glycophorin A	erythrosin	reconstituted into lipid vesicles (dimyristoyl-phosphatidylcholine)	1-2 μs; 10-30 μs	10-34	Jovin et al. (1981); Bartholdi et al., Matayoshin et al, in prepn
IgE receptor	erythrosin	basophil carcinoma cells (2H3)	ca. 10 μs	25	Zidovetzki et al., in preparation
AChR, acetylcholine receptor	eosin	postsynaptic membranes (Torpedo marmorata)	(20, 26; 13)* μs	20; 39	Bartholdi et al. (1981)
	eosin	reconstituted into lipid vesicles (soybean, DMPC)	47-56 μs	30	Criado et al., unpublished data
EGF (epidermal growth factor) receptor	erythrosin	epidermoid carcinoma cells (A-431)	25-90 μs; 200-350 μs	4; 37	Zidovetzki et al. (1981)
		membrane fragments	20 μs; 90 μs	4; 37	Matayoshi et al. (1983)
$(Na^+ + K^+)$-ATPase	eosin	membranes (Squalus acanthias)	immobile	37	Matayoshi et al. (1983)
H-2^k histocompatibility antigen	eosin	lymphoma cells (T-41)	15-30 μs	25	Damjanovich et al. (1983)

*after reduction with dithiothreitol or alkaline extraction; otherwise no anisotropy decay up to 0.5 ms.

combined effects lead to non-exponential decay modes, which have been
the subject of intense theoretical (e.g. Barkley and Zimm, 1979) and
experimental study. The latter include triplet-state depolarization
by linear dichroism (Hogan et al., 1982). Challenges to the experi-
mentalist increase in the case of protein-nucleic acid complexes,
particularly those involving the regulatory and enzymatic factors
responsible for genetic expression and replication.

The methods described in this chapter have been applied to two
systems in which the hydrodynamic properties of DNA-protein complexes
were revealed by anisotropic rotational behaviour in the μs and sub-
μs time domains. In one case, the disposition of DNA in the nucleo-
some was investigated (Wang et al., 1982). In another study, the
complexes of Escherichia coli RNA polymerase and DNA were examined
using Rose Bengal noncovalently bound to the protein (Austin et al.
1983). The free enzyme exhibited anisotropic rotation in linear
dichroism measurements with rotational correlation times of 2 μs and
about 0.7 μs, corresponding to the dimeric (0.1 M NaCl) and monomeric
(1 M NaCl) species (Fig. 4A). Binding of the polymerase to the synthetic polynucleotide poly[d(A-T)] or calf thymus DNA led to a slower,
more complex depolarization with at least two components, one about
0.5 μs and the other 10-20 μs (Fig. 4B). Similar results were
obtained when the covalently bound probe eosin isothiocyanate was
employed. The slower component was interpreted to arise from the
overall tumbling of the complex which was dominated by the DNA, while
the faster time was proposed to report the spinning motions about the
long axis of the DNA. The dependence of the decay processes on DNA
chain length, protein concentration, and temperature have been
examined and discussed in connection with the possible influences of
conformational changes, sliding motions of the protein along the DNA,
and the torsional-bending deformations predicted by an elastic filamentous model of DNA.

Coupling of photophysical decay and chemical kinetics: Acridine-DNA complex

The interaction of the acridine dyes with DNA has been the subject of
numerous investigations as these dyes exhibit a variety of biologically significant properties. These planar chromophores demonstrate

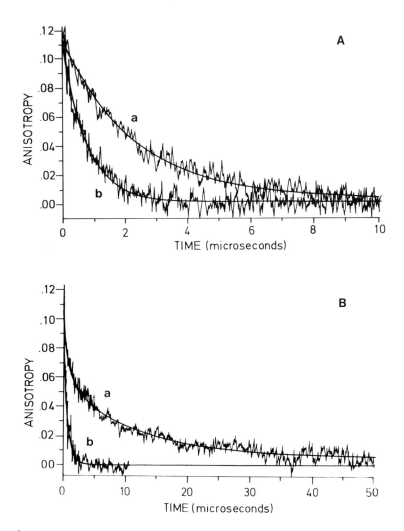

FIG. 4 Anisotropy decay of *E. coli* RNA polymerase (A) free, and (B) bound to poly[d(A-T)] . The probe was Rose Bengal bound to the enzyme and the signal was derived from the linear dichroism of the triplet-triplet absorption. a) 0.1 M NaCl. b) 1 M NaCl. The high salt concentration monomerizes the enzyme and disrupts the complex with DNA. Temperature = 5°C. (Data from Austin *et al.*, 1983).

antibacterial and mutagenic activity and their binding properties to polynucleotide polymers serve as models for the interaction with DNA of other more complex biologically active molecules such as certain carcinogens, antibiotics or proteins (Waring, 1981). The structures of two members of this class of dyes, proflavin and acridine orange, are shown below.

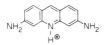

Proflavin cation

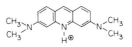

Acridine orange cation

A number of three-dimensional structures of the acridine dyes co-
crystallized with self-complementary dinucleotide mono- and diphos-
phates have shed light on the nature of the intercalating and nonintercalating interactions of such drugs bound to these mini-helices
(reviewed by Neidle and Berman, 1983). However, it might be dangerous
to generalize from the structural information found for the crystallo-
graphic structures of these systems to the interactions in solution
of intercalating agents with much longer DNA polymers. One obvious
difference between the complexes formed with such mini-helices versus
oligonucleotides is that the contribution of "end" effects is far less
important for polymers of >100 base pairs. Such end effects in the
crystal structures entail nonintercalating drug species which can
involve dye-dye and dye-base pair stacking arrangements. An example
of the latter configuration, dye stacked with a base pair, is observed
in the crystal structure of a 3:2 complex of proflavin with the
dinucleoside phosphate cytidylyl-3'5'-guanosine (Neidle et al., 1977).

Another feature of acridine-DNA interactions which is very useful
in cell cytology is the metachromatic staining of single-vs. double-
stranded nucleic acids. Acridine orange (AO) complexed to double-
stranded DNA emits a green fluorescence ($\lambda \sim 520$ nm), while complexation to single-stranded nucleic acids results in a decrease in this
fluorescence and an increase in red luminescence ($\lambda \sim 650$ nm). Saturation of single-stranded nucleic acid with AO results in a precipitate
that exhibits red luminescence. In a model proposed by Kapuscinski
et al. (1982), AO molecules are sandwiched between consecutive nucleo-
tides along the single stranded structure, and at dye-phosphate ratios
(D/P) in the range 0.6 - 1, single-stranded nucleic acids saturated
with dye associate laterally and precipitate. In addition, the
appearance of the long wavelength luminescence is accompanied by a
ten-fold increase in emissive lifetime compared with free dye and a
four-fold increase compared with dye intercalated into double-helical

nucleic acids (Borisova and Tumerman, 1965). As suggested by
Kapuscinksi *et al.* (1982), agglomeration may protect the AO chromophores from frequent collisions with O_2 and solvent, hence permitting
radiative decay which might otherwise be quenched. At lower P/D
ratios, where the concentration of the sandwich intercalation structure is substantial, such red luminescence is not observed.

Li and Crothers (1969), using the temperature jump method with
absorption detection, showed that the binding of proflavin to DNA
exhibits two different types of binding, the properties of which
depend upon the base composition. Subsequent studies on the binding
of proflavin to both natural and synthetic DNA polymers (Ramstein *et
al.*, 1980) have been interpreted by a model that describes intercalation of the drug to poly[d(G-C)] as a simple one-step bimolecular
process and to poly[d(A-T)] as a two-step process in which the proflavin first binds externally and rapidly to the polymer and then
intercalates:

$$\text{proflavin + poly[d(G-C)]} \underset{k_{-1}}{\overset{k_1}{\rightleftharpoons}} C_i \tag{11}$$

$$\text{proflavin + poly[d(A-T)]} \underset{\text{fast}}{\overset{K_1}{\rightleftharpoons}} C_e \underset{k_{-2}}{\overset{k_2}{\rightleftharpoons}} C_i \tag{12}$$

where C_e denotes the complex of DNA and externally bound dye and C_i
refers to the case of the intercalated species. At $17^{\circ}C$ with P/D
>100, and an ionic strength of 0.11, $k_1 = 1.6 \times 10^7$ M^{-1} s^{-1}, $k_{-1} = 170$
s^{-1}, $K_1 = 0.2$ mM, $k_2 = 3.6 \times 10^3$ s^{-1}, and $k_{-2} = 80$ s^{-1}. The relaxation
times in the first case (Equation 11) are >6 ms and in the second
case (Equation 12) span the range 0.3 - 12 ms.

In a manner analogous to a temperature jump we have applied a perturbation to such drug-DNA complexes by applying an instantaneous
light pulse which excites the various proflavin species to their
respective triplet states. The formalism for such a light perturbation technique is briefly dealt with in a review by Rigler and Ehrenberg (1973). Applied to the models in Equations (11) and (12), this
leads to the chemical equilibria (Case I and Case II respectively):

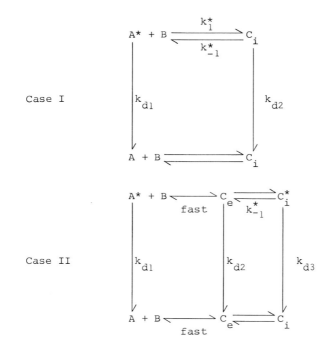

Solutions of the resulting differential equations for the two chemes yield expressions for the relaxation rates as functions of the kinetic rate constants and concentrations of the excited-state species. For direct intercalative binding with no preceding complexation step, Case I:

$$\tau_1^{-1} + \tau_2^{-1} = k_1^*[B] + k_{-1}^* + k_{d1} + k_{d2}$$

$$\tau_1^{-1} \cdot \tau_2^{-1} = k_1^*[B]k_{d2} + k_{d1}(k_{-1}^* + k_{d2})$$

(13)

For intercalation preceded by complex formation in steady state equilibrium, Case II:

$$\tau_1^{-1} + \tau_2^{-1} \quad \{1/(1 + K^*[B])\}\{k_{d1} + K^*[B](k_{d2} + k_1^*)\}$$

$$+ k_{-1}^{'Q} + k_{d3}$$

(14)

$$\tau_1^1 \cdot \tau_2^1 = \{1/(1 + K^*[B])\}\{k_{d1}(k_{-1}^* + k_{d3})$$

$$+ K^*[B](k_1^*k_{d3} + k_{-1}^*k_{d2} + k_{d2}k_{d3})\}$$

where $K*=C*/A*B$ and all k_{di} (i=1,2,3) refer to the rate constants for
radiative decay of the excited-state species from the triplet state
to the ground state. A, C_i and C_e are proflavin species in the excited
(*) and ground (unstarred) states, and B is a binding site on the DNA.
If the rates of chemical re-equilibration are on the same time scale
as the triplet lifetime decay, then the chemical kinetic rates will
be coupled to the rates of luminescence decay from the triplet state.
In addition, the expressions for the relaxation times derived for
each model are relatively simple functions of the DNA concentrations.
In Case I the sum and product of the two reciprocal relaxation times
shown in Equation (13) predict a linear dependence on the DNA concentration, while the analogous expressions for model II (Equation 14)
predict a behaviour which is linear at low DNA concentration and
plateaus at higher concentrations. With a more substantial depletion
of the ground state encountered at higher excitation intensities, the
ground-state and excited-state photophysical and chemical kinetics
are coupled, allowing, in principle, an investigation of the ground-
state equilibria (Rigler and Ehrenberg, 1973).

Both acridine orange and proflavin exhibit room temperature phosphorescence and delayed fluorescence free or bound to DNA in the μs-
ms time range (Parker and Joyce, 1973). Upon binding of proflavin to
poly[d(A-br^{5}U)] (dbr^{5}U is the brominated analogue of thymine), the
phosphorescence intensity measured at 77°K is enhanced over that
observed under identical conditions for binding of the drug to poly
[d(A-T)] (Galley and Purkey, 1972). In addition, at this low temperature, the decay shows two components, one with a lifetime of 2 s and
a second with a lifetime of ∿0.6 s. The latter has been attributed
to proflavin species bound close to the bromine atom. Experimentally,
this heavy atom effect results in a decrease in fluorescence and
phosphorescence lifetimes and an enhancement of the phosphorescence
quantum yield.

We have examined the proflavin-DNA interaction by time-resolved
delayed luminescence in fluid media, i.e. at room temperature. Under
conditions of low salt concentration (10 mM Tris-HCl, 0.1 mM EDTA,
pH 7.0) and at room temperature, proflavin bound to poly[d(A-br^{5}U)]
exhibits delayed luminescence in the μs-ms time domain (Figs. 5 and 6).

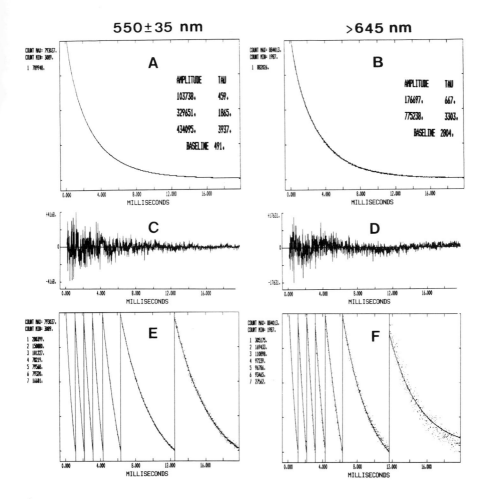

FIG. 5 Total decay curves and analysis for the proflavin (10 μM)-poly[d(A-br⁵U)] (50 μM) complex observed in two regions of the spectrum. Buffer: 10 mM Tris-HCl pH 7.5, 10 mM NaCl, 0.1 mM EDTA. (A),(B) - Superposition of the fit and the total decay of the emitted light (the step-like nature of the data at later times is purely a function of the limited resolution of the graphics display unit). The amplitudes and lifetimes (in μs) of the fit are listed in the right hand corners. The maximum and minimum number of counts of the decay and the total range of counts are listed to the left of the graph. (C),(D) - residuals of the fit - units of the ordinate are the absolute differences between the unweighted fit and the raw data. (E), (F) - expanded display of a superposition of the three component fit (solid line) and the data (dots). The maximum and minimum number of counts and the range of counts for each expansion panel are listed to the left of the graph.

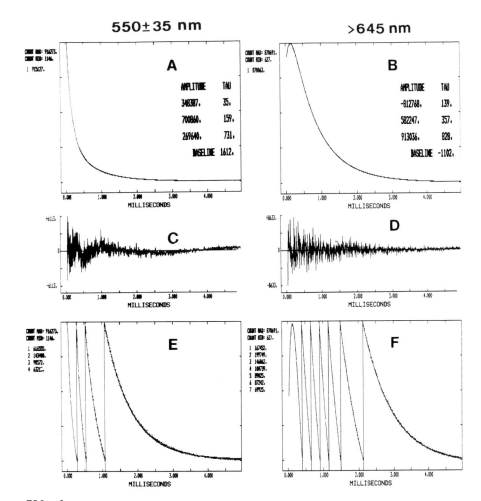

550±35 nm

>645 nm

AMPLITUDE	TAU
3483387,	35,
700860,	159,
269640,	731,
BASELINE	1612,

AMPLITUDE	TAU
-812768,	139,
582247,	357,
913038,	828,
BASELINE	-1102,

FIG. 6 Total decay curves and analyses for the proflavin (0.5 μM) – poly[d(A-br^{5}U)] (50 μM) complex observed in two regions of the spectrum. All other conventions are stated in the legend to Fig. 5.

As shown in Fig. 5B, at a phosphate/drug (P/D) ratio of 5 the emission at λ >645 nm shows a three-component decay. The shortest lifetime (∼140 μs) has a negative amplitude while the two longer lifetimes (0.36 and 0.83 ms) have positive amplitudes. This result demonstrates that there are at least two long-lived excited-state species of proflavin, one being produced from the other. The delayed emission observed through a broad bandpass filter centered at (550 ± 35) nm was also best analysed in terms of a three component decay (Fig. 5A).

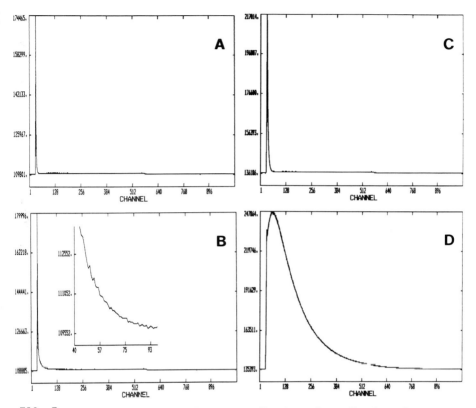

FIG. 7 Effect of O_2 depletion on proflavin and proflavin-poly-[d(A-br^{5}U)] complex. All curves represent 1024 channels at 10 μs/channel of the emission at λ >645 nm. Buffer: as in Fig. 5. (A),(B) - 5 μM proflavin unpurged and argon-purged, respectively (inset in (B) is an expansion of the earlier time channels). (C),(D) - 9.4 μM proflavin and 50 μM DNA unpurged and argon-purged, respectively.

The effect of O_2 quenching on the red emission can be seen for free dye alone in solution (Figs. 7A and 7B) and for the proflavin-DNA complex (Figs. 7C and 7D). Comparison of Figs. 7A and 7B with Fig. 7D shows that any free dye that may exist in the presence of the poly-nucleotide is not expected to contribute appreciably to the red luminescence. Similar results are observed for the luminescence centred about 550 nm (data not shown).

At the higher P/D ratio of 100 (lower proflavin concentration, Fig. 6) the decay times are considerably longer and the negative component at λ >645 nm is no longer observed. A series of experiments

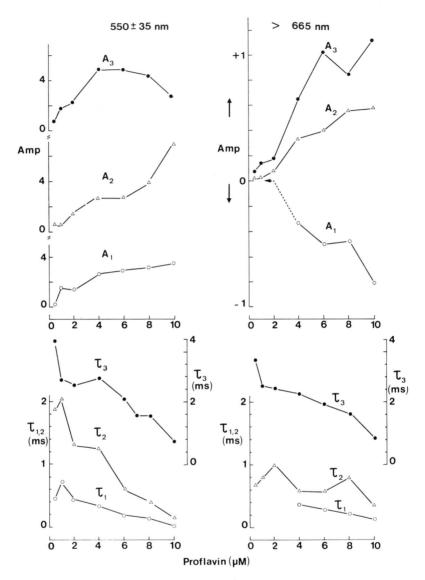

FIG. 8 Dependence of the radiative lifetimes and amplitudes of pro-flavin-poly[d(A-br^{5}U)] complex on the total drug concentration. Concentration of poly[d(A-br^{5}U)] is 50 μM throughout yielding a P/D range of 5 - 100. Data collected at (550 ± 35) nm were taken with the combination of a Corion BB 5500 broad band pass filter and a Schott glass KV470 cutoff filter, and data at λ >645 nm were collected with a Schott glass KV550 cutoff filter in combination with a long wavelength pass RG645 filter. A_i and τ_i represent relative amplitudes and decay times respectively. The amplitudes at >645 nm have not been corrected relative to measurements at 550 nm for the considerably reduced instrumental sensitivity. Buffer: as in legend to Fig. 5. The solid lines have no theoretical significance.

exploring the effect of drug concentration are summarized in Fig. 8. In general, as the concentration of proflavin is reduced, the decay times increase and the amplitudes decrease. A detailed interpretation of these and related data will be presented elsewhere. We restrict ourselves here to some qualitative observations. The initial rise in the red emissive decay suggests that a chemical kinetic process occurs on the time scale of the triplet lifetime, as is further supported by the concentration dependencies of the lifetimes on the total dye concentration as seen in Fig. 8. In order to explain a negative amplitude component we propose a dynamic exchange between intercalated dye and stacked dye located on the outside of the helix. For example, consider the case of a proflavin molecule bound close to a bromine atom and another bound at a site too far away to experience the heavy atom effect ($>$ ~5 $\overset{\circ}{A}$). Since it is likely that the emission quantum yield differs between the two environments, exchange between the two sites during the lifetime of the excited state could result in an initial increase in the phosphorescence. The wavelength dependence of all three time constants and accompanying amplitudes at the low salt concentration with a P/D of 5 is shown in Fig. 9. The variations with wavelength demonstrate the existence of several species and possibly even more complex kinetic schemes than those presented in Cases I and II above. The negative amplitudes observed for proflavin at early times in the luminescence decay are also found in the binding of acridine orange to poly[d(A-br^{5}U)] in 0.16 M salt (Fig. 10).

A complete understanding of the luminescence decay behaviour observed for the drug-DNA complexes such as those presented here clearly requires a more detailed characterization of the physical and chemical properties of the system including equilibrium states and kinetic pathways. We are currently conducting more detailed investigations on proflavin and acridine orange complexed to poly[d(A-br^{5}U)] as well as to other heavy atom substituted polymers such as: poly-[d(G-br^{5}C)], poly[d(G-io^{5}C)] (Malfoy et al., 1982; McIntosh et al.,

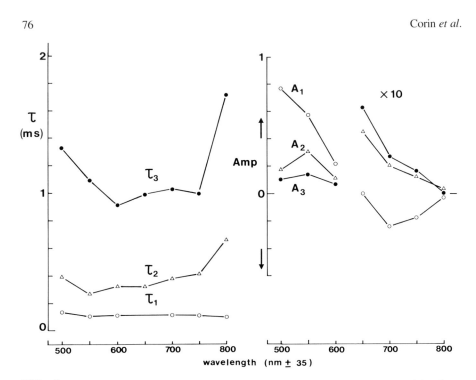

FIG. 9 Wavelength dependence of the relaxation times and amplitudes of proflavin (10 μM) - poly[d(A-br^{5}U)] (50 μM) complex, ionic strength 0.015. Emission was observed through a series of broad band pass filters: Corion BB (5000 - 8000) series. Buffer: as in legend to Fig. 5).

1983), and chemically brominated poly[d(G-C)], all of which are constitutively in a left-handed Z-DNA helical configuration, and to poly[d(A-io^{5}U)]. Preliminary measurements of the latter iodinated polymer complexed with proflavin indicate that the lifetimes are even longer than those measured in the case of the brominated polynucleotide.

REFERENCES

Austin, R.H., Chans, S.S. and Jovin, T.M. (1979). *Proc. Natl. Acad. Sci. USA* 76, 5650-5654.

Austin, R.H., Karohl, J. and Jovin, T.M. (1983). *Biochemistry* 22, 3082-3090.

Barkley, M.D. and Zimm, B.H. (1979). *J. Chem. Phys.* 70, 2991-3007.

Bartholdi, M., Barrantes, F.J. and Jovin, T.M. (1981). *Eur. J. Biochem.* 120, 389-397.

Borisova, O.F. and Tumerman, L.A. (1965). *Biofizika* 10, 32.

Damjanovich, S., Trón, L., Szöllősi, J., Zidovetzki, R., Vaz, W.L.C., Regateiro, F., Arndt-Jovin, D.J. and Jovin, T.M. (1983). *Proc. Natl. Acad. Sci. USA* 80, 5985-5989.

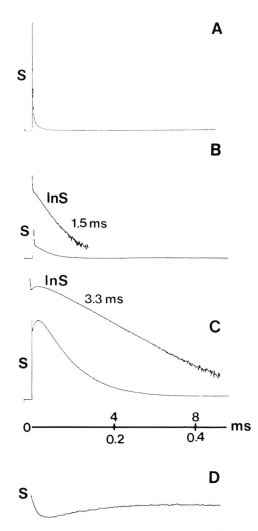

FIG. 10 Complex of acridine orange and DNA. (A) acridine orange
(0.9 μM) - total decay S(t) at >645 nm; (B) acridine orange (0.9 μM)
and poly[d(A-T)] (50 μM) - S(t) and lnS(t); (C) acridine orange
(0.9 μM and poly[d(A-br^{5}U)] (51 μM) - S(t) and lnS(t); (D) expanded
portion of S(t) in (C). Buffer: 8 mM Tris-HCl, pH 7.6, 0.28 M KCl.

Galley, W.C. and Purkey, R.M. (1972). *Proc. Natl. Acad. Sci. USA*
    69, 2198-2202.
Garcia de la Torre, J. and Bloomfield, V.A. (1981). *Quart. Rev. Bio-
    phys.* 14, 81-139.
Hogan, M., Wang, J., Austin, R.H., Monitto, C.L. and Hershkowitz, S.
    (1982). *Proc. Natl. Acad. Sci. USA* 79, 3518-3522.

Jovin, T.M., Bartholdi, M., Vaz, W.L.C. and Austin, R.H. (1981).
    Ann. N.Y. Acad. Sci. 366, 176-196.
Kapuscinski, J., Darzynkiewicz, Z. and Melamed, M.R. (1982). *Cyto-
    metry* 2, 201-211.
Kawato, S. and Kinosita, K.,Jr. (1981). *Biophys. J.* 36, 277-296.
Kinosita, K.,Jr., Kawato, S. and Ikegami, A. (1977). *Biophys. J.*
    20, 289-305.
Li, H.J. and Crothers, D.M. (1969). *J. Mol. Biol.* 39, 461-477.
Malfoy, B., Rousseau, N. and Leng, M. (1982). *Biochemistry* 21, 5463-
    5467.
Matayoshi, E.D., Corin, A.F., Zidovetski, R., Sawyer, W.H. and Jovin,
    T.M. (1983). *In* "Mobility and Recognition in Cell Biology" (Eds
    H. Sund and C. Veeger), pp. 119-134. Walter de Gruyter and Co.,
    New York.
Moore, C., Boxer, D. and Garland, P. (1979). *FEBS Lett.* 108, 161-166.
McIntosh, L., Grieger, I., Eckstein, F., Zarling, D.A., van de Sande,
    J.M. and Jovin, T.M. (1983). *Nature* 304, 83-86.
Neidle, S., Achari, A., Taylor, G.L., Berman, H.M., Carrell, H.L.,
    Glusker, J.P. and Stallings, W.C. (1977). *Nature* 269, 304-307.
Neidle, S. and Berman, H.M. (1983). *Prog. Biophys. Molec. Biol.* 41,
    43-66.
Parker, C.A. and Joyce, T.A. (1973). *Photochem. Photobiol.* 18, 467-
    474.
Ramstein, J., Ehrenberg, M. and Rigler, R. (1980). *Biochemistry* 19,
    3938-3948.
Rigler, R. and Ehrenberg, M. (1973). *Quart. Rev. Biophys.* 6, 139-199.
Saffman, P.G. and Delbruck, M. (1975). *Proc. Natl. Acad. Sci. USA*
    72, 3111-3113.
Striker, G. (1982). *In* "Deconvolution and Reconvolution of Analytic
    Signals" (Ed. M. Bouchy), pp. 329-354. ENSIC-INPL, Nancy.
Turro, N.J., Lui, K.-C., Chow, M.-F. and Lee, P. (1978). *Photochem.
    Photobiol.* 27, 523-529.
Wang, J., Hogan, M. and Austin, R.H. (1982). *Proc. Natl. Acad. Sci.
    USA* 79, 5896-5900.
Waring, M.J. (1981). *Ann. Rev. Biochem.* 50, 159-192.
Zidovetzki, R., Yarden, Y., Schlessinger, J. and Jovin, T.M. (1981).
    Proc. Natl. Acad. Sci. USA 78, 6981-6985.

Transient Dichroism and Rotational Diffusion of Macromolecules: Applications to Membrane Proteins

R.J. CHERRY

INTRODUCTION

Optical spectroscopy has long proved useful for investigating rotational motion of macromolecules in aqueous solution through the technique of fluorescence depolarization. The method is successful because fluorescence lifetimes, which are typically $\sim 10^{-8}$ s, are not too different from the rotational relaxation times to be measured. When the rotational relaxation time is slower than about 10^{-6}s, however, the method fails because the emission decays before any detectable rotation can occur. Such slow rotational diffusion occurs for macromolecular assemblies when their diameter becomes greater than about 20 nm, for example ribosomes and virus particles. Alternatively, slow diffusion may be a consequence of a viscous environment, as in the case of proteins in biological membranes.

In general, measurement of rotation in the microsecond time range, or longer, requires signals from a spectroscopic state of correspondingly long lifetime. Such long-lived states are provided, for example, by the complex light-induced changes which occur in the retinal-containing membrane proteins rhodopsin and bacteriorhodopsin or by the triplet state of probe molecules (Cone, 1972; Razi Naqvi et al., 1973; Cherry et al., 1976; Cherry and Schneider, 1976; Cherry, 1978; Hoffmann et al., 1979). As far as basic principles are concerned, it is immaterial whether such states are detected by

absorption or by emission, although there may of course be important
practical considerations which determine the methodology employed
(Austin et al., 1979; Moore et al., 1979; Johnson and Garland, 1981).

Spectroscopic methods of measuring rotation depend on photoselection, whereby an oriented population of excited molecules is
optically selected from an initially random distribution. This is
most usefully achieved by excitation with linearly polarized light,
and those molecules whose transition dipole moment for absorption is
parallel or at a small angle to the electric vector of the incident
light are preferentially excited. Signals arising from the excited
molecules in general reflect their anisotropic distribution so that
emission signals are polarized and absorption signals are dichroic.
When excitation is by a brief pulse of light, the initial emission or
absorption anisotropy decays with time as the selected molecules
become randomized by Brownian rotation. From the rate (or rates) of
decay, rotational relaxation times may be determined.

Transient dichroism may be measured by a laser flash photolysis
apparatus, full details of which have been described elsewhere
(Cherry, 1978). Data are analysed by calculating the absorption
anisotropy r(t) given by:

$$r(t) = \frac{A_{||}(t) - A_{\perp}(t)}{A_{||}(t) + 2A_{\perp}(t)} \tag{1}$$

where $A_{||}(t)$, $A_{\perp}(t)$ are the absorbance changes at time t after excita-
tion for light polarized parallel and perpendicular with respect to
the electric vector of the exciting flash. Frequently r(t) is dependent only on rotational motion and geometrical factors, although
complications can arise when the absorption transient contains more
than one component. Theoretical expressions for r(t) for macromolecules in isotropic solution were originally derived to interpret
fluorescence depolarization experiments (Rigler and Ehrenberg, 1973).
Although transient dichroism has sometimes been used in solution
studies (Lavalette et al., 1977; Hogan et al., 1983), its major
application has been to investigate rotational diffusion of proteins
in membranes. The first such measurement was performed with rhodopsin and utilized the spectroscopic properties of its intrinsic retinal

chromophore (Cone, 1972). Subsequently, triplet probes were introduced to provide a general method for measuring rotational diffusion
in the microsecond to millisecond time range (Razi-Naqvi et al., 1973;
Cherry et al., 1976; Cherry and Schneider, 1976). The diffusion of
membrane proteins has already been the subject of a number of reviews
(Cherry, 1979, 1982a; Edidin, 1981; Hoffmann and Restall, 1983).
This article summarizes some of our recent studies on protein diffusion in two different membrane systems. In the first of these, a
simple model membrane is used to test the theoretical basis for interpreting diffusion experiments in membranes. In the second, the
dynamic properties of viral glycoproteins are investigated.

BACTERIORHODOPSIN

Bacteriorhodopsin contains an intrinsic chromophore, retinal, which
in the light adapted state is in the all-trans form. Excitation of
bacteriorhodopsin produces a complex cycle of spectroscopic intermediates (Stoeckenius et al., 1979). Rotation is usually measured
by observing the transient dichroism of ground-state depletion signals
at 570 ımı (Cherry, 1982b).

Although bacteriorhodopsin forms a crystalline lattice in the
native purple membrane, it exists as a monomer when reconstituted into
lipid vesicles, provided the lipids are in the liquid-crystalline
phase and the lipid:protein mole ratio (L/P) is greater than about
100:1 (Cherry et al., 1978). In early investigations of the rotation
of membrane proteins, it was proposed that a simple model for interpreting the data would be one in which the protein rotated strictly
only around the membrane normal, as illustrated in Fig. 1 (Cherry et
al., 1976). Bacteriorhodopsin reconstituted into lipid bilayers has
proved to be an excellent system for testing this model. Figure 2
shows the experimental anisotropy decay for monomeric bacteriorhodopsin in a liquid-crystalline lipid bilayer. The theoretical expression
for r(t) derived for the model in Fig. 1 is:

$$r(t) = r_0 (A_1 [\exp -D_{||}t] + A_2 \exp[-4D_{||}t] + A_3) \qquad (2)$$

where $A_1 = 3\sin^2\theta\cos^2\theta$, $A_2 = (3/4)\sin^4\theta$, $A_3 = (1/4)(3\cos^2\theta-1)^2$, θ is
the angle between the transition dipole moment and the membrane

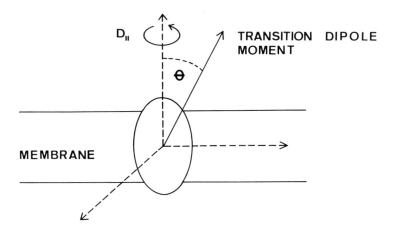

FIG. 1 Model used for interpretation of transient dichroism measurements with bacteriorhodopsin. It is assumed that rotation occurs only around the membrane normal. $D_{\parallel}$ is the rotational diffusion coefficient, while θ defines the direction of the transition dipole moment.

normal, $D_{\parallel}$ the diffusion coefficient for rotation about the membrane normal and r_0 the experimental anisotropy at time zero (Cherry, 1978; Kawato and Kinosita, 1981).

The ability of Equation (2) to fit the experimental data has been tested by curve fitting with $D_{\parallel}$, θ and r_0 as adjustable parameters (Cherry and Godfrey, 1981). An excellent fit is obtained as illustrated by the example in Fig. 2, provided bacteriorhodopsin is monomeric. Under conditions which favour protein aggregation, namely cooling below the lipid phase transition or low lipid:protein ratios, Equation (3) no longer provides a good fit to the data.

The relaxation times obtained in the above experiments have recently been combined with lateral diffusion coefficients to test the validity of the Saffmann-Delbrück equations (Peters and Cherry, 1982). Lateral diffusion coefficients were measured by the method of fluorescence

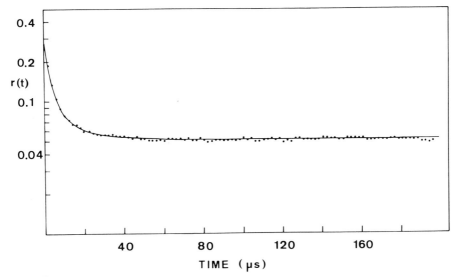

FIG. 2 Anisotropy decay curve for bacteriorhodopsin in dimyristoyl-phosphatidylcholine vesicles (L/P=220) at 28°. The solid line is the best fit of Equation (2) to the experimental points with the parameters r_0=0.28, $\phi_{||}$=12.6 µs and θ=77°.

TABLE 1 Rotational and lateral diffusion coefficients of bacterio-rhodopsin in dimyristoylphosphatidylcholine vesicles at different L/P (phospholipid:protein mole ratio) and temperatures

L/P ratio	T (°C)	D_L ($\mu m^2/s$)	$D_{||}$ ($10^4\ s^{-1}$)	a (nm)
140	28.5	1.8	7.8	1.97
140	32	2.3	7.6	2.35
210	24.5	1.4	6.4	1.89
210	28.5	2.4	7.9	2.39

microphotolysis, also known as fluorescence photobleaching recovery (FPR) and fluorescence recovery after photobleaching (FRAP), in the same reconstituted system after labelling bacteriorhodopsin with the probe eosin-5-isothiocyanate. Table 1 lists some lateral and rotational diffusion coefficients obtained with different lipid:protein ratios and at different temperatures. The Saffman-Delbruck equations are

$$D_{||} = \frac{kT}{4\pi a^2 h\eta} \tag{3}$$

$$D_L = \frac{kT}{4\pi\eta h} [\ln\left(\frac{\eta h}{n_w a}\right) - 0.5772]$$ (4)

where the protein is modelled as a cylinder of radius a spanning a
membrane of width h (Saffman and Delbrück1975). The membrane is
treated as a fluid of viscosity η, while η_w is the viscosity of the
surrounding aqueous phase. Equations (3) and (4) are approximations
which are accurate when $\eta_w \ll \eta$. Exact expressions for $D_{\parallel}$ and D_L in
the case of an arbitrary relationship between η and η_w have also been
obtained (Hughes et al., 1981). Eliminating η from Equations (3) and
(4) gives:

$$\frac{D_L}{D_{\parallel}} = a^2 [\ln\left(\frac{kT}{4\pi a^3 \eta_w D_{\parallel}}\right) - 0.5772]$$ (5)

Values of a calculated by inserting the measured diffusion coefficients for bacteriorhodopsin into Equation (5) are listed in the last
column in Table 1. On averaging these values, the diameter (2a) of
bacteriorhodopsin is found to be (4.3 ± 0.5) nm. This is in reasonable agreement with the 0.7 nm resolution structural model of Henderson and Unwin (1975) in which the widest cross-sectional diameter in
the plane of the membrane is 3.5 nm.

   A further test of the validity of Equation (4) was performed by
adding sucrose to the vesicle preparation to increase the viscosity
of the aqueous phase. It was found that the predicted logarithmic
dependence of D_L on η_w was closely obeyed. In other experiments, Vaz
et al. (1982) have found that Equation (4) also correctly predicts
the dependence of D_L on protein size. Thus it may be concluded that
the Saffman-Delbruck equations do provide a good description of
protein diffusion in membranes.

   This conclusion gives rise to two important consequences. First,
measurements of D_L and $D_{\parallel}$ (under conditions in which there is free
diffusion) enable the protein diameter to be calculated and should
therefore provide an unequivocal demonstration of whether a given
protein exists in a monomeric or oligomeric state in the membrane.
Second, Equation (3) enables D_L to be calculated from measurement of
$D_{\parallel}$ alone, provided the protein diameter can be estimated. In cell
membranes, this value of D_L which applies to local (Brownian)

diffusion, may be significantly different from values measured by the technique of fluorescence microphotolysis, which apply to long-range (i.e. several μm) diffusion. Such differences give information on restrictions to long-range diffusion, while the local D_L may be of particular importance in relation to collision-controlled reactions.

SENDAI-VIRUS GLYCOPROTEINS

The lipid envelope of Sendai virus contains two different glycoproteins known as HN and F. The HN protein is involved in the initial adsorption of the virus to cellular receptors, while the F protein mediates fusion between the envelope and the target membrane (Choppin and Scheid, 1980). In view of the large structural rearrangements which must occur during fusion, it is likely that the dynamic properties of the glycoproteins contribute towards the fusion event. It is therefore of considerable interest to attempt to characterize their molecular motion. To this end we have labelled virus particles with two triplet probes, eosin-5-isothiocyanate (eosin-SCN) and eosin-5-thiosemicarbazide (eosin-TSC) which bind covalently to amino groups and galactose residues (after oxidation with galactose oxidase) respectively (Lee et al., 1983). Using sodium dodecyl sulphate-polyacrylamide gel electrophoresis and fluorescence detection of the probes, it was established that both HN and F proteins were labelled by the two eosin derivatives.

Rotational motion of the glycoproteins was investigated by observing transient dichroism following excitation of the eosin triplet state. Figure 3 shows r(t) curves obtained with eosin-SCN labelled virus at different temperatures. In (a), the measurements are recorded over 2 ms, while in (b) the shorter time range of 0.5 ms is used. Over the longer time range it can be seen that the anisotropy decay curves have a similar shape to that observed with bacteriorhodospin. There is a marked temperature dependence with the mobility increasing with increasing temperature. Qualitatively similar results were obtained with eosin-TSC labelled virus (Fig. 3c).

To quantify the results the data were fitted by the following equation:

$$r(t) = (r_0 - r_\infty) \exp[-t/\phi] + r_\infty \tag{6}$$

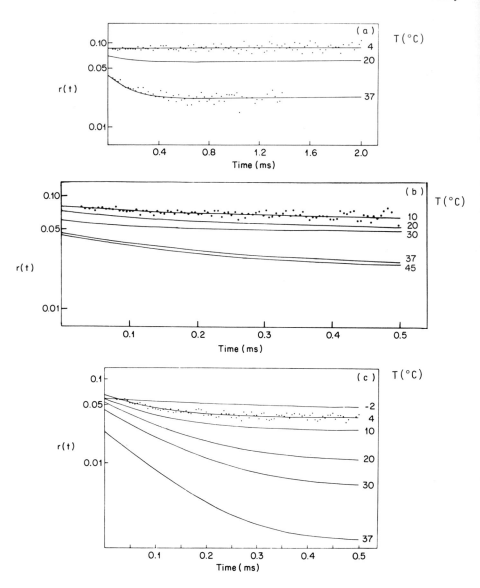

FIG. 3 Effect of temperature on the time-dependent absorption anisotropy, r(t), of eosin-labelled Sendai virus. (a) and (b) eosin-SCN
labelled virus. (c) eosin-TSC labelled virus. Samples were in PBS
buffer containing 50% glycerol. Fresh samples were used at each temperature. The laser intensity of the exciting flash at each temperature was adjusted to approximately the same magnitude. Solid lines
were obtained by curve fitting Equation (6) to the data. For the purpose of clarity, the experimental points are omitted on some curves;
the signal/noise ratio for all the curves was similar.

where r_0 is the initial anisotropy, r_∞ the constant anisotropy
observed at long times and ϕ an apparent correlation time. This
single exponential equation was used instead of Equation (2) because
a double exponential does not significantly improve the goodness of
fit. Moreover, Equation (2) is only valid for a single rotating
species and the data are not sufficiently accurate to test whether
this condition is valid. The parameters obtained from the curve
fitting are plotted as a function of temperature in Fig. 4. The time
constant ϕ was about 200 µs at $37^{\circ}C$ and did not vary significantly
with temperature from $10^{\circ}-45^{\circ}C$. Since the rotational correlation
time for the whole virus under the experimental conditions is deduced
to be about 5 ms, the observed anisotropy decays must arise from
rotational motion of the glycoproteins in the membrane.

To investigate whether any phase change of the viral lipids
occurred over the measured temperature range, the steady state fluorescence polarization of diphenylhexatriene incorporated into the
viral membrane was measured as a function of temperature (Shinitzky
and Barenholz, 1978). The fluorescence polarization varied smoothly

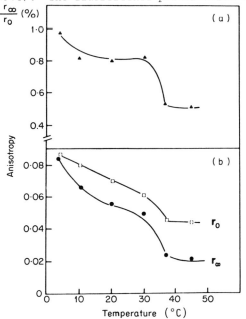

FIG. 4 Temperature dependence of parameters r_0, r_∞ and $(\frac{r_\infty}{r_0})$ for
eosin-SCN labelled Sendai virus determined by curve fitting Equation
(6) to the r(t) curves in Fig. 3.

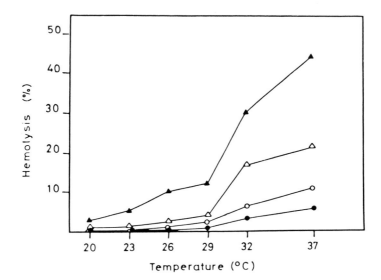

FIG. 5 Haemolysis of human erythrocytes incubated with Sendai virus for 3 (●), 6 (O), 12 (Δ) and 24 (▲) min at various temperatures. Haemolysis is expressed as a percentage of the total lysis induced by distilled water.

with temperature in the range 4-40°C indicating that there are no abrupt changes in lipid motion in this temperature range.

To investigate the relationship between the mobility of the glycoproteins and their biological function, we studied the temperature-dependence of the haemolytic activity with human erythrocytes. The kinetics of the haemolytic interaction were determined by measuring lysis induced by submaximal concentrations of virus (i.e. virus concentrations induced 50% haemolysis during 30 min of incubation with erythrocytes at 37°C) at various temperatures and incubation times. Cells incubated with virus at temperatures below 20°C were not lysed even after prolonged (1 h) times of incubation. As shown in Fig. 5, the extent of haemolysis at temperatures between 20 and 29°C was still less than 5% after 3 to 12 min of incubation and reached only 12% after 24 min of incubation. The temperature range above 29°C was characterized by a marked increase in haemolytic activity which was particularly pronounced after 12 and 24 min of incubation, the latter producing a curve with the steepest rise between 29 and 33°C.

Interpretation of anisotropy decays

The time constant of 200 μs which characterizes the anisotropy decays may be compared with the rotational relaxation time $(D_{\parallel}^{-1}$ of bacteriorhodopsin (∿15 μs) in fluid lipid bilayers. Since the molecular weight of bacteriorhodopsin is 26000 - compared with about 5500 for that of the hydrophobic anchor of the viral glycoproteins (Scheid et al., 1978) - it would be expected that the viral glyco-proteins would have faster rotation if they existed as freely rotating monomers. The relatively slow rotation time of viral glycoproteins is therefore a strong indication that their rotational mobility is restricted, probably by the formation of small aggregates (Kim et al., 1979; Markwell and Fox, 1980). In addition, interaction of the densely packed spikes on the viral surface or interaction with internal proteins could also restrict rotational motion.

The low values of r_0 indicate that the probes attached to the viral glycoproteins undergo some restricted motion which is fast compared with the time resolution of the experiment (20 μs). The theoretical maximum value of r_0 is 0.4, and this can be reduced typically to about 0.25, by instrumental factors. The large difference between this value and the observed r_0 must be due to motion. The present data are remarkably similar to those previously obtained with Ca^{2+}-ATPase in sarcoplasmic reticulum vesicles where it was deduced that internal flexibility of the protein must at least in part account for the low values of r_0 (Bürkli and Cherry, 1981). We consider that this is the most probable explanation in the case of the viral glyco-proteins too. The main alternative is independent probe motion. Although this almost certainly occurs, it is generally highly restricted in angular amplitude, so that r_0 is only reduced to about 0.10-0.15 by this mechanism alone (Kawato and Kinosita, 1981). Moreover, cross-linking of the viral glycoprotein by glutaraldehyde substantially increases r_0, as would be expected for segmental motion, but not for independent probe motion. Finally, measurements of the steady-state fluorescence anisotropy of the eosin probe yielded high values which is a further indication that independent probe motion is highly restricted.

The present results may be compared with previous studies of
erythrocyte glycoproteins, where both restricted independent probe
motion and segmented motion were deduced to occur (Cherry *et al.*,
1980).

Effects of temperature on motion

As can be seen in Fig. 3, the r(t) curves become flatter as the
temperature is decreased. By 4°C the glycoproteins have largely
become immobile on the time scale of the experiment. A similar
behaviour has been observed with a number of other membrane proteins
and ascribed to self-aggregation (Nigg and Cherry, 1979; Bürkli and
Cherry, 1981; Müller *et al.*, 1981). As discussed above, the viral
glycoprotein probably already exists as small aggregates at 37°C.
The effect of decreasing temperature can be explained by the formation
of larger relatively immobile aggregates. In Fig. 4, the normalized
residual anisotropy ($\frac{r_{\infty}}{r_0}$) which is sensitive to the fraction of
immobile glycoproteins, is plotted against temperature. It can be
seen that there is a sharp change in r_{∞}/r_0 between 30 and 35°C.

The value of r_0 is a measure of the contribution of fast motions
to the anisotropy decay. Figure 4a shows that r_0 decreases with
increasing temperature and, like r_{∞}, decreases most rapidly between
30 and 35°C. The dependence of r_0 on temperature can be interpreted
as a decrease in the amplitude of segmental motion with decreasing
temperature, which could result from an increase in steric hindrance
arising from self-aggregation of proteins.

Overall there is clear evidence for a substantial increase in
mobility which occurs in the temperature range of about $30-35^{\circ}$C. It
should be emphasized that the parameters r_0 and r_{∞} which reflect this
change correspond to amplitudes rather than rates of motion. The
time constant α does not vary significantly with temperature.

Relationship between glycoprotein mobility and function

Adsorption of virus to cell membranes mediated by the HN glycoprotein
is temperature indpendent while the glycoprotein mobility is strongly
temperature dependent. This indicates that the electrostatic binding
of the HN protein to its receptor is independent of the degree of
rotational freedom of HN glycoproteins. This is reminiscent of

temperature independent binding of a lectin to glycophorin (Lee and
Grant, 1980). In contrast, haemolysis and fusion are strongly temperature dependent. It is of particular interest that the temperature profile of the haemolytic activity (Fig. 5) exhibits similarities
to that of glycoprotein mobility.

The fluorescence depolarization studies we performed with diphenylhexatriene incorporated in the virus, as well as those of others with
erythrocyte membranes (Shinitzky and Inbar, 1976; Kinosita et al.,
1981), do not reveal any distinct changes in the fluidity of the
lipid phase over the range of 4 to 40°C. Therefore the abrupt
increase in haemolytic activity from 29 to 37°C cannot be attributed
to any lipid phase changes in the erythrocyte membrane, nor the viral
lipids. For the same reason, the sudden increase in glycoprotein
mobility in this temperature range is also unrelated to a lipid phase
change.

It is noteworthy that in other studies, the extent of transfer of
viral phospholipids to erythrocytes as a result of fusion showed a
dramatic increase at about 25°C (Maeda et al., 1975) mediated by the
F protein (Maeda et al., 1981). Since the ability of the virus to
fuse with a target membrane and thereby, in the case of erythrocytes,
induce haemolysis depends on molecular properties of the F protein
and its activation by a proteolytic cleavage (Scheid and Choppin,
1974; Gething et al., 1978; Richardson et al., 1980), the molecular
mobility and flexibility could well represent an important functional
feature of this molecule. Furthermore, the conformational changes of
spike molecules which were recently recognized to accompany viral
fusion (Hsu et al., 1982) are also likely to be reflected in an
altered freedom of mobility. Thus the interrelation of biochemical
and structural properties of the viral glycoprotein spikes with their
rotational mobility and/or flexibility may well be an important new
factor in the understanding of the molecular basis of viral fusion.

ACKNOWLEDGEMENTS

I thank Carmen Zugliani for expert technical assistance and the Fritz
Hoffman-La Roche Foundation, the Swiss National Science Foundation
and the Science and Engineering Research Council for financial support.

REFERENCES

Austin, R.H., Chan, S.S. and Jovin, T.M. (1979). *Proc. Natl. Acad. Sci. USA* 76, 5650-5654.
Bürkli, A. and Cherry, R.J. (1981). *Biochemistry* 20, 138-145.
Cherry, R.J. (1978). *Methods Enzymol.* 54, 47-61.
Cherry, R.J. (1979). *Biochim. Biophys. Acta* 559, 289-327.
Cherry, R.J. (1982a). *In* "Membranes and Transport" (Ed. A.N. Martonosi), Vol. I. pp. 145-152. Plenum Publishing Co., New York.
Cherry, R.J. (1982b). *Methods Enzymol.* 88, 248-254.
Cherry, R.J., Bürkli, A., Busslinger, M., Schneider, G. and Parish, G.R. (1976). *Nature* 263, 389-393.
Cherry, R.J. and Godfrey, R.E. (1981). *Biophys. J.* 36, 257-276.
Cherry, R.J., Müller, U., Henderson, R. and Heyn, M.P. (1978). *J. Molec. Biol.* 121, 283-298.
Cherry, R.J., Nigg, E.A. and Beddard, G.S. (1980). *Proc. Natl. Acad. Sci. USA* 77, 5899-5903.
Cherry, R.J. and Schneider, G. (1976). *Biochemistry* 15, 3657-3661.
Choppin, P.W. and Scheid, A. (1980). *Rev. Infect. Dis.* 2, 40-61.
Cone, R.A. (1972). *Nature New Biol.* 236, 39-43.
Edidin, M. (1981). *In* "Membrane Structure" (Eds J.B. Finean and R.H. Michel), pp. 37-82. Elsevier/North-Holland Biomedical Press, Amsterdam.
Gething, M.J., White, J.M. and Waterfield, M.D. (1978). *Proc. Natl. Acad. Sci. USA* 75, 2737-2740.
Henderson, R. and Unwin, P.N.T. (1975). *Nature* 257, 28-31.
Hoffmann, W., Sarzala, M.G. and Chapman, D. (1979). *Proc. Natl. Acad. Sci. USA* 76, 3860-3864.
Hoffmann, W. and Restall, C.J. (1983). *In* "Topics in Molecular and Structural Biology" (Ed. D. Chapman), pp. 1-113. MacMillan, London.
Hogan, M., Wang, J. and Austin, R.H. (1983). *In* "Mobility and Function in Proteins and Nucleic Acids", Ciba Foundation Symposium 93, pp. 226-245. Pitman, London.
Hsu, M.C., Scheid, A. and Choppin, P.W. (1982). *Proc. Natl. Acad. Sci. USA* 79, 5862-5866.
Hughes, B.D., Pailthorpe, B.A. and White, L.R. (1981). *J. Fluid. Mech.* 110, 349-372.
Johnson, P. and Garland, P.B. (1981). *FEBS Lett.* 132, 252-256.
Kawato, S. and Kinosita, K. Jr. (1981). *Biophys. J.* 36, 277-296.
Kim, J., Harna, K., Miyaka, Y. and Okodo, Y. (1979). *Virology* 9, 523-535.
Kinosita, K. Jr., Kataoka, R., Kimura, Y., Gotoh, O. and Ikegami, A. (1981). *Biochemistry* 20, 4270-4277.
Lavalette, D., Amand, B. and Pochon, F. (1977). *Proc. Natl. Acad. Sci. USA* 74, 1407-1411.
Lee, P.M. and Grant, C.W. (1980). *Can. J. Biochem.* 58, 1197-1205.
Lee, P.M., Cherry, R.J. and Bächi, T. (1983). *Virology* 128, 65-76.
Maeda, T., Asano, A., Ohki, A., Okada, Y. and Ohnishi, S. (1975). *Biochemistry* 14, 3736-3741.
Maeda, T., Asano, K., Toyana, S. and Ohnishi, S. (1981). *Biochemistry* 20, 5340-5345.
Markwell, M.A.K. and Fox, C.F. (1980). *J. Virol.* 331, 152-166.
Moore, C., Boxer, D. and Garland, P. (1979). *FEBS Lett.* 108, 161-166.

Müller, M., Krebs, J.J.R., Cherry, R.J. and Kawato, S. (1981). *J. Biol. Chem.* 257, 1117-1120.
Nigg, E. and Cherry, R.J. (1979). *Biochemistry* 16, 3457-3465.
Peters, R. and Cherry, R.J. (1982). *Proc. Natl. Acad. Sci. USA* 79, 4317-4321.
Razi Naqvi, K., Gonzalez-Rodriguez, J., Cherry, R.J. and Chapman, D. (1973). *Nature New Biol.* 245, 249-251.
Richardson, C.D., Scheid, A. and Choppin, P.W. (1980). *Virology* 105, 205-222.
Rigler, R. and Ehrenberg, M. (1973). *Quart. Rev. Biophys.* 6, 139-199.
Saffman, P.G. and Delbrück, M. (1975). *Proc. Natl. Acad. Sci. USA* 72, 3111-3113.
Scheid, A. and Choppin, P.W. (1974). *Virology* 57, 475-490.
Scheid, A., Graves, M.C., Silver, S.M. and Choppin, P.W. (1978). *In* "Negative Strand Viruses and the Host Cell" (Eds B.W.J. Mahy and R.D. Barry) pp. 183-191. Academic Press, New York.
Shinitzky, M. and Barenholz, Y. (1978). *Biochim. Biophys. Acta* 515, 367-394.
Shinitzky, M. and Inbar, M. (1976). *Biochim. Biophys. Acta* 433, 133-149.
Stoeckenius, W., Lozier, R.H. and Bogolmoni, R.A. (1979). *Biochim. Biophys. Acta* 505, 215-278.
Vaz, W.L.C., Criado, M., Madeira, V.M.C., Schoellmann, G. and Jovin, T.M. (1982). *Biochemistry* 21, 5608-5612.

Optical Methods for Measuring the Rotational Diffusion of Membrane Proteins

P.B. GARLAND and P. JOHNSON

INTRODUCTION

The use of time-resolved optical methods for studying the rotational
and other less extensive diffusional movements of membrane proteins
has been reviewed recently by several authors (Cherry, 1979; Jovin
et al., 1981; Garland, 1982; Hoffmann and Restall, 1983). Furthermore the principles and practice of these methods have been fully
described for linear dichroism (Cherry, 1978), phosphorescence (Austin
et al., 1979; Moore et al., 1979), delayed fluorescence (Austin et
al., 1979; Greinert et al., 1979; Garland and Moore, 1979) and fluorescence depletion (Johnson and Garland, 1981,1982). Our aim in this
chapter is to indicate the nature of the biologically important questions that one might usefully answer with the aid of time-resolved
measurements of the non-translational diffusional movements of membrane proteins, and how the scope of such questions can be greatly
increased by appropriate technical or methodological advances.

Modes of membrane protein movements studied by photoselection methods

Figure 1 illustrates the physical modes of movement that in theory
would be available to a photosensitive molecule (probe) attached to a
membrane protein molecule. Movement of the probe throughout any of
these ranges would, because it is a movement, result in time-dependent decay of the anisotropy of the probe induced by a photoselection
flash at zero time. Several modes of movement can be recognized:

SPECTROSCOPY AND THE DYNAMICS
OF MOLECULAR BIOLOGICAL SYSTEMS

95

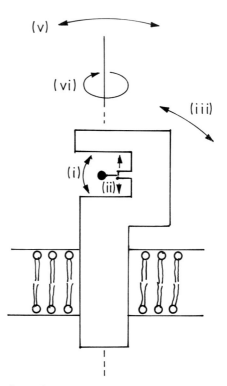

FIG. 1 Various modes of motion of a probe molecule attached to a membrane protein, and of the protein itself. The motional modes shown are: (i) free motion of the probe about its point of attachment; (ii) motion of the attachment point e.g. an ε-NH$_2$ of a lysine residue; (iii) segmental motion, or flexion of part of the protein; (iv) rotational motion of the whole protein regarded as a rigid body, proceeding about an axis normal to the plane of the membrane; (v) wobbling of the whole protein due to tilting of the axis around which rotation can occur.

motion of the probe at its attachment point, movement of the attachment point, flexion of the polypeptide segment or domain of protein to which the probe is attached, uniaxial rotation of the whole protein about an axis normal to the plane of the membrane, and movement of the rotation axis itself resulting in wobbling or rocking of the whole protein. The extent to which each of these can proceed is a static or structural parameter, and can be expected to vary with the nature of the probe (i.e. intrinsic or extrinsic), with its means of attachment, and with the structural properties of the protein and its neighbouring molecules. Because the extent of these movements reflects

structural properties of the protein, a change in such an extent indicates a structural (conformational) change of the protein. These
differing modes of probe movement can to some extent be time resolved
from each other, as indicated below.

Fuller accounts of the depolarizing effects of segmental motions
of proteins are given by Kawato and Kinosita (1981), Kinosita et al.
(1982), Harvey and Cheung (1981), and Hanson et al. (1981).

Time domains of modes of membrane protein movement

The viscosity of the hydrophobic interior of a phospholipid bilayer
limits the relaxation times for uniaxial diffusional rotation of
transmembrane protein to values of 10-20 μs or longer (Cone, 1972;
Saffmann and Delbrück, 1975). By contrast, the relaxation times for
movement of extrinsic probes about the site of attachment, or of the
immediate attachment site itself, are in the nanosecond region
(Badea and Brand, 1979; Cherry et al., 1980). These two types of
motion can therefore be time-resolved from each other with ease; the
nanosecond motions are best measured with conventional prompt fluorescence methods, whereas the microsecond motions require use of a triplet probe method (Cherry, 1979; Austin et al., 1979; Moore et al.,
1979; Johnson and Garland, 1981). These different time domains are
illustrated in Fig. 2.

Motions of segments or domains of proteins can occur with relaxation times as brief as 33 ns (Yguerabide et al., 1970) or even less
(Hanson et al., 1981), as measured for immunoglobulin G. There is no
upper limit for the relaxation times of segmental motion of proteins.
It follows therefore that separation of segmental motions of protein
from motion of the probe itself or from rotation of the whole protein
is possible only if the relaxation time for segmental flexion lies
within the range of a few nanoseconds to a few microseconds. Tech-
nically, this is a difficult time domain to explore by photoselection
methods. Prompt fluorescence for most probes expires beyond about
100 ns, while signal-to-noise considerations complicate the use of
triplet probes at high time resolution. Nevertheless, direct measurements of segmental motion in the microsecond region have been made on
membrane bound skeletal muscle sarcoplasmic Ca^{2+}-dependent ATPase
labelled with erythrosin by time-resolved measurement of the

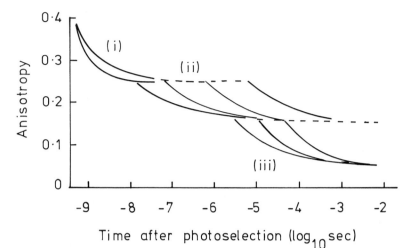

FIG. 2 Time domains of the motions of a probe attached to a membrane
protein. The figure is for a photoselection giving a maximal zero
time value of 0.4 for the anisotropy parameter. It is assumed that
the angle between the dipole moment and any given axis of rotation is
such that rotation always has a depolarizing effect (i.e. of Equations (6)-(8) is greater than zero). More than one aisotropy decay
course is shown for each time domain, to emphasize that within any
one time domain there is a range of possible relaxation times. The
time domains shown are: (i) 10^{-9} to 10^{-8} s, for rotation of the probe
and its point of attachment; (ii) 10^{-8} to $>10^{-2}$ s, for segmental
flexion; (iii) 10^{-5} to $>10^{-2}$ s, for rotation or tilting of the whole
protein.

depolarization of its laser-flash induced phosphorescence (Spiers *et
al.*, 1983). The response of this segmental motion to a temperature-
dependent conformational change at 11-13°C (Lippert *et al.*, 1981) was
also detected.

Problems to be addressed

As outlined above, time-resolved measurements of the rotational motion
of appropriate photosensitive probe molecules attached to the proteins
of membranes yields data that convey information on both the static
and dynamic properties of the proteins. It is doubtful if the dynamic
properties are finely tuned for biological function: for instance, a
submicrosecond segmental flexion of an enzyme has several orders of
magnitude in hand before it could become rate limiting for catalytic
processes. On the other hand, the structural information that can be
obtained from measurements of rotational motion can lead to biologi-

cally significant interpretations. For example, changes in the
extent and rate of segmental flexion would indicate an underlying
conformational change of the protein (Spiers et al., 1983). Another
example of such use of triplet probes is to be found with the proton-
translocating ATPase of chloroplasts (Wagner and Junge, 1982; Wagner
et al., 1981,1982).

Changes (or lack of changes) in the global rotation of a membrane
protein in response to ligand binding or to the presence of other
proteins in the membrane can indicate whether the ligand-binding protein has undergone oligomerization or not, e.g. binding of epidermal
factor to its receptor (Zidovetzki et al., 1981), or whether one protein interacts with another, e.g. the effect of an alkali-extractable
peripheral protein on the rotational behaviour of the acetylcholine
receptor of the electric organ of *Torpedo marmorata* (Lo et al., 1980;
Bartholdi et al., 1981). It should be noted here that the rotational
relaxation time of a membrane protein is proportional to the square
of the radius of the transmembranous part of the protein (Saffmann
and Delbrück, 1975). Table 1 summarizes studies made in recent years
on the rotation of membrane proteins.

These brief comments and examples indicate some of the ways in
which measurements of rotational diffusion can be used in seeking to
advance knowledge and understanding of membranous systems. However,
there remains the question of which method of measurement should be
used, which is the subject of the following section.

DEVELOPMENT OF OPTICAL METHODS AND EXTRINSIC PROBES

Photoselection methods for measuring rotational diffusion of molecules
require that the molecule concerned should be capable of being converted by light to a new state or species that is spectroscopically
detectable. The light-activated state must have a lifetime not significantly less than the rotational relaxation time that is to be
measured. The molecule may have intrinsic light-sensitive properties
as in the case of rhodopsin or the photodissociable CO-complexes of
oxidases. Failing that, light sensitivity can be conferred by attachment of an extrinsic probe. The use of fluorescent probes as extrinsic probes is familiar. The number of membrane-bound proteins with
intrinsic light sensitivity of long (μs-ms) lifetime is limited,

TABLE 1 Summary of studies made on the rotational diffusion of membrane proteins

Protein and membrane	Probe	Question(s) posed	References
Rhodopsin of frog retinal rods	Intrinsic	How rapidly can a membrane protein rotate?	1
Cytochrome a_3 of mitochondria	Intrinsic (CO-complex)	Free to rotate or in complexes with other mitochondrial proteins?	2-5
Cytochrome o and a_1 of E. coli	Intrinsic (CO-complex)	Free to rotate or in respiratory assemblies?	6
Cytochrome P450 of microsomes	Intrinsic (CO-complex)	Association with its NADPH-dependent reductase?	7-9
Bacteriorhodopsin in liposomes	Intrinsic	Is the Saffman-Delbruck model correct?	10,11
Ca^{2+}-dependent ATPase of sarcoplasmic reticulum	Eosin Erythrosin*	What is the oligomeric state of the enzyme?	12-15
Band 3 of erythrocytes	Eosin	Why is its lateral diffusion so slow? What are its contacts with other proteins?	16-18
Acetylcholine receptor of electric organ	Erythrosin* Eosin*	What is the role of peripheral membrane proteins in immobilizing the receptor?	19,20
Epidermal growth factor receptor of cell surface	Erythrosin*	Does growth factor binding cause microaggregation of receptors?	21
H^+-translocating ATPase of chloroplasts	Eosin	What are the effects of "membrane energization" on ATPase conformation?	22,23

References: (1) Cone, 1972. (2) Junge and DeVault, 1975. (3) Kunze and Junge, 1975. (4) Kawato et al., 1980, 1981. (5) Reid, 1981. (6) Garland et al., 1982. (7) Richter et al., 1979. (8) Kawato et al., 1982. (9) Gut et al., 1982. (10) Cherry and Godfrey, 1981. (11) Peters and Cherry, 1982. (12) Hoffman et al., 1979. (13) Hoffman et al., 1980. (14) Burkli and Cherry, 1981. (15) Spiers et al., 1983. (16) Cherry et al., 1976. (17) Nigg and Cherry, 1979,1980. (18) Matayoshi et al., 1983. (19) Lo et al., 1980. (20) Bartholdi et al., 1981. (21) Zidovetzki et al., 1981. (22) Wagner and Junge, 1980,1982. (23) Wagner et al., 1981,1982.

(*Studies made with phosphorescence. All other studies used linear dichroism.)

unless one includes the triplet state of tryptophan (Strambini and
Galley, 1976) to rhodopsins and carbon monoxide-binding photodissociable cytochromes. The possibility of using the triplet state of
flavin in flavoproteins has apparently not been explored.

Cherry and his colleagues introduced tetrabromofluorescein-(i.e.
eosin-) isothiocyanate as an extrinsic probe that was suitable for
measuring slow rotation (μs-ms) by virtue of the long lifetime of its
triplet state (Cherry and Schneider, 1976; Cherry, Burkli et al.,
1976; Cherry, Cogoli et al., 1976). Dibromofluorescein-isothiacyanate
has also been used as a triplet probe (Lavalette et al., 1977; Pochon
et al., 1978). The choice of these halogen-substituted fluoresceins
was dictated by several considerations. First, the quantum yield for
triplet formation is moderate to high. Second, they can be excited
in their visible absorption wavelengths around 510-530 nm, a region
readily accessed either by flash-lamp pumped dye lasers or frequency-
doubled Nd-YAG lasers. Third, they have high extinction coefficients
for their visible region absorption peaks, facilitating both triplet
formation and its measurements by ground-state depletion. Eosin can
also be measured by its phosphorescence (Garland and Moore, 1979;
Austin et al., 1979) with a considerable gain in sensitivity over
ground-state depletion measured as an extinction change. Even greater
sensitivity for the phosphorescence method was obtained by the introduction of tetraiodofluorescein-(i.e. erythrosin)-isothiocyanate
(Moore and Garland, 1979). The molecular and spectroscopic requirements for extrinsic triplet probes as just described are relatively
strict. These constraints can be relaxed considerably when the triplet state is generated and measured by the laser-microscope instrument
that we described in 1981 as part of the so-called fluorescence depletion method (Johnson and Garland, 1981). This method combines the
sensitivity of prompt fluorescence methods with the long lifetime of
the triplet state as will be further discussed below. Apart from the
great gain in sensitivity over previous methods, the fluorescence
depletion method relaxes considerably the constraints otherwise
imposed on the probe properties and on the choice of laser. Thus, it
is possible to use fluorescent triplet probes of relatively low
quantum yield for triplet formation, particularly where slow rotations

are under study (Johnson and Garland, 1982,1983). Following this
introduction on the development of triplet probes, we describe the
optical methods presently available for measuring the decay of flash-
induced triplet probe anisotropy.

Linear dichroism

Descriptions of laser-flash photolysis equipment for measuring linear
dichroism and its time-dependent decay due to molecular rotation have
been given by Austin *et al.*, 1979; Cherry, 1978; Garland and Moore,
1979; Hoffman *et al.*, 1979; Junge, 1972; Junge and DeVault, 1975;
Lavalette *et al.*, 1977; Lessing and von Jena, 1979; Matayoshi *et al.*,
1983 and Spiers *et al.*, 1983. There is considerable variation among
these designs, but some generalizations can be drawn:

(i) signal averaging from repetitive flashes is essential for most
purposes to obtain adequate sensitivity and signal-to-noise ratio;

(ii) approximately 20-30% flash-induced ground-state depletion is
required: more than that reduces anisotropy, less reduces sensitivity;

(iii) the laser flash must deliver 10-50 mJ cm^{-2} in the 520-540 nm
region in order to obtain satisfactory ground-state depletion,
unless

(iv) the laser-flash beam is focussed to a narrow diameter (Lessing
and von Jena, 1979), when relatively less powerful lasers are
required;

(v) a flash-lamp pumped dye laser with a coaxial flash tube and a
flash width of 0.3 µs at half-peak height (Lo *et al.*, 1980) provides a very satisfactory excitation source at relatively modest
price;

(vi) measurement in the sub-microsecond time domain requires either
a frequency-doubled Nd-YAG laser or a N_2-pumped dyelaser;

(vii) focussing a monochromator-selected measuring beam from an
arc or filament source into the laser-flash illuminated region of
the cuvette is problematical, and in any case, such sources are not
optimal for fast kinetic work;

(viii) a CW gas laser has several advantages as the measuring beam:
it can be easily passed through the flash-illuminated region of
the sample, and it is more powerful than a conventional lamp source,

so reducing both photomultiplier shot noise and the relative size of electrical and optical artifacts accompanying the excitation flash.

With signal averaging and eosin as triplet probe, the linear dichroism method requires that there be not less than about 10^{-6} M probe in the membrane suspension under study. Not more than 1 mg of membrane protein per ml of suspension is allowable, otherwise light scattering degrades polarization. Consequently, the linear dichroism method is limited to proteins of relatively high abundance in the membranes. The signal-to-noise ratio falls markedly as the sample path length is decreased. Consequently, linear dichroism is unsuited for microscopic use except in the unusual case of naturally-occurring membrane stacks (Cone, 1972).

Phosphorescence and delayed fluorescence

Delayed fluorescence is of the same wavelength as prompt fluorescence. As a consequence of this, flash photolysis equipment for measuring delayed fluorescence must have a means of protecting photomultiplier tubes against overload and damage from the high intensity prompt fluorescence arising during the excitation flash. Greinert et al. (1979) used a mechanical shutter with a dead-time of 75 µs, too long for the rotational relaxation times of many membrane proteins. Photomultiplier gating is preferable (Austin et al., 1979; Matayoshi et al., 1983).

Phosphorescence is emitted at much longer wavelengths than fluorescence. For eosin, fluorescence emission is maximal at about 550-560 nm whereas phosphorescence is maximal at 650-670 nm. A high-pass blocking filter is sufficient to protect the photomultiplier tube from overload (Moore et al., 1979; Spiers et al., 1983). Alternatively, photomultiplier gating can be used (Austin et al., 1979; Matayoshi et al., 1983). The sensitivity of the phosphorescence method is much greater than that of linear dichroism, and satisfactory results can be obtained with as little as 10^{-8} M probe in the sample. A microscopic version using an acousto-optically modulated argon ion laser as the excitation source (Garland, 1981) has a sensitivity to erythrosin of about 10^8-10^9 molecules (unpublished observations).

Generally, eosin phosphorescence measurements are several-fold less sensitive than those of erythrosin phosphorescence.

In comparing various optical methods of measuring rotational relaxation times it is necessary to introduce the orientation of the dipole moments used for photoselection and for measurement. Reference to Fig. 3 shows that the dipole moment for photoselection, $\vec{\mu}_p$, makes an angle θ_p with respect to the membrane normal around which unaxial rotation of the whole molecule is considered to occur. Similarly, $\vec{\mu}_m$ and θ_m refer to the dipole moment used for measurement and the angle it makes with the normal. For a randomly orientated suspension of membranes in a photoselection experiment, the time-dependent anisotropy parameter $r(t)$ is defined (Jablonski, 1961) by:

$$r(t) = \frac{S_{\parallel}(t) - S_{\perp}(t)}{S_{\parallel}(t) + 2S_{\perp}(t)} \tag{1}$$

where $S_{\parallel}(t)$ and $S_{\perp}(t)$ are the measuring signals at time t after the photoselection flash measured in polarization planes either parallel $(S_{\parallel})$ or perpendicular $(S_{\perp})$ to the plane of polarization of the photoselection flash. If D is the uniaxial rotational diffusion coefficient then (Rigler and Ehrenberg, 1973):

$$r(t) = A_1 e^{-Dt} + A_2 e^{-4Dt} + A_3 \tag{2}$$

where:

$$A_1 = 1.2\sin\theta_p \sin\theta_m \cos\theta_p \cos\theta_m \cos\phi \tag{3}$$

$$A_2 = 0.3\sin^2\theta_p \sin^2\theta_m (\cos^2\phi - \sin^2\phi) \tag{4}$$

$$A_3 = 0.1(3\cos^2\theta_p - 1)(3\cos^2\theta_m - 1) \tag{5}$$

in which ϕ is the angle in the plane of the membrane between the projections of $\vec{\mu}_p$ and $\vec{\mu}_m$ onto the plane (Fig. 3).

For linear dichroism measurements the same absorption band can be used both for photoselection and measurement, in which case $\theta_p = \theta_m = \theta$ and $\phi = 0$, whereupon:

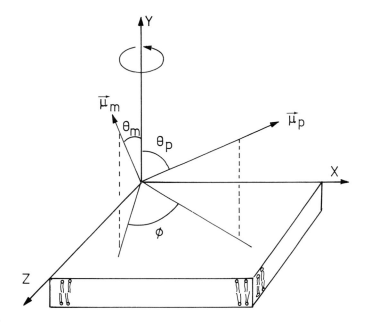

FIG. 3 Orientation of dipole moments in photoselection experiments. Of the orthogonal axes shown, the X and Z-axes lie in the plane of the membrane and Y-axis is normal to that plane. θ_p is the angle between the dipole moment for photoselection and the Y-axis, θ_m is the angle between the dipole moment for measurement and the Y-axis, and ϕ is the angle between their projections onto the membrane plane. θ_m and θ_p are interchangeable.

$$A_1 = 1.2\sin^2\theta\cos^2\theta \tag{6}$$

$$A_2 = 0.3\sin^4\theta \tag{7}$$

$$A_3 = 0.1(3\cos^2\theta-1)^2 \tag{8}$$

In these circumstances, r_0, the zero-time anisotropy, has its maximum value of 0.4 (Albrecht, 1961,1970), and with uniaxial molecular rotation decays towards a time-independent limiting value (r_∞) given by A_3 (Equations (2) and (8)). Because A_1, A_2 and A_3 depend on θ which is not usually known and may have a range of values for an extrinsic probe, the interpretation of anisotropy decay curves for uniaxial rotation is not straightforward. Moreover, the situation is even worse for phosphorescence where θ_p and θ_m are not necessarily equal. As a result the r_0 value for phosphorescence may lie anywhere between -0.2 and 0.4, and $r(t)$ may either grow or decay within these limits, depending on the values of θ_p, θ_m and ϕ. A further complication arises from the possibility that these angles may change with temperature and probe environment.

Fluorescence depletion

It will be clear from our comments above that the linear dichroism method is relatively insensitive, that phosphorescence is less so, but that neither offers in general sufficient sensitivity to work at the microscopic or single cell level. Furthermore, the analysis of anisotropy decay curves is complicated by uncertainties about the angles θ_p, θ_m, ϕ (for phosphorescence) or θ (for linear dichroism). This uncertainty means that the time-dependent relaxation of anisotropy due to rotation may not readily be interpretable unequivocally in terms of components in D, 4D or of a mixture of such components, particularly in the former case. Moreover, this drawback is less troublesome than uncertainty in A_3, the time-independent anisotropy term, because it is important to know if an observed value for r_∞ is that anticipated from the A_3 term alone or if it also contains a contribution from molecules that have not rotated at all during the probe lifetime.

In view of the relatively low sensitivity of the then available methods, we developed a new method of triplet-state detection, which we call fluorescence depletion (Johnson and Garland, 1981,1982). In principle it is similar to linear dichroism, but in practice differs in that the absorption of the measuring beam of light is not measured by its attenuation on passing through the sample, but by the fluorescence that such absorption excites. Ground-state depletion is therefore detected as a depletion of fluorescence. We also used an acousto-optically modulated continuous argon ion laser as the source both for photoselection and measurement. The incident laser powers are low, a few milliwatts for photoselection and a few microwatts for measurement. Sufficient intensity for triplet formation in a few microseconds exposure of the sample to the photoselecting power is achieved by focussing via a fluorescence microscope on to the cell membranes under study. We have also developed a convenient way of establishing anaerobic conditions (otherwise oxygen quenches the triplet state) on a microscope slide or in a circulation chamber (Johnson and Garland, 1983).

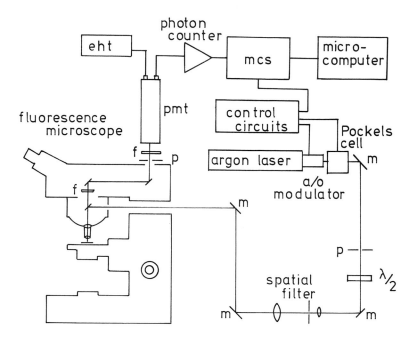

FIG. 4 Components and optical arrangements of a fluorescence depletion apparatus. The apparatus is centred around a fluorescence microscope and a 1 watt argon ion laser. The laser beam power is varied
with an analogue acousto-optic modulator (a/o modulator) and reflected by mirrors (m) through a Pockels cells, a pinhole (p), a half-
wave plate ($\lambda/2$) or polarization rotator, and a spatial filter, to a
focus at the rear image plane of the microscope objective. Fluorescence from the sample is passed by mirrors and filters (f) to a photomultiplier tube (pmt) provided with a high voltage supply (eht) and
set up for photon counting. A multichannel scaler (mcs) accumulates
photon counts, and its output is transferred to a microcomputer.
Timing and pulses to the acousto-optic modulator, Pockels cell and
multichannel scaler are provided by a hybrid analogue/digital control
circuit.

   Figure 4 shows the components of our fluorescence depletion instru-

ment. It is a natural development of the apparatus that we had pre-

viously developed for measuring, by fluorescence photobleaching

recovery, the lateral diffusion of membrane molecules (Garland, 1981).

This in its turn was based on the fluorescence photobleaching recovery

instrument described by Webb and coworkers (Axelrod et al., 1976;

Koppel et al., 1976). A full description of our fluorescence deple-

tion instrument is given in our earlier papers (Johnson and Garland,

1981,1982). Recent improvements include the addition of an on-line
microcomputer for data handling. An earlier version (Garland, 1982)
used a pulsed measuring beam in a box-car mode. We think that this
may have advantages in sensitivity and time resolution over the present continuous measuring beam. A microprocessor-controlled version
is currently under development for re-evaluation of the box-car mode.

   The sensitivity of the fluorescence depletion method is high, yielding satisfactory triplet decay curves from as little as 27000 molecules
of eosin-labelled band 3 on part of an erythrocyte ghost (Johnson and
Garland, 1981) or 4000 molecules of rhodamine-labelled albumin in a
model system (Garland, 1982). Limiting factors in sensitivity are
probe stability throughout several thousand cycles of triplet formation and decay, and fluorescence arising from microscope slides. The
choice of probe is not restricted to those with high quantum yields
for triplet formation, because the photoselection energy can be easily
increased merely by prolonging the duration of the photoselection
pulse (Johnson and Garland, 1982). There is however a trade-off
between sensitivity and time resolution.

   If the propagation axes of the photoselecting and measuring beams
are parallel to the axis around which rotation occurs (i.e. the membrane normal - see Fig. 5), then a simpler relationship than that of
Equation (2) holds for the relationship between the diffusion coefficient and the observed signals $S_{\parallel}(t)$ and $S_{\perp}(t)$. Following on from
Cone (1972), if the time-dependent polarization $p(t)$ is defined as:

$$p(t) \ = \ \frac{S_{\parallel}(t) \ - \ S_{\perp}(t)}{S_{\parallel}(t) \ + \ S_{\perp}(t)} \tag{9}$$

then:

$$p(t) \ = \ p_0 e^{-4Dt} \tag{10}$$

   Equation (10) replaces the more complex Equation (2) if the correct
geometry can be achieved, with the considerable gain that there is
only one exponential term and no time-independent term. The dipole
moments $\vec{\mu}_p$ and $\vec{\mu}_m$ are parallel because the photoselecting and measuring wavelengths are identical. Under these conditions the zero-time
polarization p_0 is independent of θ and has a value of 0.5.

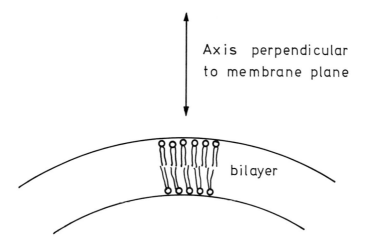

FIG. 5 Parallel propagation axes for photoselection and measurement, normal to the membrane plane. The figure shows a phospholipid bilayer and, perpendicular to it, the axis of a single laser beam used both for photoselection and for measurement in a fluorescence depletion apparatus. Parallel axes for photoselection and measurement are obtained by use of a fluorescence microscope with epi-illumination. To maximize zero-time anisotropy the numerical aperture of the microscope objective should be low, not more than 0.65 and preferably less.

We have obtained the appropriate geometry with a single red cell ghost (Johnson and Garland, 1981) and also with a single large liposome (Johnson and Garland, 1983). With rough cell surfaces this will not, of course, be possible. In Table 2, various features of the linear dichroism, phosphorescence, and fluorescence depletion methods of measuring the triplet state and decay of its anisotropy are summarized.

We should mention for comparison an aerobic version of the fluorescence depletion method developed independently by Smith et al. (1981). This method utilizes irreversible oxidative photobleaching of a fluorescence probe as the photoselection step, and is therefore a single-shot method as far as any single photobleached area of membrane is concerned. The quantum yield for aerobic photobleaching is probably in the range 10^{-5} - 10^{-6} (Britt and Moniz, 1973), whereas that for triplet formation ranges from about 10^{-1} - 10^{-2} (Lessing and von Jena, 1979). Proportionately higher laser energies must therefore be delivered to obtain adequate photobleaching, and these powers must not be so high that unacceptable temperature increases occur at the

TABLE 2 Summary of optical methods for measuring rotational diffusion in the μs-ms range, using extrinsic triplet probes.

Feature	Linear dichroism[1-3]	Phosphorescence[2-5]	Fluorescence depletion[6]
Photoselection	50-100 mJ laser flash in 15 ns-2 μs	0.1-5 mJ flash is adequate	Modulated Argon ion laser
Sample	Cuvette	Cuvette	Microscope slide
Sample volume (illuminated)	0.5-1.0 ml	0.5-1.0 ml	Area approx. 10 μm^2
Measuring beam	Lamp or laser	None	Same laser, alternated
Instrument geometry	$90°$ between photoselection and measuring beams	"T"-format	Fluorescence microscope
Signal averaging rate	0.2 - 10 Hz	5-50 Hz	50-200 Hz
Photomultiplier mode	Current	Current	Photon-counting
Detection limit (approx.)	10^{-10} mole (eosin)	10^{-12} mole (erythrosin)	10^{-20} mole (eosin)
Probe requirement for triplet quantum yield	Strict	Less strict	Relaxed
Constraints on probe chemistry	Moderate	Severe (must phosphoresce)	Relaxed
Time resolution	10-20 μs	0.3-1 μs	10-50 μs

References: (1) Cherry, 1978. (2) Jovin et al., 1981. (3) Garland and Moore, 1979. (4) Spiers et al., 1983. (5) Matayoshi et al., 1983. (6) Johnson and Garland, 1981,1982.

membrane sample (Axelrod, 1977). In practice this means that the
method of Smith *et al.* (1981) has a photoselection period of at least
several milliseconds, and cannot be used for detection of other than
relatively slow rotation. However, because of the infinitely long
lifetime of the photobleached state, the method is uniquely well-
suited for measuring slow rotation on a time scale of many milliseconds or more. The sensitivity of the method cannot benefit from
signal averaging, but this is compensated for, in part at least, by
the much higher measuring beam powers that can be employed. So far
this method has been applied to measuring the very slow lateral diffusion   curved surface of rotation of lipid probes in phospholipid vesicles or multilayers (Smith *et al.*, 1981), but not to proteins.

FUTURE DEVELOPMENT OF THE FLUORESCENCE DEPLETION METHOD

Lipid rotation

The presence of gel-phase microdomains of lipid in an otherwise
liquid-crystalline membrane would result in at least two greatly
differing rates of lipid rotational diffusion. There would be a fast
rate in the nanosecond region corresponding to the liquid-crystalline
phase, and a slower rate corresponding to the rotation in the gel-
phase regions. The slower rotations in the gel-phase region can have
two origins: either slow rotation of individual molecules in an otherwise immobile domain, or rotation of the whole domain containing
otherwise immobile molecules. In order to facilitate the study of
such rotations in the μs-ms time scale, we have explored the triplet-
formation of N,N'-di(octadecyl)indocarbocyanine (diI_{18}) and N,N'-di-
(tetradecyl)indocarbocyanine (diI_{14}), using the fluorescence depletion method (Johnson and Garland, 1983). We found that these dyes
can be converted to the triplet state with reasonable ease, and that
they have a relatively oxygen-insensitive triplet lifetime of several
milliseconds. Using a probe-to-phospholipid ratio of approx. 1:50000,
it was possible to make measurements with a time resolution of 20 μs
on a spot 10 μm in diameter focussed on a single large paucilamellar
liposome. Rotational relaxation times of 0.1 - 0.4 ms were deleted
as the temperature of the sample passed through the phase-transition
temperature of the liposome phospholipid.

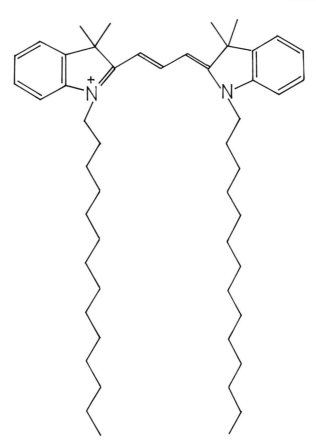

FIG. 6 Structure of N,N'-di(tetradecyl)indocarbocyanine

An unusual feature of fluorescent lipid probes such as diI_{18} (Fig. 6) is that they retain the dipole axes of their fluorescent chromophore in a relatively fixed orientation, parallel to the membrane surface (Badley et al., 1973; Yguerabide and Stryer, 1971; Axelrod, 1979; Smith et al., 1981). It follows that the only effective depolarizing motion of these probes in a phospholipid bilayer is that due to uniaxial rotation around the membrane normal (Fig. 7). Rotation of the fluorescent chromophore round the bilayer-penetrating acyl chains is not possible. This restriction of depolarizing motion to uniaxial rotation of the whole probe is not the case with other previously used lipid probes such as diphenylhexatriene (Dale et al., 1977; Lakowicz et al., 1979) or anthroyl-substituted fatty acids

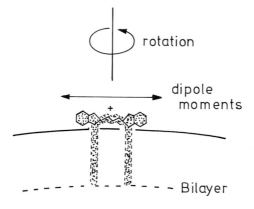

FIG. 7 Orientation of di-acyl carbocyanine dye incorporated into a bilayer. Incorporation into only the outer leaflet of the bilayer is shown. Comparison with Fig. 6 will identify the chromophore and acyl-parts of the molecule (after Axelrod, 1979).

(Thulborn and Sawyer, 1978; Tilley et al., 1979) where depolarization can arise from several motions: rotation of the whole molecule, wobbling, and independent motion of the chromophore if the chromophore and lipid are not one and the same molecule.

We believe that it would be worthwhile to extend the use of carbocyanine dyes as triplet probes by synthesizing derivatives that are bromo- or iodo-substituted in the chromophore. Such derivatives would have a much larger quantum yield for triplet formation, and could therefore be used in conjunction with nanosecond laser pulses to explore the range of lipid rotation from 10^{-7} to 10^{-2} s. This region is essentially unexplored, yet should not remain so in view of the potential importance of lipid microdomains in biological membranes (Karnovsky et al., 1982).

Two-laser instrument

There are applications where the fluorescence depletion method might be more effective if two lasers were used, rather than one laser that is modulated between photoselection and measuring powers. In a two-laser instrument, one laser could be used for photoselection, and might be a high repetition rate N_2-pumped dye laser. The other laser would provide the measuring beam, and could be an argon ion laser of modest power. Such an arrangement would have two advantages. First, the method could be scaled up to larger sample volumes or areas, as might be met in a cuvette or with a confluent monolayer of cultured cells. Such scaling-up would improve the signal-to-noise ratio. Second, the photoselection pulse would be short (e.g. 0.5 - 15 ns, depending on the laser) and enable fast time resolution to be obtained. However, only probes with a high quantum yield for triplet formation could be used, such as eosin or dibromofluorescein, otherwise large temperature rises might occur at the membrane if the photoselection laser intensity was increased, or triplet formation would be insufficient if it was not.

Modulation methods

As an alternative to pulsed time-resolved methods, consideration might be given to modulation methods similar in principle to those already established for measuring the decay of prompt fluorescence in the nanosecond time region (Weber, 1969, 1981). In such methods the excitation source is sinusoidally modulated with a period similar to that of the lifetime under study. The emitted fluorescence appears as a modulated signal shifted in phase with respect to the exciting waveform. Fluorescent lifetimes can be calculated either from the phase shift or from the modulation amplitude of the emitted signal. The method was been developed for resolving heterogeneous systems with more than one lifetime (Weber, 1981; Lakowicz and Cherek, 1981; Jameson and Weber, 1981), and also for quantitation of anisotropic rotations (Weber, 1977; Mantulin and Weber, 1977; Lakowicz et al., 1979). Translation of modulation methods to the fluorescence depletion method should be simple. The appropriate modulation frequencies are about 1 KHz or less, and are readily obtained with simple electronics and an analogue acousto-optic modulator acting on an argon ion laser. Data acquisition can be performed with a digital signal averager, and all data manipulations with an on-line micro- or mini-computer. One additional data manipulation step would be required to compensate the modulation of the fluorescence signal caused solely by modulation of the excitation source irrespective of any triplet formation.

Other fluorescent triplet probes

We have already described the use of carbocyanine dyes as triplet probes for the fluorescence depletion method. In principle the method could be applied to a whole range of already available fluorescent probes by using a continuous-wave gas laser operating in the ultra-violet (e.g. He-Cd at 325 nm; argon at 351-364 nm). Some advantages that might accrue include the possible study of NADH and NADPH bound as substrates to membrane-bound enzymes, and of fluorescent choles-terol and fatty acid analogues.

CONCLUDING REMARKS

Improvements in optical methods for measuring slow rotational move-ments in the time range from 1 μs upwards have advanced greatly in the last few years. Following the first practical description of the use of a triplet probe for membrane studies by Cherry et al. (1976), the detection sensitivity has been improved 10^{10}-fold (Johnson and Garland, 1981). Time resolution has also been greatly increased, from 20 μs to 1 μs (Spiers et al., 1983) or less (Jovin et al., 1981) by use of phosphorescence and/or improved instrumentation. The range of probes has been extended to the use of carbocyanine dyes as probes of slow lipid rotation (Johnson and Garland, 1983). Instrumentation has moved away from powerful flash-lamp pumped dye lasers of low repetition rate towards N_2-pumped dye lasers of high repetition rate. It is also possible to use microscopic methods via a laser-microscope combination (Fig. 4) that can also measure lateral diffusion coeffi-cients. Nevertheless, despite these considerable advances, and pos-sibly more to come, measurements of the rotational diffusion of mem-brane components should not be looked at as more than merely one of many powerful techniques that can be applied to answering questions concerning membrane structure and function.

ACKNOWLEDGEMENTS

Work referred to from our own laboratory was supported by the Royal Society, the Medical Research Council and the Science and Engineering Research Council. P.J. was in receipt of a postgraduate studentship from hte SERC, held in collaboration with Dr R.A. Clegg of the Hannah Research Institute, Ayr.

REFERENCES

Albrecht, A.C. (1961). *J. Molec. Spectrosc.* 6, 84-108.
Albrecht, A.C. (1970). *Prog. React. Kinetics* 5, 301-334.
Austin, R.H., Chan, S.S. and Jovin, T.M. (1979). *Proc. Natl. Acad. Sci. USA* 76, 5650-5654.
Axelrod, D. (1977). *Biophys. J.* 18, 129-133.
Axelrod, D. (1979). *Biophys. J.* 26, 557-574.
Axelrod, D., Koppel, D.E., Schlessinger, J., Elson, E. and Webb, W.W. (1976). *Biophys. J.* 16, 1055-1069.
Badea, M.G. and Brand, L. (1979). *Meth. Enzymol.* 61, 379-425.
Badley, R.A., Martin, W.G. and Schneider, H. (1973). *Biochemistry* 12, 286-275.
Bartholdi, M., Barrantes, J. and Jovin, T.M. (1981). *Eur. J. Biochem.* 120, 389-397.
Britt, A.D. and Moniz, W.V. (1973). *J. Org. Chem.* 38, 1057-1059.
Bürkli, A. and Cherry, R.J. (1981). *Biochemistry* 20, 138-145.
Cherry, R.J. (1978). *Meth. Enzymol.* 54, 47-61.
Cherry, R.J. (1979). *Biochim. Biophys. Acta* 559, 289-327.
Cherry, R.J., Burkli, A., Busslinger, M., Schneider, G. and Parish, G.R. (1976). *Nature (London)* 263, 389-393.
Cherry, R.J., Cogoli, A., Oppliger, M., Schneider, G. and Semenza, G. (1976). *Biochemistry* 15, 3653-3656.
Cherry, R.J. and Godfrey, R.E. (1981). *Biophys. J.* 36, 257-276.
Cherry, R.J., Nigg, E.A. and Beddard, G.S. (1980). *Proc. Natl. Acad. Sci. USA* 77, 5899-5903.
Cherry, R.J. and Schneider, G. (1976). *Biochemistry* 15, 3657-3661.
Cone, R.A. (1972). *Nature New Biol.* 236, 39-43.
Dale, R.E., Chen, L.A. and Brand, L. (1977). *J. Biol. Chem.* 252, 7500-7510.
Garland, P.B. (1981). *Biophys. J.* 33, 481-482.
Garland, P.B. (1982). *In* "Membranes and Transport" (Ed. A.N. Martonosi), Vol. 1, pp. 153-158. Plenum Press, New York.
Garland, P.B., Johnson, K. and Reid, G.A. (1982). *Biochem. Soc. Trans.* 10, 484-485.
Garland, P.B. and Moore, C.H. (1979). *Biochem. J.* 183, 561-572.
Greinert, R., Staerk, H., Stier, A. and Weller, A. (1979). *J. Biochem. Biophys. Meth.* 1, 77-83.
Gut, J., Richter, C., Cherry, R.J., Winterhalter, K.H. and Kawato, S. (1982). *J. Biol. Chem.* 257, 7030-7036.
Hanson, D.C., Yguerabide, J. and Schumaker, V.N. (1981). *Biochemistry* 20, 6842-6852.
Harvey, S.C. and Cheung, H.-C. (1980). *Biopolymers* 19, 913-930.
Hoffmann, W. and Restall, C. (1983). *In* "Topics in Molecular and Structural Biology" (Ed. D. Chapman), pp. 1-113. MacMillan, London.
Hoffman, W., Sarzala, M.G. and Chapman, D. (1979). *Proc. Natl. Acad. Sci. USA* 76, 3860-3864.
Hoffman, W., Sarzala, M.G., Gomez-Fernandez, J.C., Goni, F.M., Restall, C.J., Chapman, D., Heppeler, G. and Kreutz, W. (1980). *J. Molec. Biol.* 141, 119-132.
Jablonski, A. (1961). *Z. Naturforsch.* A16, 1-4.
Jameson, D.M. and Weber, G. (1981). *J. Phys. Chem.* 85, 953-958.
Johnson, P. and Garland, P.B. (1981). *FEBS Lett.* 132, 252-256.
Johnson, P. and Garland, P.B. (1982). *Biochem. J.* 203, 313-321.

Johnson, P. and Garland, P.B. (1983). (1983). *FEBS Lett.* 153, 391-394.

Jovin, T.M., Bartholdi, M., Vaz, W.L.C. and Austin, R.M. (1981). *Ann. N.Y. Acad. Sci. USA* 78, 389-397.

Junge, W. (1972). *FEBS Lett.* 25, 109-112.

Junge, W. and DeVault, D. (1975). *Biochim. Biophys. Acta* 408, 200-214.

Karnovsky, M.J., Kleinfeld, A.M., Hoover, R.L. and Klausner, R.D. (1982). *J. Cell Biol.* 94, 1-6.

Kawato, S., Gut, J., Cherry, R.J., Winterhalter, K.H. and Richter, C. (1982). *J. Biol. Chem.* 257, 7023-7029.

Kawato, S. and Kinosita, K. Jr. (1981). *Biophys. J.* 36, 277-296.

Kawato, S., Sigel, E., Carafoli, E. and Cherry, R.J. (1980). *J. Biol. Chem.* 255, 5508-5510.

Kawato, S., Sigel, E., Carafoli, E. and Cherry, R.J. (1981). *J. Biol. Chem.* 256, 7518-7527.

Kinosita, K.Jr., Ikegami, A. and Kawato, S. (1982). *Biophys. J.* 37, 461-464.

Koppel, D.E., Axelrod, D., Schlessinger, J., Elson, E.L. and Webb, W. W. (1976). *Biophys. J.* 16, 1313-1329.

Kunze, U. and Junge, W. (1977). *FEBS Lett.* 80, 429-434.

Lakowicz, J.R. and Cherek, H. (1981). *J. Biol. Chem.* 256, 6348-6353.

Lakowicz, J.R., Prendergast, F.G. and Hogen, D. (1979). *Biochemistry* 18, 508-519.

Lavalette, D., Amand, B. and Pochon, F. (1977). *Proc. Natl. Acad. Sci. USA* 74, 1407-1411.

Lessing, H.E. and von Jena, A. (1979). *In* "Laser Handbook" (Ed. M.L. Stitch), pp. 753-846. North-Holland, Amsterdam.

Lippert, J.L., Lindsay, R.M. and Schultz, R. (1981). *J. Biol. Chem.* 256, 12411-12416.

Lo, M.M.S., Garland, P.B., Lamprecht, J. and Barnard, E.A. (1980). *FEBS Lett.* 111, 407-412.

Mantulin, W.M. and Weber, G. (1977). *J. Chem. Phys.* 66, 4092-4099.

Matayoshi, E.D., Corin, A.F., Zidovetzki, R., Sawyer, W.H. and Jovin, T.M. (1983). *In* "Mobility and Recognition in Cell Biology" (Eds H. Sund and C. Veeger), pp. 119-134. Walter de Gruyter and Co., New York.

Moore, C., Boxer, D. and Garland, P. (1979). *FEBS Lett.* 108, 161-166.

Moore, C.H. and Garland, P.B. (1979). *Biochem. Soc. Trans.* 7, 945-946.

Nigg, E.A. and Cherry, R.J. (1979). *Biochemistry* 18, 357-365.

Nigg, E.A. and Cherry, R.J. (1980). *Proc. Natl. Acad. Sci. USA* 77, 4702-4706.

Peters, R. and Cherry, R.J. (1982). *Proc. Natl. Acad. Sci. USA* 79, 4317-4321.

Pochon, F., Amand, B. and Lavalette, D. (1978). *J. Biol. Chem.* 253, 7496-7499.

Reid, G.A. (1981). Ph.D. Thesis, University of Dundee.

Richter, C., Winterhalter, K.H. and Cherry, R.J. (1979). *FEBS Lett.* 102, 151-154.

Rigler, R. and Ehrenberg, M. (1973). *Quart. Rev. Biophys.* 6, 139-199.

Saffman, P.G. and Delbrück, M. (1975). *Proc. Natl. Acad. Sci. USA* 72, 3111-3113.

118

Garland and Johnson

Spiers, A., Moore, C.H., Boxer, D.H. and Garland, P.B. (1983). *Biochem. J.* 213, 67-74.
Smith, L.M., Weiss, R.M. and McConnell, H.M. (1981). *Biophys. J.* 36, 73-91.
Strambini, G.B. and Galley, W.C. (1976). *Nature* 260, 554-556.
Thulborn, K.R. and Sawyer, W.H. (1978). *Biochim. Biophys. Acta* 511, 125-140.
Tilley, L., Thulborn, K.R. and Sawyer, W.H. (1979). *J. Biol. Chem.* 254, 2592-2594.
Wagner, R., Carrillo, N., Junge, W. and Vallejos, R.H. (1981). *FEBS Lett.* 136, 208-212.
Wagner, R., Carrillo, N., Junge, W. and Vallejos, R.H. (1982). *Biochim. Biophys. Acta* 680, 317-330.
Wagner, R. and Junge, W. (1980). *FEBS Lett.* 114, 327-333.
Wagner, R. and Junge, W. (1982). *Biochemistry* 21, 1890-1899.
Weber, G. (1969). *Meth. Enzymol.* 16, 380-394.
Weber, G. (1977). *J. Chem. Phys.* 66, 4081-4091.
Weber, G. (1981). *J. Phys. Chem.* 85, 949-953.
Yguerabide, J. and Stryer, L. (1971). *Proc. Natl. Acad. Sci. USA* 68, 1217-1221.
Yguerabide, J., Epstein, H.F. and Stryer, L. (1970). *J. Molec. Biol.* 51, 573-590.
Zidovetzki, R., Yarden, Y., Schlessinger, J. and Jovin, T. (1981). *Proc. Natl. Acad. Sci. USA* 78, 6981-6985.

Phosphorescence Depolarization Techniques in the Study of Membrane Protein Rotational Mobility

E.K. MURRAY, C.J. RESTALL and D. CHAPMAN

INTRODUCTION

The study of diffusion, both rotational and lateral, in membranes is
of interest because such movements may affect, or be affected by, the
specific functions of the protein under study, and because information
may be gained about the micro-environment and state of aggregation of
the protein. Methods of examining and quantifying protein rotational
mobility have been developed by several groups of workers. This
motion occurs on a micro- to milli-second timescale and is suitable
for investigation by phosphorescence depolarization and transient
absorption dichroism. These techniques utilize the preferential
absorption, by a chromophore attached to the protein, of polarized
light and monitor the anisotropy decay, either in emission or absorption, as the protein rotates. The chromophore may be intrinsic, such
as the retinal component of bacteriorhodopsin (Cherry, 1978; Hoffman
et al., 1980; Cherry and Godfrey, 1981), or it may be necessary to
covalently bind a suitable phosphorescent probe to a site on the protein under study. Such probes include derivatives of the heavy halogen-substituted fluorescein analogues eosin and erythrosin (Garland
and Moore, 1979). These can be utilized to observe transient absorption dichroism decays (Naqvi et al., 1973; Cherry, 1979; Chapman and
Restall, 1981), or in the more sensitive measurements of flash-induced
phosphorescence anisotropy decay (Garland and Moore, 1979; Austin et

al., 1979) or delayed fluorescence anisotropy decay (Greinert *et al.*,
1982).

The rotational mobilities of several proteins have been investigated using these techniques, including cytochrome oxidase (Junge and
Devault, 1975; Kawato *et al.*, 1980), the band 3 protein of human
erythrocytes (Cherry and Nigg, 1979; 1981; Austin *et al.*, 1979; Jovin
et al., 1981), the acetylcholine receptor of the Torpedo electric
organ (Lo *et al.*, 1980; Bartholdi *et al.*, 1981), glycophorin reconstituted in dimyristoylphosphatidylcholine vesicles (Jovin *et al.*, 1981),
the Ca^{2+} ATPase of sarcoplasmic reticulum (Hoffman *et al.*, 1979;
Moore *et al.*, 1979; Burkli and Cherry, 1981; Speirs *et al.*, 1983),
zinc cytochrome *c* bound to mitochondrial membranes (Dixit *et al.*, 1982)
and rabbit liver cytochrome P450 (Greinert *et al.*, 1982).

The time-resolved methods of monitoring membrane protein rotational
mobility can provide detailed information about protein movement, but
they require sophisticated and expensive apparatus and the analysis of
the results may be a complex procedure. In some circumstances it may
not be necessary to perform such detailed measurements. It is possible to make qualitative assessments of such rotational mobilities
using the much simpler technique of time-averaged phosphorescence
depolarization (Murray *et al.*, 1983), which may be carried out on
relatively inexpensive, commercially available luminescence spectro-
photometers. Some of the limits and qualifications to the interpretation of such data will be discussed and the time-resolved and time-
averaged phosphorescence depolarization methods will be compared with
regard to the kind of information they can provide about rotational
mobility.

TIME-RESOLVED PHOSPHORESCENCE ANISOTROPY DECAY

Phosphorescence of a triplet probe, covalently bound to a membrane
protein, is induced by a short pulse of vertically polarized laser
light. In general, the resulting (singlet) absorption and (triplet)
emission are initially dichroic, and this dichroism is depleted in
time by the rotational motion of the membrane protein. The experiment
is designed to observe the phosphorescence emitted at right angles to
the excitation beam. Components polarized parallel and perpendicular
to the electric vector of the excitation beam are selected and monitored simultaneously as a function of time, characteristically over a

period of around 0.5 ms. Signal averaging is routinely used to
improve the signal to noise ratio of the data. The phosphorescence
anisotropy, defined as the difference between parallel and perpendicularly polarized emission intensities weighted by the total phosphorescence intensity, may then be computed as a function of time. The
response of the measuring system to the orthogonally polarized emission components is determined by rotating a quarter-wave plate in the
excitation beam path through $45°$ to produce a horizontally polarized
exciting pulse which should give rise to identical "parallel" and
"perpendicular" intensities. The optics are adjusted to produce visu-
ally approximately identical signals for both polarized emission components. Inspection alone, however, cannot guarantee that a true
balance over the whole time-span has been achieved, and it is necessary to compute a correction ratio (G-factor). This is used to normalize the orthogonally polarized intensities excited by vertically
polarized light. The values of this ratio may vary slightly, but
significantly with time and it is then necessary to use its time dependence when computing anisotropy values. This effect and other small
instabilities in the system are best eliminated by alternating the
collection of anisotropy data with the collection of G-factor data,
and calculating the time-dependent anisotropy as:

$$r(t) = [G(t)I_{\parallel}(t) - I_{\perp}(t)]/s(t) \qquad (1)$$

where $s(t)$, the total emission intensity is given by

$$s(t) = G(t)I_{\parallel}(t) + 2I_{\perp}(t) \qquad (2)$$

Examples of parallel and perpendicularly polarized emission component signals and their computed anisotropy and total emission decays
are shown in Fig. 1. These were obtained in our laboratory from the
Ca^{2+}-Mg^{2+} ATPase of sarcoplasmic reticulum covalently labelled with
erythrosin isothiocyanate. Both the total emission and phosphorescence anisotropy appear to decay in a multiexponential manner:

$$s(t) = \Sigma_i \alpha_i \exp[-t/\tau_i] \qquad (3)$$

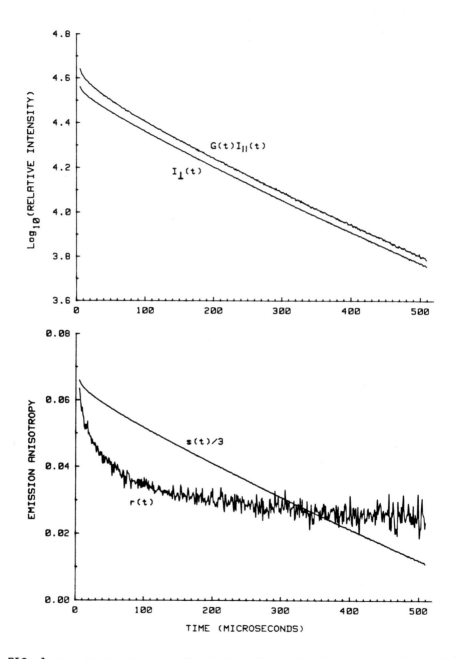

FIG. 1 Parallel and perpendicularly polarized emission signals, and the resultant anisotropy and emission decay curves, obtained with erythrosin labelled Ca^{2+}-Mg^{2+} ATPase of sarcoplasmic reticulum at $25^{\circ}C$.

and:

$$r(t) = \sum_j \beta_j \exp[-t/\phi_j] \tag{4}$$

where τ_i is the lifetime of the i'th component of the phosphorescence decay and ϕ_j is the rotational correlation time of the j'th component of the anisotropy decay. The anisotropy usually has a residual non-decaying value on the time-scale of the measurement at long times (r_∞) due to the protein being unable to rotate about axes parallel to the lipid bilayer. Any protein which is stationary in the bilayer on the time-scale of the experiment also contributes to the value of the residual anisotropy.

Although an anisotropy may be fitted arbitrarily to a series of independent exponential terms plus a constant, this does not provide direct quantitative information relating to the physical motion of the protein. For that, appropriate relationships between the various correlation times ϕ_j and pre-exponential terms β_j according to theoretical considerations of possible models (Kinosita et al., 1977, 1982; Lipari and Szabo, 1980; Kawato and Kinosita, 1981; Zannoni, 1981) are required. For example, the simplest model for membrane protein rotation, which only considers one-dimensional rotation of the protein as a whole about the normal to the lipid bilayer, gives rise to an anisotropy decay of the form:

$$r(t) = \beta_1 \exp(-D_m t) + \beta_2 \exp(-4D_m t) + \beta_3 \tag{5}$$

where D_m is the rotational diffusion coefficient of the protein and β_3 the residual anisotropy. If the phosphor is rigidly attached to the protein, the pre-exponential terms are given (see e.g. Rigler and Ehrenberg, 1973) by:

$$\beta_1 = (6/5)\sin\theta_a \cos\theta_a \sin\theta_e \cos\theta_e \cos\xi$$

$$\beta_2 = (3/10)\sin^2\theta_a \sin^2\theta_e (\cos^2\xi - \sin^2\xi) \tag{6}$$

$$\beta_3 = (1/10)(3\cos^2\theta_a - 1)(3\cos^2\theta_e - 1)$$

where θ_a and θ_e are the polar angles made by the absorption and emission transition moments respectively with the axis of rotation, i.e. with the bilayer normal, and ξ is the dihedral angle made by their projections onto the bilayer plane. The expression for β_3 assumes all labelled proteins to be identical and rotating. These coefficients are related to the angle θ_{ae} between absorption and emission moments by:

$$\beta_1 + \beta_2 + \beta_3 = 0.4[(3/2)\cos^2\theta_{ae} - (1/2)] = r_f \qquad (7)$$

where r_f is known as the fundamental anisotropy.

Unless the absorption and emission transitions are coincident $(\theta_a = \theta_e = \theta, \xi = 0)$, Equations (5) and (6) can already lead to quite complicated decay courses, even in this simple case. A number of further complications may arise, however:

(i) dimerization or, in general, oligomerization equilibria for the protein (or association equilibria with other proteins),

(ii) as one limit of the above, aggregation of part of the protein population leading to an immobile fraction,

(iii) restricted segmental motion of the phosphor itself and/or a segment or segments to which it is attached,

(iv) wobbling of the protein about the bilayer normal,

(v) heterogeneity of orientation of the phosphor on the protein either within a unique binding site or between different binding sites.

In (iii), part or all of the segmental motion may occur on a time-scale much faster than can be resolved in the experiment and the transitions may, or may not, be azimuthally randomized about the axis of rotation. Effects (iii) - (v) lead, in general, to ill-defined averaging of the pre-exponential factors given in Equation (6), and additional decay terms in Equation (5). In some cases, however, such averaging may lead to considerably simpler anisotropy decay expressions. Thus, for a homogeneous system in which rapid segmental motion over a restricted range occurs with full azimuthal averaging of the transition moments about the axis of this motion, Equation (6) reduces (cf. Wahl *et al.*, 1970) to:

$$\beta_1 = 3r_0 \sin^2\theta_m \cos^2\theta_m$$

$$\beta_2 = (3/4)r_0 \sin^4\theta_m \qquad\qquad (8)$$

$$\beta_3 = r_0 [(3/2)\cos^2\theta_m - (1/2)]^2$$

where θ_m is the angle made by the axis of rapid rotation with the bilayer normal. The zero-point anisotropy, r_0, is decreased from its limiting fundamental value r_f given in Equation (7), by a factor depending on the range of the segmental motion. If, on the other hand, complete static randomization of the transition moments were to result from binding site heterogeneity, then (cf. Brochon et al., 1972):

$$\beta_1 = \beta_2 = 0.4r_f; \qquad \beta_3 = 0.2r_f \qquad\qquad (9)$$

If fast azimuthal randomization of the transition moments occurs, and, in addition, the orientations of the axes for this motion are statistically randomized, r_f in Equation (9) will be replaced simply by r_0.

A further complication that may need to be considered arises if the decay of the phosphorescence itself is complex, i.e. not monoexponential. A biexponential decay, for instance, may signify protein heterogeneity (it might also arise kinetically in a homogeneous system), and if this is linked to rotational heterogeneity, complex anisotropy decay curves may arise even when the individual depolarizations are simple. Such, for instance, has been the interpretation of an initial rise in anisotropy observed for the Torpedo electric organ acetylcholine receptor labelled with an erythrosin derivative of α-bungarotoxin (Lo et al., 1980). On the other hand, the effect can also be observed in a homogeneous system for appropriate combinations of θ_a, θ_e and ξ, e.g. when the absorption and emission transitions lie on either side of the bilayer normal, rotation of the protein thereby giving rise initially to an increasing alignment of the transitions, as invoked in another study of this system using eosin rather than erythrosin as the phosphor (Bartholdi et al., 1981). Since the observed anisotropy decay in the first interpretation is an average one weighted by the phosphorescence decays, its course will change if the emission lifetimes are varied, e.g. by quenching with oxygen, whereas this will not be true of the homogeneous situation, however complex the emission kinetics may be, and as pointed out by Jovin (1983), this provides a reasonable criterion for separating these two possibilities.

Initial increases in anisotropy with time have also been observed in eosin-labelled erythrocyte ghosts (Austin *et al.*, 1979) and for erythrosin-labelled epidermal growth factor-receptor complexes (Zidovetzki *et al.*, 1981).

The observation of an r_0 value considerably reduced from the value of r_f, which can be measured in highly viscous or solid media (Garland and Moore, 1979), is tentative evidence that rapid reorientational motion of the probe, or a small segment to which it is attached, may have resulted in essentially complete azimuthal averaging of the absorption and emission transition moments. The converse is not true, however, since r_0 will be identical with r_f if one of these transition moments happens to be aligned along the axis of rapid rotation resulting in this rapid motion not being detected by depolarization measurements.

If the axis for such rapid reorientational motion is located in a segment of the protein that also has restricted reorientational freedom on a time-scale long enough to be experimentally detectable, and this motion in turn is fully azimuthally averaged about some axis fixed in the protein as a whole (Fig. 2), the overall anisotropy decay, which represents the product of the normalized anisotropy decay functions for the independent motions (cf. Kawato and Kinosita, 1981) may be well approximated by:

$$r(t) = (\Sigma_i \beta_i \exp[-t/\phi_i]) + \beta_{m1} \exp[-D_m t] + \beta_{m2} \exp[-4D_m t]) + \beta_{m3} \quad (10)$$

with i at most 1 or 2. The correlation times ϕ_i contain information about both the rates and extent of restricted motion (Kinosita *et al.*, 1977, 1982). The β_m reflect the anisotropy of the axis fixed in the macromolecule and are reduced from their fundamental values by the product of factors describing the extents of depolarization due to both the fast and slower segmental motions.

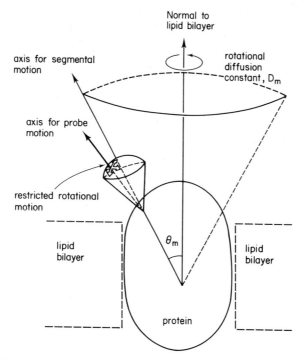

FIG. 2 Model of the rotational relaxation processes for erythrosin isothiocyanate-labelled sarcoplasmic reticulum Ca^{2+}-Mg^{2+} ATPase described fully in Restall *et al.* (1984).

In the above circumstances, the angle θ_m between the segmental reorientation axis and the normal to the lipid bilayer is obtainable from the ratio between β_{m1} and β_{m2}:

$$\theta_m = \arctan[2\sqrt{(\beta_{m2}/\beta_{m1})}] \qquad (11)$$

The residual anisotropy, β_{m3} is also related to these by:

$$\beta_{m3} = [(3/2)(\frac{\beta_{m1}}{\beta_{m1}+4\beta_{m2}}) - (1/2)]^2 \qquad (12)$$

These relationships can be utilized to test for the presence of an immobile, e.g. aggregated, fraction of protein (Greinert *et al.*, 1982; Restall *et al.*, 1984). If β_{m3} thus calculated does not correspond, within experimental error, to the residual anisotropy actually observed in the experiment, the difference will represent a residual anisotropy due to the stationary fraction:

$$\beta_{stat} = \beta_{3obs} - \beta_{m3} \qquad (13)$$

Its quantitative interpretation in terms of the fraction of stationary protein depends on whether or not fast and slower segmental motions attributed to the mobile fraction are affected by the immobilizing process, and also by whether or not there is any difference in the decay of the phosphorescence itself between the mobile and immobile forms.

Our own results, as well as others presented more qualitatively in the literature, for depolarization of the phosphorescence of specifically erythrosin-labelled Ca^{2+}-Mg^{2+} ATPase of sarcoplasmic reticulum (Restall *et al.*, 1984) indicate that the above considerations, without any appreciable change in the complex decay of the phosphorescence itself between the mobile and immobile fractions, may be appropriate to rotation of this protein isolated in its native membrane. Interestingly enough, although the observed residual anisotropy in this case is rather high, application of Equations (11) and (12) indicate that β_{m3} is close to zero ($\theta_m \simeq 60^{\circ}$, close to the "magic" angle of $\sim 55^{\circ}$ for which β_{m3} is identically zero), so that virtually all of the observed residual anisotropy is attributable in this case to immobilized protein.

TIME-AVERAGED PHOSPHORESCENCE ANISOTROPY

Commercial spectrofluorimeters may be used in the phosphorescence mode to observe the average phosphorescence anisotropy of a sample over a chosen time interval. The sample is illuminated by a pulse of vertically or horizontally (G-factor) polarized light, and the polarized components of phosphorescence emission are recorded as time-averaged signals over the interval t_1 to t_2. The time-averaged phosphorescence anisotropy thus derived represents the integral of the anisotropy decay over the measuring interval, weighted by the decay of the phosphorescence over this interval:

$$\langle r \rangle_{t_1,t_2} = [G \int_{t_1}^{t_2} I_{\parallel}(t)\,dt - \int_{t_1}^{t_2} I_{\perp}(t)\,dt]/\int_{t_1}^{t_2} s(t)\,dt =$$

$$= \int_{t_1}^{t_2} r(t)\,s(t)\,dt / \int_{t_1}^{t_2} s(t)\,dt \qquad (14)$$

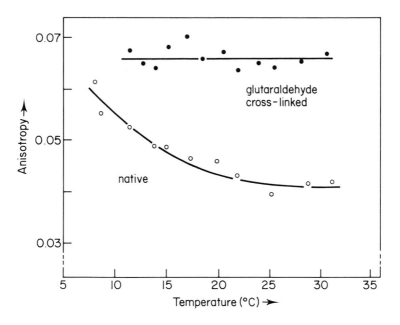

FIG. 3 Time-averaged phosphorescence anisotropy of native Ca^{2+}-Mg^{2+} ATPase labelled with erythrosin isothiocyanate (o), compared with that of the same protein after cross-linking using glutaraldehyde (●), the latter showing an absence of temperature dependence (redrawn from data presented in Murray et al., 1983).

If the lower limit t_1 is chosen to be as close to zero on the millisecond time-scale as possible - it is limited to 10 µs by the design of the spectrometer - and the upper limit is about ten times the longest phosphorescence decay time, the anisotropy measured will effectively be the "steady-state" value $<r>$, i.e. it will be essentially fully averaged over the decay course. Since the longest rotational correlation times D_m^{-1} are usually of the same order as the longest decay times, $<r>$ will be heavily weighted towards β_3, the limiting anisotropy, customarily designated r_∞. In attempting to monitor changes in mobility per se, it is thus useful to set t_2 to a value comparable with D_m^{-1}. Examples of the temperature dependence of the time-averaged phosphorescence anisotropy of mobile and immobile proteins measured under such conditions (Murray et al., 1983) are shown in Fig. 3. If aggregation resulting in complete immobilization is the phenomenon of interest, the lower limit t_1 should be set three- or four-fold higher than D_m^{-1} and t_2 about ten-fold higher than the

longest decay time so that only changes in r_∞ are effectively monitored. Such quick and simple measurements for the detection of rotational relaxation of membrane proteins under different conditions may
provide a useful qualitative monitor of mobility in experiments
designed to investigate the interrelationship of mobility and function
in such systems.

A semi-quantitative estimation of mobility may be made as follows.
The anisotropy decay may be approximated to:

$$r(t) = r_\infty + (r_0 - r_\infty)\exp[-t/<\phi>] \qquad (15)$$

where $<\phi>$ represents an average correlation time and r_0 represents
approximately the limiting anisotropy on a time-scale of a few tens
of microseconds, that is, the value obtained after sub-microsecond
and microsecond depolarization processes. By also approximating the
phosphorescence decay to a single exponential, it is possible by
taking three time intervals t_1, t_2 to obtain an average value for the
phosphorescence lifetime $<\tau>$ and the values of r_0, r_∞ and $<\phi>$ defined
in Equation (15). One of the time intervals may be chosen to give r_∞
directly as indicated above.

Another approach to semi-quantitative evaluation of these parameters is suggested from the work of Lakowicz and his colleagues on
the effects of oxygen-quenching on fluorescence anisotropies of small
membrane-bound fluorophores (Lakowicz et al., 1979a,b; Lakowicz and
Knutson, 1980). Here, the parameters entering into Equation (15) are
obtained by fitting the true steady-state anisotropy values obtained
at various concentrations of oxygen which, at high pressures, quenches
the singlet excited state to give lifetime-resolved anisotropies.
Since the triplet state giving rise to phosphorescence is about three
orders of magnitude longer-lived than the singlet which gives rise to
fluorescence, quite low oxygen tensions will quench it and reduce the
phosphorescence lifetime. Reasonable semi-quantitative estimates of
mobility may thus also be expected from the application of this technique to the phosphorescence case. Moreover, the opportunity to
estimate r_∞ directly by combining the late time-window anisotropy
measurement with such measurements potentially renders the estimates
of r_0 and $<\phi>$ more accurate than in the fluorescence case.

CONCLUSIONS

Both time-resolved and time-averaged phosphorescence depolarization techniques may be useful in the investigation of protein rotational mobility. A detailed model of protein motion can be constructed using time-resolved techniques but this type of analysis is a time-consuming and complex procedure and, as we have discussed, there are many possible complications. In situations where bulk estimates of the relative mobilities of proteins are all that is required, time-averaged phosphorescence anisotropy measurements are adequate, much quicker and simpler to perform. Particularly in combination with oxygen quenching, the time-averaged technique is also capable of providing reasonable semi-quantitative estimates of protein rotational mobility in membranes.

ACKNOWLEDGEMENTS

The authors wish to thank R.E. Dale for helpful advice in the preparation of this article. We are grateful to the S.E.R.C., the Muscular Dystrophy Group of Great Britain, the Wellcome Trust and the Royal Society for financial support.

REFERENCES

Austin, R.H., Chan, S.S. and Jovin, T.M. (1979). *Proc. Natl. Acad. Sci. USA* 76, 5650-5654.
Bartholdi, M., Barrantes, F.J. and Jovin, T.M. (1981). *Eur. J. Biochem.* 120, 389-397.
Brochon, J.-C., Wahl, Ph. and Auchet, J.-C. (1972). *Eur. J. Biochem.* 25, 20-32.
Burkli, A. and Cherry, R.J. (1981). *Biochemistry* 20, 138-145.
Chapman, D. and Restall, C.J. (1981). *Biochem. Soc. Symp.* 46, 139-145.
Cherry, R.J. (1978). *Methods in Enzymology* 54, 47-61.
Cherry, R.J. (1979). *Biochim. Biophys. Acta* 559, 289-327.
Cherry, R.J. and Godfrey, R.E. (1981). *Biophys. J.* 36, 257-276.
Cherry, R.J. and Nigg, E.A. (1979). *Progr. Clin. Biol. Res.* 30, 475-481.
Cherry, R.J. and Nigg, E.A. (1981). *Progr. Clin. Biol. Res.* 51, 59-77.
Dixit, B.P.S.N., Waring, A.J., Wells, K.O.,III, Wong, P.S., Woodrow, G.V.,III and Vanderkooi, J.M. (1982). *Eur. J. Biochem.* 126, 1-9.
Garland, P.B. and Moore, C.H. (1979). *Biochem. J.* 183, 561-572.
Greinert, R., Finch, S.A.E. and Stier, A. (1982). *Biosci. Repts.* 2, 991-994.
Hare, F. and Lussan, C. (1977). *Biochim. Biophys. Acta* 467, 262-272.
Hoffman, W., Sarzala, M.G. and Chapman, D. (1979). *Proc. Natl. Acad. Sci. USA* 76, 3860-3864.

Hoffman, W., Restall, C.J., Hyla, R. and Chapman, D. (1980). *Biochim. Biophys. Acta* 602, 531-538.

Jovin, T.M. (1983). personal communication.

Jovin, T.M., Bartholdi, M., Vaz, W.L.C. and Austin, R.H. (1981). *Ann. N.Y. Acad. Sci.* 365, 176-196.

Junge, W. and Devault, D. (1975). *Biochim. Biophys. Acta* 408, 200-214.

Kawato, S. and Kinosita, K.,Jr. (1981). *Biophys. J.* 36, 277-296.

Kawato, S., Sigel, E., Carafoli, E. and Cherry, R.J. (1980). *J. Biol. Chem.* 255, 5508-5510.

Kinosita, K.,Jr., Kawato, S. and Ikegami, A. (1977). *Biophys. J.* 20, 289-305.

Kinosita, K.,Jr., Ikegami, A. and Kawato, S. (1982). *Biophys. J.* 37, 461-464.

Lakowicz, J.R., Prendergast, F.G. and Hogen, D. (1979a). *Biochemistry* 18, 508-519.

Lakowicz, J.R., Prendergast, F.G. and Hogen, D. (1979b). *Biochemistry* 18, 520-527.

Lakowicz, J.R. and Knutson, J.R. (1980). *Biochemistry* 19, 905-911.

Lipari, G. and Szabo, A. (1980). *Biophys. J.* 30, 489-506.

Lo, M.M.S., Garland, P.B., Lamprecht, J. and Barnard, E.A. (1980). *FEBS Lett.* 111, 407-412.

Moore, C., Boxer, D. and Garland, P.B. (1979). *FEBS Lett.* 108, 161-166.

Murray, E.K., Restall, C.J. and Chapman, D. (1983). *Biochim. Biophys. Acta* 732, 347-351.

Naqvi, R.K., Gonzales-Rodrigruez, J., Cherry, R.J. and Chapman, D. (1973). *Nature* 245, 249-251.

Restall, C.J., Dale, R.E., Murray, E.K., Gilbert, C.W. and Chapman, D. (1984). *Biochemistry* in press.

Rigler, R. and Ehrenberg, M. (1973). *Quart. Rev. Biophys.* 6, 139-199.

Spiers, A., Moore, C.M., Boxer, D.H. and Garland, P.B. (1983). *Biochem. J.* 213, 67-74.

Wahl, Ph., Meyer, G., Parrod, J. and Auchet, J.-C. (1970). *Eur. Polymer J.* 6, 585-608.

Zannoni, C. (1981). *Mol. Phys.* 42, 1303-1320.

Zidovetzki, R., Yarden, Y., Schlessinger, J. and Jovin, T.M. (1981). *Proc. Natl. Acad. Sci. USA* 78, 6981-6985.

Dynamic Light Scattering and Fluorescence Photobleaching Recovery: Application of Complementary Techniques to Cytoplasmic Motility

B.R. WARE

INTRODUCTION

Among the greatest challenges to the modern biophysicist is the physical characterization of dynamic biological phenomena. Clearly, time resolution of the physical parameter to be measured is an inherent requisite. The ultimate interpretation of the data then requires a rigorous understanding of the physical meaning both of the parameter and of its time variation. In this article I shall review the principles and methodologies of two physical techniques, namely dynamic light scattering and fluorescence photobleaching recovery, for the characterization of the most basic types of motion. I shall illustrate the application of these techniques to dynamic biological phenomena by use of the example of cytoplasmic motility.

In solutions that are concentrated and/or partially ordered, the distinction between tracer transport coefficients and transport coefficients that characterize mutual motion becomes significant. If the solution contains many species, the tracer transport coefficients of each species reflect the mobility of that species, while transport parameters of the bulk medium characterize the overall hydrodynamics and rheology of the system. Dynamic light scattering techniques measure mutual transport coefficients and characterize the rheology of the bulk medium, whereas fluorescence photobleaching recovery is a tracer technique that determines the mobility of a single species within a medium. These two fundamentally complementary techniques have been applied to the study of both the rheological characteristics and molecular mobilities in the cytoplasm of living cells and in solutions of cytoplasmic proteins. The objective of these investigations is the elucidation of the molecular and mechanical details of the mechanism by which the motive force is generated.

Dynamic light scattering is a general term applied to any technique that utilizes the time variation of scattered light to infer dynamic properties of the scattering medium. The phenomenon characterized by such measurements typically include translational and rotational diffusion, vibrational motions, and any of a number of types of directed flow. Fluorescence photobleaching recovery (FPR) is a tracer technique utilized primarily for the characterization of the translational motion of specific molecular species. The species of interest is tagged with a fluorophore, and its motion is detected by photobleaching a portion of the sample and measuring the characteristic time for the bleached region to fade away as bleached and unbleached species re-randomize their distributions by translational diffusion. Although these two techniques can often be applied to the measurement of the same types of motion, the respective parameters derived are fundamentally different and often complementary. Thus the combination of the two techniques can be particularly powerful.

As an example of the complementary nature of the parameters measured by these two techniques, we consider first the quantitative descriptions of translational diffusion. The diffusive motion of a particle dissolved or suspended in a solvent may be characterized by the mean-square displacement $<x^2>$ after a period of time (t). It was demonstrated by Einstein in 1905 that this relationship may be expressed in one dimension (see Tanford, 1961) by the equation:

$$<x^2> = 2\ Dt \tag{1}$$

where D is a constant, called the translational diffusion coefficient, whose magnitude is given by the simple relationship:

$$D = kTf^{-1} \tag{2}$$

where k is Boltzmann's constant, T is absolute temperature, and f is

the frictional coefficient of the particle, which is determined generally by the linear dimension of the particle and the solvent shear
viscosity (η). For a sphere of radius R:

$$f = 6\pi\eta R \qquad (3)$$

A second means of characterizing diffusion is based on the realization that a concentration gradient of particles will relax in time
due to the random nature of diffusive motion. The relevant equation,
often called the diffusion equation or Fick's second law, for one
dimension is:

$$\frac{\partial C}{\partial t} = \frac{\partial}{\partial x}\left(D'\frac{\partial C}{\partial x}\right) \qquad (4)$$

where C is concentration. The constant D' is properly called the
mutual diffusion coefficient. It differs from the tracer diffusion
coefficient D because the gradient in concentration of solute particles may affect the average motion of a particle at a particular
position in the gradient. A general relation for the mutual diffusion
coefficient has been written (Chu, 1974) as:

$$D' = f^{-1}\frac{\partial\mu}{\partial \ln c}(1-\phi) \qquad (5)$$

where μ is the chemical potential and φ is the volume fraction of the
diffusing species. In the limit of infinite dilution, the mutual
diffusion coefficient becomes equal to the tracer diffusion coefficient, but for the conditions of many experiments, the distinction
between the two is significant.
    Fluorescence photobleaching recovery is a tracer technique that
measures the tracer diffusion coefficient of the labelled species,
provided it can be assumed that the bleached and unbleached species
are thermodynamically equivalent. Dynamic light scattering, as we
shall see, reflects the dynamics of the relaxation of microscopic
concentration fluctuations in solution and consequently must be interpreted in terms of the mutual diffusion coefficient. Thus in the
simplest case of the diffusion of a single macromolecular species,

these two techniques report related but fundamentally distinct parameters that can be substantially different, particularly in cases for
which solute-solute or solute-solvent interactions are strong. The
distinction between the two techniques becomes even greater as the
complexity of the system increases. In a complex mixture such as
cell cytoplasm, the data from dynamic light scattering are well
defined only in terms of optical parameters; each molecular species
contributes to the scattering in an undetermined proportion dependent
upon a number of factors. Light scattering data are thus useful in
such systems for the determination of bulk, average parameters of the
optics and rheology. FPR data, on the other hand, are specific to
the labelled species only and hence permit molecular characterizations
even in complex media. The two techniques thus complement each other
in an ideal way to provide a meaningful description both of the bulk
properties and of the molecular components. Before we proceed with
the illustration of this point, it will be useful to describe in more
detail the physical principles of these two techniques and the background of the biological problem to which their application will be
illustrated.

DYNAMIC LIGHT SCATTERING

Light is scattered by optical inhomogeneities in the medium of propagation. No real medium has a perfectly uniform polarizability, so
light scattering may be observed from any phase of matter to study
the origin of optical fluctuations in space and time. The sources of
these optical fluctuations may be of interest for a diverse range of
investigations. Variations in local order, temperature, pressure,
density, and composition will generally give rise to scattered light.
The intensity of the scattering will be proportional to the magnitude
of the optical fluctuations, and the time (or frequency) dependence
of the scattered light will be determined by the lifetime and velocity
of the optical fluctuations.

For biological applications, the scattering material is generally
a solution or suspension of biological macromolecules or assemblies
in an aqueous solvent. These macromolecules do not have the same
polarizability as the solvent and therefore constitute optical inhomogeneities. In most cases it is best to consider the scattering to

arise from fluctuations in the concentration of these macromolecules. The magnitude of these fluctuations is proportional to the concentration of the macromolecules, and the optical inhomogeneity resulting from the concentration fluctuation is proportional to the mass (molecular weight) of the macromolecule and to the extent to which the optical properties of the macromolecules are different from those of the solvent (usually expressed as the refractive index increment dn/dc). Measurement of the scattering intensity from a solution of known concentration of macromolecules with known refractive index increment can thus be interpreted to determine the molecular weight of the macromolecules. Variation of the scattered intensity with concentration can be used to study the extent of solution non-ideality, usually expressed as the second virial coefficient of the expansion of the osmotic compressibility in terms of concentration. For samples of large macromolecules, variation of the scattered intensity with the experimental scattering angle can be interpreted in terms of the size (mean-square radius of gyration) and shape of the macromolecules. Such experiments to deduce molecular weight, radius of gyration, and virial coefficients from the intensity of scattered light are known collectively as elastic light scattering and have been a useful component of the arsenal of the biophysicist for over three decades (see Tanford, 1961; van de Hulst, 1957; Huglin, 1972).

The scattering process described above is referred to as Rayleigh scattering, after Lord Rayleigh, who first formulated its quantitative description. Truly inelastic scattering (the Raman effect) results from an exchange of energy between the scatterer and the scattered photon. However, even the elastic scattering mechanism described by Rayleigh may have slight shifts in energy associated with it when, as is generally true, the scatterers are in motion. If there is no net transport in the observed system, the spectrum will be symmetric about the incident frequency. Thus there is no net exchange of energy between the scatterer and the incident radiation, but any individual photon may be scattered with a change in frequency. This process is often termed quasi-elastic light scattering (QELS), or it may be known by a number of other names including dynamic light scattering, laser light scattering, photon correlation spectroscopy, and light scattering spectroscopy.

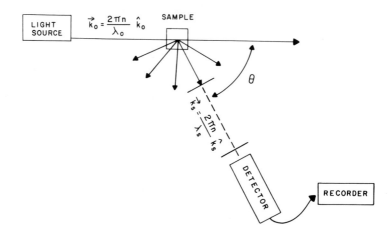

FIG. 1 Schematic diagram of a light scattering measurement, indicating the geometric parameters defined in the text.

In order to define the fundamental relationships of dynamic light scattering, we must make reference to the geometry of the measurement which is defined in Fig. 1. The incident light is characterized by wave vector $\vec{k}_0$ and the scattered light by $\vec{k}_s$. The angle between these two vectors is called the scattering angle θ. If two points in the scattering medium are separated by a distance and direction expressed by the vector $\vec{r}$, then it can be shown that the scattered light from these two points will arrive at the detector with a difference in phase Δ expressed by the relation:

$$\Delta = \vec{k}_0 \cdot \vec{r} - \vec{k}_s \cdot \vec{r} = (\vec{k}_0 - \vec{k}_s) \cdot \vec{r} \qquad (6)$$

The quantity $(\vec{k}_0 - \vec{k}_s)$ occurs so frequently in scattering processes that it is defined to be the scattering vector $\vec{k}$:

$$\vec{k} = \vec{k}_0 - \vec{k}_s \qquad (7)$$

with magnitude given by:

$$K = \frac{4\pi n}{\lambda_0} \sin\left(\frac{\theta}{2}\right) \qquad (8)$$

The scattering process may be viewed as a space-time Fourier transform of the scattering medium. At each scattering angle θ, the scattering intensity is proportional to the Fourier component of the medium with wavelength $2\pi/K$, and the time dependence of that Fourier component may be determined by measuring the intensity fluctuations of the scattered light (Pecora, 1964). This time dependence is generally expressed in terms of the correlation time τ_c of the intensity, i.e. the characteristic time taken for a fluctuation in intensity to decay to $1/e$ of its initial fluctuation magnitude. If we assume that the microscopic concentration fluctuations δC that give rise to the intensity obey the macroscopic diffusion equation (Equation 4), then we have:

$$\frac{\partial \delta C}{\partial t} = \begin{cases} D' \dfrac{\partial^2 \delta C}{\partial x^2} & \text{(one dimension)} \\[2mm] D' \nabla^2 \delta C & \text{(three dimensions)} \end{cases}$$

(9a)

(9b)

where D' is assumed space invariant. Fourier transformation of Equation (9b) to $\vec{K}$ space gives:

$$\frac{\partial \delta C}{\partial t} = D'K^2 \delta C \tag{10}$$

the solution of which is:

$$\delta C(t) = \delta C(0)\, e^{-D'K^2 t} \tag{11}$$

Thus:

$$\tau_c = (D'K^2)^{-1} \tag{12}$$

Note the correspondence between Equations (12) and (1), with the exception that the mutual diffusion coefficient must be employed since we have described the relaxation of a concentration gradient. Thus, in practice, the intensity autocorrelation function of light scattered from a solution of diffusing macromolecules is an exponential with time constant given by $1/D'K^2$. The corresponding frequency spectrum of the scattered light (the Fourier transform of the autocorrelation function) is a Lorentzian peak centred at the incident frequency with half-width at half-height (in angular frequency) given by $D'K^2$. Rigorous derivations of the theory and descriptions of the

experimental methodology and data analysis may be found in an extensive literature, which includes several fine books (Chu, 1974; Berne
and Pecora, 1976; Cummins and Pike, 1974, 1977; Chen et al., 1981).

If the scattering macromolecules are optically anisotropic, then
the scattering from each macromolecule will fluctuate as the orientation of the macromolecule fluctuates. Thus the correlation time
includes a contribution related to the rotational diffusion coefficient D which in the simplest case is given (Pecora and Steele, 1965;
Pecora, 1968a,b) by:

$$\tau_c = (D'K^2 + 6D_R)^{-1} \tag{13}$$

Unfortunately, the anisotropy of the macromolecule must be extreme in
order for the rotational contribution to the dynamic data to be
resolvable.

Another type of motion which could give rise to dynamic light
scattering is the intramolecular (segmental) motion of a flexible
coil macromolecule. If the domain of a polymer in solution is comparable to $2\pi/K$, then the reorientation of various segments of the
polymer by thermal diffusion will produce time-dependent interferences
of the scattered radiation and will thus modulate the observed intensities. The basic theory of this effect (Pecora, 1965, 1968c) predicts that the spectrum of the scattered light consists of a sum of
Lorentzian terms whose half-widths would be related to the decay
times of the normal (statistically independent) modes of vibration of
the polymer in solution. The weighting factors of these terms would
be related to the relative changes in the scattering interference
factor, which is determined by the spatial amplitude of each mode and
by the magnitude of K. Applications of this principle have included
determinations of dynamic scaling relationships and entanglements,
principally of synthetic polymers (e.g. Munch et al., 1977; Lin and
Chu, 1980; Nishio and Wada, 1980; Mathiez et al., 1980). The primary
biological system studied by this approach is the nucleic acid family
(e.g. Parthasarathy and Schmitz, 1980; Lin et al., 1981).

A common complication in QELS measurements is sample polydispersity.
Each component in solution contributes a term to the autocorrelation

function with a time constant proportional to its size, so the
measured data form is an indeterminate sum of exponentials. The simplest method of dealing with this problem is to analyse the data in
a manner that determines a defined average mutual diffusion coefficient with some information regarding the scatter of the distribution
(Koppel, 1972; Bezot *et al.*, 1978; Selser, 1979; Brehm and Bloomfield,
1975). A more sophisticated approach may permit direct interpretation of the data in terms of the size distribution function of the
sample, but extremely precise data are required for meaningful results
to be achieved (Gulari *et al.*, 1979; Provencher *et al.*, 1978).

A final case important for our consideration involves the dynamic
light scattering measurement from a sample which is undergoing directed flow. In this case the measurement can be understood as a Doppler
shift measurement, not different in principle from highway radar
velocimetry. It is easy to show (Yeh and Cummins, 1964; Ware, 1977)
that the magnitude of the Doppler shift ($\Delta\nu$, in Hz) is given by the
projection of the measured velocity onto the scattering vector:

$$\Delta\nu = (2\pi)^{-1} \ \vec{K}\cdot\vec{V} \tag{14}$$

The Doppler-shifted peak at frequency $\Delta\nu$ will be broadened by diffusion ($DK^2/2\pi$) as well as by dispersion in $\vec{V}$. The detection of the
Doppler shift, which can be as small as a fraction of Hz, requires a
"beating" method in which the scattered light is mixed with a reference beam, the local oscillator, at the surface of a photodetector.
The output of the photodetector contains a low-frequency beat component at the frequency difference between the local oscillator and the
scattered beam, i.e. the Doppler shift frequency. This technique is
known as laser Doppler velocimetry (LDV), and its many applications
are described in an extensive literature (e.g. Cummins and Pike,
1977; Ware, 1977; Drain, 1980). The important biological applications
have been in electrophoresis (Ware and Flygare, 1971; Ware, 1974;
Uzgiris, 1981; Ware and Haas, 1983) and the detection of natural biological motions, including, as we shall see, cytoplasmic motility.

FLUORESCENCE PHOTOBLEACHING RECOVERY (FPR)

An FPR measurement is simple in concept. The molecules or particles
of interest are labelled by the addition of a fluorescent dye, usually
bound covalently to a specific site. A small region of the sample
containing the labelled species is illuminated by a low-power laser
beam. After a steady-state fluorescence level has been determined, a
much more intense laser beam is allowed to illuminate the same region
of the sample for a short time. The result is the photobleaching of
some of the fluorophores in the illuminated region. This photobleaching is generally an irreversible photochemical reaction that alters
the structure of the fluorophore to a non-fluorescent configuration.
Following this bleaching pulse, the fluorescence level is again monitored by the same dim beam on the same region, so that the extent of
bleaching is reflected by the reduction in fluorescence intensity.
If the fluorescence level is then monitored for a time it will be seen
to increase as bleached fluorophores diffuse out of the region and
unbleached fluorophores diffuse in. If a single circular spot is
bleached in the sample and the fluorophores are in motion by diffusion, then the characteristic recovery time is proportional to the
square of the spot diameter and inversely proportional to the diffusion coefficient of the fluorophore-macromolecule entity. If some of
these complexes are restricted in their motion so as to be effectively
immobile, then the recovery will not be complete, and the fractional
recovery determines the fraction of mobile fluorophores.

The FPR technique was developed primarily for the study of motion
of specific components of cell membranes (Peters et al., 1974;
Jacobson et al., 1976; Axelrod et al., 1976). The numerous successful applications of FPR to the study of motion in membranes has led
to widespread adoption of the technique, but has also pointed up some
limitations in the standard methodology. In particular the problem of
adventitious bleaching by the monitoring beam introduced an unCharac-
terized experimental uncertainty. In addition, standard techniques
required exact superposition of bleaching and monitoring beam and a
completely stationary specimen so that the region bleached and the
region monitored would be exactly the same. We (Lanni and Ware, 1982)
have developed a new approach that circumvents these problems by

combining the advantages of periodic pattern photobleaching (Smith and McConnell, 1978) with scanning detection (Koppel, 1979). To summarize our approach, a periodic photobleaching pattern is produced in the sample by brief, intense illumination through a grating. The grating is then translated at constant speed through the reference beam to produce a moving pattern on the specimen. A modulation of fluorescence emission is produced as the bleached pattern and the moving illumination pattern fall into and out of phase. The resulting photocurrent contains an ac component whose frequency is determined by the spacing and velocity of the grating and whose amplitude relative to the dc component is determined by the extent of photobleaching. In effect the modulation ac envelope $E(t)$ is a measure of the spatial Fourier component of the fluorescence intensity with wave vector K' given by:

$$K' = 2\pi/L$$

where L is the spacing between the lines of the grating. The ac component decays by diffusion as a simple exponential by analogy with Equation (11):

$$E(t) = E(0)(e^{-DK'^2t} + f_{LM}) \tag{15}$$

where $E(0)$ is the magnitude of the modulation envelope immediately after photobleaching. The fraction f_{LM} of persistent modulation indicates the fraction of labelled species that are not mobile on the time scale of the measurement. A schematic diagram of the FPR modulation detection method is shown in Fig. 2. This method does not require a stationary specimen or exact superposition of bleaching and monitoring beam. Any bleaching during the monitoring process can be corrected for since the average fluorescence level (i.e. the dc photocurrent) can be measured independently of the modulation envelope.

CYTOPLASMIC MOTILITY

The capacity of single cells for active motion has been an object of fascination for centuries. All cells exhibit some form of directed force generation (e.g. cell division), and some cells show a remarkable motility that enables them to translocate for selective advantage. For the purpose of this discussion we will omit the locomotion

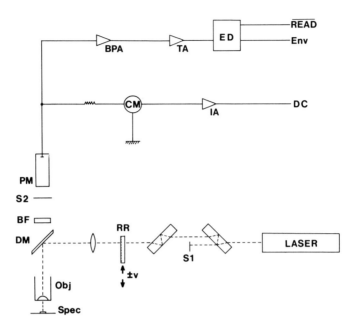

FIG. 2 Schematic diagram of the modulation detection system. A standard epifluorescence microscope equipped with dichroic mirror (DM) and barrier filter (BF) is modified so as to have a moveable grating (Ronchi Ruling, RR) in the plane of the illuminator field stop. The light source is an argon ion laser with associated beam control optics for photobleaching (SI). A photomultiplier (PM) receives fluorescence emission from the specimen. The photocurrent is measured by a dc microammeter (CM) with an isolated recorder output (IA). The ac components of the photocurrent in the range 10-300 Hz are amplified by the bandpass amplifier (BPA). The resulting signal can be processed by a spectrum analyser used as a power envelope detector. More simply, the broadband output can be further amplified by a tuned stage (TA) and processed by an amplitude envelope detector (ED). The output of the envelope detector is the fluorescence recovery signal (E(t). The signal READ is used to trigger a data recorder when a stable value of E(t) is available during each cycle of the modulation signal. From Lanni and Ware (1982).

mechanisms that rely on flagellary propulsion and direct our attention to active processes within the cytoplasm. Examples of cytoplasmic motility to be considered include amoeboid motion and various other types of protoplasmic streaming.

Protoplasmic streaming was first observed in 1774 (Corti, 1774), shortly after the development of the optical microscope, and the vast amount of work on this subject has relied almost exclusively on optical imaging of the motion of large particles in the protoplasm to characterize the flow. The velocities, velocity distributions, and more detailed properties of the flow inside these cells, including the response of the flow to various external perturbations, are important clues to the nature of the generation of the motive force. A comprehensive review of the classical work in this field prior to 1959 has been given by Kamiya (1959). More recent work in the area has focussed on ultrastructural details and on the application of techniques that can measure new details of the flow characteristics (Allen and Allen, 1978; Kamiya, 1981). As we shall illustrate later, one of the best new techniques for this purpose is laser Doppler velocimetry.

Although rheological measurements are essential for a complete understanding of the physical mechanism of cytoplasmic motility, the most fundamental level of understanding can only be approached by resolving the molecular structural and dynamic details of the motile apparatus. Over the past two decades, there has been substantial progress in the isolation and identification from a number of cell types of the many proteins that seem to be involved in cytoplasmic motility (Korn, 1978, 1982). As this effort continues, there has also been a requirement to develop physical assays for the polymerization, crosslinking, gelation, ordering, and contraction of supramolecular assemblies of contractile proteins in the reconstituted systems in which the activities of the cytoplasmic components can best be assayed. The most sophisticated current objective is the development of *in vivo* observation and experimentation using molecular techniques at the level of the motile organism.

The primary component of the motile apparatus is actin, which comprises as much as 15-20% of total cytoplasmic protein (Pollard and Weihing, 1974; Korn, 1978). Like skeletal muscle actin, cytoplasmic actin is a single-chain globular protein (G-actin) that self-assembles upon addition of salt or ~ 1 mM levels of divalent cations to form long filaments (F-actin) which are present in muscle as the thin filaments and are dispersed in cytoplasm as microfilaments.

The assembly and disassembly of actin filaments and networks is one of the primary processes and key puzzles of the mechanism of

cytoplasmic motility. It is known that the cation-induced nucleation
of G-actin to form the nucleus of a filament is followed by rapid
addition of monomers to the growing filament (Oosawa and Asakura,
1975), and that net addition is more rapid at one end of the filament
than at the other (Woodrum *et al.*, 1975). It is also known that the
assembly, crosslinking, and disassembly of actin *in vivo* are regulated
by numerous specialized cytoplasmic proteins (Korn, 1982; Craig and
Pollard, 1982; Lin *et al.*, 1982; Weeds, 1982). Some of the regulatory functions can also be accomplished by smaller molecules which
serve well for model studies probing the mechanism of actin assembly.
Particularly important in this regard are the cytochalasins (Tanenbaum, 1978), which inhibit assembly by binding to the growing end of
filaments (Brenner and Korn, 1979; Flanagan and Lin, 1980; MacLean-
Fletcher and Pollard, 1980) and possibly by cleaving intact filaments
(Hartwig and Stossel, 1979; Maruyama *et al.*, 1980). Many of the
details of the fundamental mechanisms of regulatory proteins and other
factors in altering the state of actin assembly are now subjects of
active investigation.

The assembly of actin and the other proteins of the supramolecular
complexes of motile cytoplasm can be measured by the increase in viscosity, flow birefringence, light scattering, or turbidity, or by
direct electron-microscopic visualization of filaments. G-actin concentrations are generally estimated from the proportion of non-sedi-
mentable protein. Changes in fluorescence of attached dyes (Detmers
et al., 1981; Kouyama and Mihashi, 1981; Tait and Frieden, 1982a) or
of efficiency of energy transfer between chromophores (Taylor *et al.*,
1981; Pardee *et al.*, 1982) have recently been adapted as new assays
of the proportion of assembled actin. The whole of the arsenal of
applied physical techniques, ranging from the relatively crude to the
highly sophisticated, is still as yet unable to provide us with some
of the parameters of fundamental interest, and most of these techniques cannot be applied *in vivo*. We will show in the following section, that dynamic light scattering provides a valuable new approach
to cytoplasmic rheology and, in the section subsequent to it, that
FPR can measure the most meaningful molecular parameters of supramolecular assembly both *in vitro* and *in vivo*.

APPLICATION OF DYNAMIC LIGHT SCATTERING TO CYTOPLASMIC MOTILITY

Because of the high frequency and short wavelength of optical light (e.g. in comparison to radar), laser Doppler velocimetry can be applied to the measurement of very small velocities in very small regions. Thus, it is possible to measure the flow characteristics of streaming protoplasm in localized regions within a single living cell. The LDV technique has several advantages over microscopy for these studies: it is automatic; it is objective; it can detect the motion of submicroscopic particles; and the light scattering angular factors can be used to infer details of the motion. Mustacich and Ware (1974) were the first to measure protoplasmic streaming velocities by LDV; these first studies were on the internodal and leaf cells of *Nitella*, a fresh-water alga. In subsequent publications (Mustachich and Ware, 1976a, 1977a), we have described an extensive characterization of the flow of protoplasm in the *Nitella* internodal cells.

A typical laser Doppler spectrum from an internodal cell of *Nitella* is presented in Fig. 3. The distribution of intensity is peaked at 93 Hz, which corresponds to 72 µm/s, approximately the velocity that would be measured by optical microscopy. A study of the angular dependence of the spectrum verified that both the peak position and the peak width were proportional to the first power of K (Equation 14), indicating that the laser Doppler spectrum reflects the velocity distribution of the protoplasm. Only at the highest scattering angles (where smaller particles are relatively more dominant) was there a component of the width that varied as K^2 and was thus attributable to diffusion. A major result of these angular studies was that all particles, large to submicroscopic, are streaming with the same average velocity. The low-frequency peak was shown to be a modulation component due to the chloroplast array and does not represent low velocities. The velocity histogram was shown to be consistent with a spatial flow profile that is somewhat flatter than parabolic but clearly not a flat plug flow.

The response of the *Nitella* cell to external stimuli and perturbations has been used extensively to gather more information on the mechanism of protoplasmic streaming. Using LDV technology, we were

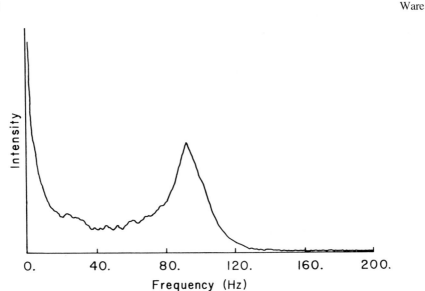

FIG. 3 Frequency spectrum of the scattered light from an internodal cell of *Nitella flexilis*. The scattering angle was 28.8° and the temperature was 25°C. The peak frequency at 93 Hz corresponds to a streaming velocity of 72 μm/s. There is a narrow distribution of frequencies about the peak frequency (93 Hz), with considerable intensity at lower frequencies. Little intensity is observed at much higher frequencies. From Mustacich and Ware, 1977a).

able to measure an entire velocity histogram following stimulation in less time than it would have taken to track a single particle by optical microscopy. Stimuli that were employed included light, temperature, and a variety of drugs and chemical agents. The photoinhibition of the streaming was utilized in a double-beam experiment to determine the spatial limits of inhibition. The two beams were of different colour; one was an inhibiting beam and the other, at lower intensity, was used to probe the flow about the point of inhibition. The inhibition was found to extend to about 100 μm in the upstream direction and about 400 μm in the downstream direction, indicating a spatial limit and directionality that could be related to the length and orientation of actin filaments in the cell.

A quite different type of protoplasmic streaming is present in the myxomycetes (plasmodial slime moulds). A particularly common example is the plasmodial phase of the slime mould *Physarum polycephalum*. The plasmodium of this organism takes the form of a fan, with the

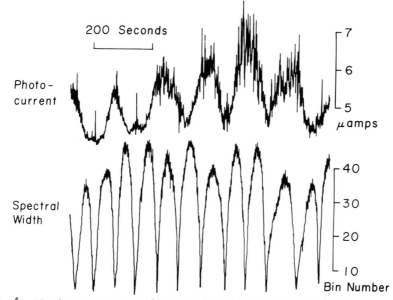

FIG. 4 Simultaneous recording to the photocurrent and Doppler spectral width of light scattered from a plasmodial vein of *Physarum*. The Doppler detection scheme used in these experiments was not sensitive to streaming direction so the spectral width recordings are "rectified". Both photocurrent and velocity can be seen to oscillate with approximately the same frequency but with a different phase. The photocurrent oscillations correlate closely with the changing diameter of the plasmodial tube. From Mustacich and Ware (1977b).

advancing front in the form of a thin sheet and with a reticular structure leading to the sheet. Protoplasm streams back and forth in the network of tubes in the plasmodium in a regular oscillatory fashion with a period of about one minute. The peak velocities of the flow are greater than 1 mm/s.

Mustacich and Ware (1976b, 1977b) have designed a special tracking circuit to interface with a real-time spectrum analyser and track these flow velocities in time. A record of the protoplasmic streaming in *Physarum* obtained with this circuit is shown in Fig. 4. The lower trace shows the output of the tracking circuit, which locks into a particular intensity level on the Doppler spectrum and reports the highest bin number of the spectrum analyser in which that intensity occurred. This provides a rectified representation of the velocities. In this recording a bin number of 40 corresponds to a Doppler shift of 1.01 kHz, which in turn corresponds to a velocity of 1.82 mm/s.

The upper trace is a simultaneous record of the intensity at the
photodetector, which in this case was shown to be primarily a function
of tube size. The changes in velocity and tube size occur at approximately the same frequency but, because of the complex tube structure
of the plasmodium, do not have the same phase.

Because of the oscillations in *Physarum* streaming velocities,
spectra had to be collected in the minimum possible time and interpreted without signal averaging. Mustacich and Ware (1977b) have
discussed the several potential complications of the spectral line-
shapes and concluded that the Doppler spectrum may be interpreted to
a first approximation as the velocity histogram of the flow in the
tube. From the Doppler spectrum the median velocity at the maximum
point in the velocity cycle was calculated to be 0.65 mm/s, and the
velocity at the 90th percentile was about 1.95 mm/s. Detectable
intensity was present at Doppler shifts corresponding to velocities
well in excess of 3 mm/s. These higher velocities exceed any velocities reported by optical microscopy (Gray and Alexopoulos, 1968). The
distribution of velocities within the organism had previously been
thought to be either parabolic or flattened parabolic profiles, but
this was based on optical microscopy performed at the lowest velocities in the oscillation cycle. We have shown that the Doppler spectrum at the maximum in the velocity cycle was consistent only with a
much sharper velocity profile, and found that a hyperbolic profile
agreed quantitatively with the data. However, at lower velocities,
Doppler spectra were observed which were consistent with parabolic
profiles. From these data, we surmise that the velocity profile
begins as a very flat distribution across the tube at low velocities,
becomes parabolic as the velocity increases, and goes to a hyperbolic
profile at the highest velocities, presumably as a result of the non-
Newtonian character of the complex protoplasm. Another surprising
result of the *Physarum* studies was the observation of components of
velocity transverse to the tube, which were found to be nearly as
high as the longitudinal velocities. The depolarized spectrum in
both orientations was found to be of the same form. The temperature
dependence of the streaming was measured, and it was found that
although the magnitudes of the velocities did not change greatly with

temperature, the frequency of oscillation was a linear function of temperature, increasing by more than a factor of four from 10 to 30°C.

Although all of the examples cited are from our own laboratory, I am pleased to note that there have been a number of interesting and informative LDV studies of protoplasmic streaming from other groups (e.g. Langley et al., 1976; Sattelle and Buchan, 1976; Newton et al., 1977; Sattelle et al., 1979). As we have emphasized, these studies yield hydrodynamic and rheological data that place important limits on plausible mechanisms for the generation of the motive force. Details of the molecular mechanism of cytoplasmic motility must be derived from techniques with molecular specificity.

APPLICATION OF FLUORESCENCE PHOTOBLEACHING RECOVERY TO CYTOPLASMIC MOTILITY

Actin may be labelled with fluorescent dyes using reagents that react with either the sulfhydryl or amino groups on the protein. Our labelling procedures have been described in detail (Lanni et al., 1981) and the biological activity of fluorescently-labelled actin has been established by our collaborators (Taylor and Wang, 1980; Wang et al., 1982). The data shown in this contribution were all obtained using actin which had been labelled at the sulfhydryl site (Cys 374) by the reagent 5-iodoacetamido fluorescein (5-IAF). The monomeric labelled actin is thus called G-AF-actin. An FPR trace of the diffusion of G-AF-actin is shown in Fig. 5. The decay of the modulation signal, indicating the fading of the striped pattern bleached into the solution, is well fit by a single exponential. The decay constant can be used to calculate accurately the tracer diffusion coefficient of the single diffusing species. The ability to measure the tracer diffusion coefficient is a distinct advantage, since that coefficient is most directly related to the molecular friction constant (Equation 2). This advantage may be particularly important for G-actin, since the measurement must be made at low salt concentrations, a condition for which the mutual diffusion coefficient (measured by dynamic light scattering and boundary spreading techniques) may be difficult to interpret (Equation 5). Our best value for the tracer diffusion coefficient of G-actin ($D_{20,w}$) is $7.15 \pm 0.1 \times 10^{-7}$ cm^2/s. This value is consistent with the accepted values of molecular weight (42,000),

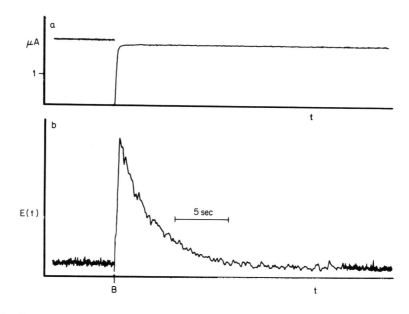

FIG. 5 FPR trace of G-AF-actin. E(t) is the envelope of the sine-wave modulation that decays exponentially after the bleach pulse (B). The top trace shows the corresponding level of dc photocurrent. From Ware and Lanni (1983).

sedimentation coefficient (3.3 S), and partial specific volume (0.732 ml/g), but it is somewhat lower than the diffusion coefficients of several other globular proteins of approximately the same molecular weight. The difference may reflect molecular asymmetry or increased hydration. Our number is much less anomalous than the classical value of 5.3×10^{-7} cm^2/s (Mihashi, 1964), and historical, erroneous speculations based on that number can be laid to rest.

A sequential series of FPR measurements which monitor the progress of actin assembly is shown in Fig. 6. From the plateau levels of such data we can determine the kinetics of appearance of filamentous material in a quantitative fashion, and from the decay constants we have a measure of the size of the species that are in the mixture at the same times. By a slight variation of experimental parameters we can measure the much lower diffusion coefficients of the filaments and thus provide an estimate of their length. Ability to measure both filament length and state of association of non-filamentous material

Actin + KCl (50 mM)

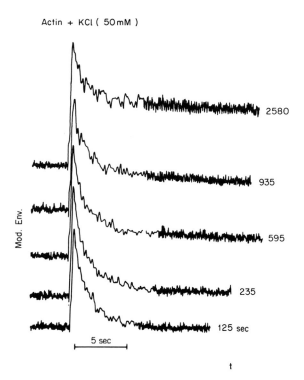

FIG. 6 Sequence of FPR traces demonstrating assembly of actin (18 μM) following addition of KCl to a concentration of 50 μM. At each successive time, there is an increase in the immobile fraction of label (filaments), indicated by the distance of the plateau of E(t) above the initial level. The decay constant of each exponential reflects the average diffusion coefficient of the mobile species at that time. From Ware and Lanni (1983).

is a unique and powerful advantage of our methodology. An example of such a determination through the course of polymerization is shown in Fig. 7. The D_{LM} (low mobility) decrease throughout the approach to steady state, indicating a lengthening of filaments up to lengths of between 10 and 100 μm, the latter being the short dimension of the capillaries in which the experiments are carried out. The decrease is faster and the plateau value lower at higher salt concentrations. For both experiments the D_{HM} (high mobility) stay in the range expected for monomer. This is the most direct evidence to date that the unassociated species in the presence of growing filaments is predominantly monomer.

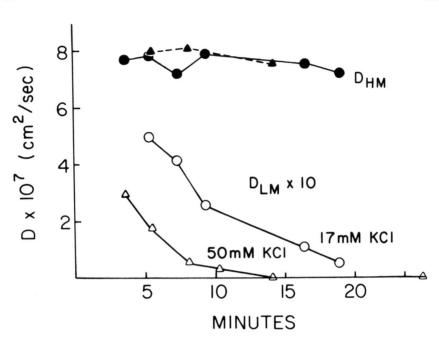

FIG. 7 A plot of the diffusion coefficients ($D_{20,w}$) of the high-mobility (D_{HM}) and low-mobility (D_{LM}) fractions of actin as a function of time through the course of self-assembly following addition of KCl. Data for two separate experiments using 17 mM KCl and 50 mM KCl, respectively are plotted. In both cases D_{LM} is multiplied by 10 to permit plotting on the same axis. From Ware and Lanni (1983).

We are pursuing several lines of experimentation to study the effects of cytochalasins (B and D) on actin assembly. By the FPR assay we detect three distinct effects:

(1) reduced fraction of actin in filaments;

(2) shortened filaments;

(3) accelerated kinetics of association from monomer upon addition of salt.

Effects (1) and (3) are illustrated in Fig. 8. Such results will eventually be explained in terms of three putative activities of cytochalasins, (i) nucleation, (ii) filament capping, (iii) filament breaking. The FPR assay is clearly a valuable tool to provide molecular definition to cytochalasin activity. The cytochalasins in turn are most useful as probes for the study of actin assembly.

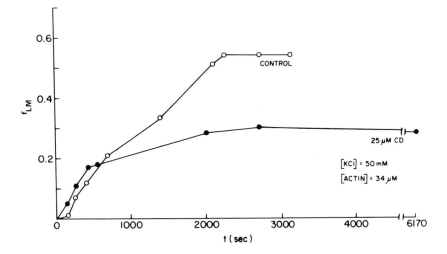

FIG. 8 Kinetics of appearance of the low-mobility fraction (f_{LM}) of actin filaments in the presence and absence of cytochalasin D, which has the effect of accelerating the initial rate of assembly but leading at longer times to a reduced fraction of assembly. From Ware and Lanni (1983).

The effects of other proteins on the assembly and crosslinking of actin filaments can also be studied in the same direct fashion using FPR. As a simple example we show in Fig. 9 the effects of aldolase, a basic protein known to associate with actin filaments. Clearly the aldolase increases the fraction of low-mobility actin. From the reduced slope of the plateau line, it is also apparent that aldolase decreases D_{LM}, probably by crosslinking filaments. The high-mobility state of aggregation (D_{HM}) judged from the time constant of initial recovery is not affected by the presence of aldolase, indicating that the monomers diffusing in the supramolecular matrix are not retarded by the network.

The highest level of complexity foreseen in this line of investigation is the cytoplasm of living cells. In a collaborative effort with the group of D.L. Taylor we recently performed the first such studies, with gratifying results (Wang *et al.*, 1982). The free-living amoeba species *Amoeba proteus* and *Chaos carolinensis* were microinjected with the fluorescein-labelled proteins G-actin, bovine serum albumin, ovalbumin, and ribonuclease A. The diffusion coefficients of the proteins in the cytoplasm were then measured using our FPR

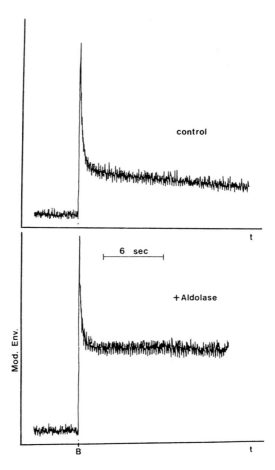

FIG. 9 FPR trace of a solution of AF-actin at low concentration, after addition of 6 mM MgCl$_2$. In the experiment represented by the lower trace, a substoichiometric quantity of aldolase was included in the solution. From Ware and Lanni (1983).

apparatus. For all proteins the diffusion coefficients measured were reduced by a factor of two or three from the values in water. This reduction is much less than the factor of 70 reported by Wojcieszyn *et al.* (1981) in a similar experiment in which labelled bovine serum albumin was injected into fibroblasts. The difference in the two observations probably reflects the great difference in the extent of rigid structure in the respective cytoplasms of amoebae and fibroblasts. We are very enthusiastic about the potential of this type of measurement for the study of the complex hydrodynamics of cytoplasm.

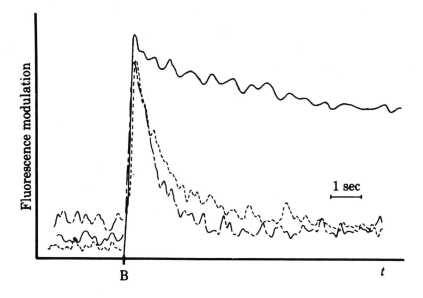

FIG. 10 Kinetics of recovery after photobleaching, recorded by using modulation detection. The curves represent the measured magnitude of the alternating current component (extent of modulation) in the photo-current. A x16 objective and ruling of 100 lines per 2.5 cm were used with Zeiss epifluorescence optics. This combination produces a striped pattern in the specimen with a period of 29.5 μm. Bleaching occurred at position B on the trace. —·—, fluorescein isothiocya-nate-labelled bovine serum albumin; ---, AF-actin; ——, AF-actin with phalloidin. From Wang *et al.* (1982).

The most interesting results of this study, however, pertained to the state of actin in the cells. The basic findings can be illustrated with the three sets of data shown in Fig. 10. From the recovery time con-stants, it appears that actin is diffusing more slowly than bovine serum albumin, which has a molecular weight about 50% greater than that of actin. These data are thus consistent with the supposition that G-actin is present in the cytoplasm only as a complex with another molecule of substantial size, presumably profilin or a similar protein. The most surprising result was the observation, apparent from Fig. 10, that the immobile fraction of actin was only about 10% of the total. If we can assume from the preponderance of previous evidence that this labelled actin distributes itself through the cell in the same manner as native actin, we must conclude that a much lower fraction of actin is present as static filaments than had

previously been suspected. The functional importance of maintaining
a large proportion of unassembled actin can be accepted but is certainly not fully understood. The solid trace in Fig. 10 shows the
effect of adding phalloidin, a drug that is known to stabilize actin
filaments. Clearly a huge increase in the proportion of low-mobility
filaments is verified. Finally, we performed a series of experiments
designed to probe the mobility of actin in different parts of the
Chaos cell. We showed that the mobility of actin was greater in the
advancing tips than in the tail and that the actin in the plasmalemma
region was characterized by a much higher fraction of low mobility
(45-85%).

The application of fluorescence techniques to biological systems
has been a productive area of research for decades, but the sophisti-
cation of the methodology and the scope of the applications has never
been greater. We believe that our results demonstrate that FPR is a
most powerful technique for the detailed study of molecular assembly
and of transport in a molecular matrix. Indeed other groups are
already applying this technique to similar problems both *in vitro*
(Tait and Frieden, 1982b; Tellam and Frieden, 1982) and *in vivo* (Kreis
et al., 1982).

CONCLUSION

Recent advances in experimental techniques have made possible a more
detailed level of characterization of transport in complex media such
as the cytoplasm of a cell. Dynamic light scattering experiments
characterize hydrodynamic and rheological properties through their
influence on the average optical refractive index of the medium over
a distance scale determined by the scattering angle and the wavelength of the incident radiation. Fluorescence photobleaching
recovery experiments characterize the tracer mobility of labelled
species in the complex medium over the characteristic distance defined
by the incident illumination pattern. We believe that these two complementary forms of dynamic information may be combined effectively
with structural studies for a rather complete characterization of
concentrated and/or partially ordered systems. We are pursuing similar studies on other complex media such as gels and low-salt concentration solutions of polyelectrolytes, both of which are of immense

practical importance as well as theoretical interest. Experiments
such as these have made us very aware of the lack of understanding of
the basic physical principles that govern transport in complex media.
We anticipate that an expanded experimental effort made possible by
the development of new techniques, coupled with heightened theoretical
interest, will lead eventually to a molecular and physical understanding of even such an intricate process as the active motion of a
living cell.

ACKNOWLEDGEMENTS

I am grateful to my coworkers, whose names are cited in the appropriate literature. Work in my laboratory on cytoplasmic motility is
supported by a grant from the National Science Foundation.

REFERENCES

Allen, R.D. and Allen, N.S. (1978). *Annu. Rev. Biophys. Bioeng.* 7,
    469-495.
Axelrod, D., Koppel, D.E., Schlessinger, J., Elson, E. and Webb, W.W.
    (1976). *Biophys. J.* 16, 1055-1060.
Berne, B.J. and Pecora, R. (1976). *In* "Dynamic Light Scattering"
    Wiley, New York.
Bezot, P., Ostrowsky, N. and Hesse-bezot, C. (1978). *Opt. Commun.*
    25, 14-18.
Brehm, G.A. and Bloomfield, V.A. (1975). *Macromolecules* 8, 663-665.
Brenner, S.L. and Korn, E.D. (1979). *J. Biol. Chem.* 254, 9982-9985.
Chen, S.-H., Chu, B. and Nossal, R. (1981). *NATO Advanced Study
    Institute,* Series B, Vol. 73. Plenum Press, New York.
Chu, B. (1974). *In* "Laser Light Scattering". Academic Press, New
    York.
Corti, B. (1774). *Lucca,* Italy.
Craig, S.W. and Pollard, T.D. (1982). *Trends Biochem. Sci.* 7, 88-92.
Cummins, H.Z. and Pike, E.R. (1974). *In* "Photon Correlation and
    Light Beating Spectroscopy". *NATO Advanced Study Institute*, Series
    B, Vol. 3, Plenum Press, New York.
Cummins, H.Z. and Pike, E.R. (1977). *In* "Photon Correlation Spectro-
    scopy and Velocimetry". NATO Advanced Study Institute, Series B,
    Vol. 23, Plenum Press, New York.
Detmers, P., Weber, A., Elzinga, M. and Stephens, R.E. (1981). *J.
    Biol. Chem.* 256, 99-105.
Drain, L.E. (1980). *In* "The Laser Doppler Technique". Wiley, New
    York.
Flanagan, M.D. and Lin, S. (1980). *J. Biol. Chem.* 255, 835-838.
Gray, W.D. and Alexopoulis, C.J. (1968). *In* "Biology of the Myxo-
    mycetes". Ronald Press, New York.
Gulari, E., Gulari, E., Tsunashima, Y. and Chu, B. (1979). *J. Chem.
    Phys.* 70, 3965-3972.
Hartwig, J.H. and Stossel, T.P. (1979). *J. Mol. Biol.* 134, 539-553.

Huglin, M.B. (1972). *In* "Light Scattering from Polymer Solutions". Academic Press, New York.

Jacobson, K., Wu, E.-S. and Poste, G. (1976). *Biochim. Biophys. Acta* $\underline{433}$, 215-222.

Kamiya, N. (1959). *In* "Protoplasmatologia, Handbuch der Protoplasmaforschung". Band VIII. Springer-Verlag, Vienna.

Kamiya, N. (1981). *Annu. Rev. Plant Physiol.* $\underline{32}$, 205-236.

Koppel, D.E. (1972). *J. Chem. Phys.* $\underline{57}$, 4814-4820.

Koppel, D.E. (1979). *Biophys. J.* $\underline{28}$, 281-291.

Korn, E.D. (1978). *Proc. Natl. Acad. Sci. USA* $\underline{75}$, 588-599.

Korn, E.D. (1982). *Physiol. Rev.* $\underline{62}$, 672-737.

Kouyama, T. and Mihashi, K. (1981). *Eur. J. Biochem.* $\underline{114}$, 33-38.

Kreis, T.E., Geiger, B. and Schlessinger, J. (1982). *Cell* $\underline{29}$, 835-845.

Langley, K.H., Piddington, R.W., Ross, D. and Sattelle, D.B. (1976). *Biochim. Biophys. Acta* $\underline{444}$, 893-898.

Lanni, F., Taylor, D.L. and Ware, B.R. (1981). *Biophys. J.* $\underline{35}$, 351-364.

Lanni, F. and Ware, B.R. (1982). *Rev. Sci. Instrum.* $\underline{53}$, 905-908.

Lin, S., Wilkins, J.A., Cribbs, D.H., Grumet, M. and Lin, D.C. (1982). Cold Spring Harbor Symp. Quant. Biol. $\underline{XLVI}$, 625-632.

Lin, S.C., Thomas, J.C., Allison, S.A. and Schurr, J.M. (1981). *Biopolymers* $\underline{20}$, 209-230.

Lin, Y.-H. and Chu, B. (1980). *Macromolecules* $\underline{13}$, 1025-1026.

MacLean-Fletcher, S. and Pollard, T.D. (1980). *Cell* $\underline{20}$, 329-341.

Maruyama, K., Hartwig, J.H. and Stossel, T.P. (1980). *Biochim. Biophys. Acta* $\underline{626}$, 494-500.

Mathiez, P., Mouttet, C. and Weisbuch, G. (1980). *J. Phys. (Paris)* $\underline{41}$, 519-523.

Mihashi, K. (1964). *Arch. Biochem. Biophys.* $\underline{107}$, 441-448.

Munch, J.P., Lemaréchal, P. and Candau, S. (1977). *J. Phys. (Paris)* $\underline{38}$, 1499-1509.

Mustacich, R.V. and Ware, B.R. (1974). *Phys. Rev. Lett.* $\underline{33}$, 617-620.

Mustacich, R.V. and Ware, B.R. (1976a). *Biophys. J.* $\underline{16}$, 373-388.

Mustacich, R.V. and Ware, B.R. (1976b). *Rev. Sci. Instrum.* $\underline{47}$, 108-111.

Mustacich, R.V. and Ware, B.R. (1977a). *Biophys. J.* $\underline{17}$, 229-241.

Mustacich, R.V. and Ware, B.R. (1977b). *Protoplasma* $\underline{91}$, 351-367.

Newton, S.A., Ford, N.C., Langley, K.H. and Sattelle, D.B. (1977). *Biochim. Biophys. Acta* $\underline{496}$, 212-225.

Nishio, I. and Wada, A. (1980). *Polymer J.* $\underline{12}$, 145-152.

Oosawa, F. and Asakura, S. (1975). *In* "Thermodynamics of the Polymerization of Protein". Academic Press, New York.

Pardee, J.D., Simpson, P.A., Stryer, L. and Spudich, J.A. (1982). *J. Cell Biol.* $\underline{94}$, 316-324.

Parthasarathy, N. and Schmitz, K.S. (1980). *Biopolymers* $\underline{19}$, 1655-1666.

Pecora, R. (1964). *J. Chem. Phys.* $\underline{40}$, 1604-1614.

Pecora, R. (1965). *J. Chem. Phys.* $\underline{43}$, 1562-1564.

Pecora, R. (1968a). *J. Chem. Phys.* $\underline{48}$, 4126-4128.

Pecora, R. (1968b). *J. Chem. Phys.* $\underline{49}$, 1036-1043.

Pecora, R. (1968c). *J. Chem. Phys.* $\underline{49}$, 1032-1035.

Pecora, R. and Steele, W.A. (1965). *J. Chem. Phys.* $\underline{42}$, 1872-1879.

Peters, R., Peters, J., Tews, K.H. and Bahr, W. (1974). *Biochim. Biophys. Acta* 367, 282-294.

Pollard, T.D. and Weihing, R.R. (1974). *CRC Crit. Rev. Biochem.* 2, 1-65.

Provencher, S.W., Hendrix, J., DeMaeyer, L. and Paulussen, N. (1978). *J. Chem. Phys.* 69, 4273-4276.

Sattelle, D.B. and Buchan, P.B. (1976). *J. Cell Sci.* 22, 633-643.

Sattelle, D.B., Green, D.J. and Langley, K.H. (1979). *Phys. Scripta* 19, 471-475.

Selser, J.C. (1979). *Macromolecules* 12, 909-916.

Smith, B.A. and McConnell, H.M. (1978). *Proc. Natl. Acad. Sci. USA* 75, 2759-2763.

Tait, J.F. and Frieden, C. (1982a). *Arch. Biochem. Biophys.* 216, 133-141.

Tait, J.F. and Frieden, C. (1982b). *Biochemistry* 21, 3666-3674.

Tanenbaum, S.W. (1968). *In* "Cytochalasins: Biochemical and Biological Aspects". Elsevier, Amsterdam.

Tanford, C. (1961). *In* "Physical Chemistry of Macromolecules". Wiley, New York.

Taylor, D.L. and Wang, Y.-L. (1980). *Nature* 284, 405-410.

Taylor, D.L., Reidler, J., Spudich, J.A. and Stryer, L. (1981). *J. Cell Biol.* 89, 362-367.

Tellam, R., and Frieden, C. (1982). *Biochemistry* 21, 3207-3214.

Uzgiris, E.E. (1981). *Adv. Coll. Interface Sci.* 14, 75-171.

van de Hulst, H.C. (1957). *In* "Light Scattering by Small Particles". Wiley, New York.

Wang, Y.-L., Lanni, F., McNeil, P.L., Ware, B.R. and Taylor, D.L. (1982). *Proc. Natl. Acad. Sci. USA* 79, 4660-4664.

Ware, B.R. (1974). *Adv. Coll. Interface Sci.* 4, 1-44.

Ware, B.R. (1977). *In* Chemical and Biochemical Applications of Lasers". (Ed. C.B. Moore), Vol. 2, PP. 199-239. Academic Press, New York.

Ware, B.R. and Flygare, W.H. (1971). *Chem. Phys. Lett.* 12, 81-85.

Ware, B.R. and Haas, D.D. (1983). *In* "Fast Methods in Physical Biochemistry and Cell Biology". (Eds R. Sha'afi and S. Fernandez), Elsevier/North-Holland Biomedical Press, Amsterdam.

Ware, B.R. and Lanni, F. (1983). *In* "Proc. Int. Conf. Photochemistry and Photobiology". (Ed. A. Zewail), Harwood Academic Publishers, New York.

Weeds, A. (1982). *Nature* 296, 811-816.

Wojcieszyn, J.W., Schlegel, R.A., Wu, E.-S. and Jacobson, K.A. (1981). *Proc. Natl. Acad. Sci. USA* 78, 4407-4410.

Woodrum, D.T., Rich, S.A. and Pollard, T.D. (1975). *J. Cell Biol.* 67, 231-237.

Yeh, Y. and Cummins, H.Z. (1964). *Appl. Phys. Lett.* 4, 176-178.

Fluorescence Photobleaching Techniques and Lateral Diffusion

D. AXELROD

INTRODUCTION

Since its introduction in 1974, the fluorescence photobleaching technique has been used extensively to study lateral motion of membrane
lipids and proteins, protein polymerization, and surface chemical
kinetics. Along with these applications, numerous variations of the
technique have been introduced, with corresponding variations in the
name: FPR (fluorescence photobleaching recovery), FRAP (fluorescence
recovery after photobleaching or fluorescence redistribution after
photobleaching), FRAPP (fluorescence recovery after pattern photobleaching), and FM (fluorescence microphotolysis). This review
places emphasis on the technical features of the variations of the
photobleaching technique.

All of the variations depend upon the irreversible photobleaching
(fading) of fluorescence in a two or three dimensional region in some
sort of spatially inhomogeneous pattern. The rate of spatial inter-
mixing of bleached and unbleached fluorophore can then be monitored
photometrically and related to a diffusion coefficient or linear flow
speed.

A typical general configuration of the optical and electronic system is shown in Fig. 1. The figure shows three major components of
the system about which this discussion is centred:

(i) the laser beam <u>intensity modulator</u> by which the rate of photobleaching can be altered;

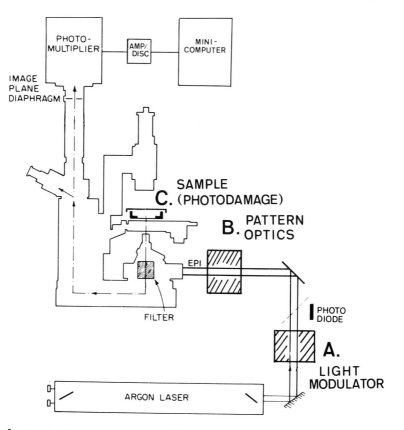

FIG. 1 Typical apparatus for fluorescence microphotolysis, utilizing the epi-illumination system of an inverted microscope. The three aspects of this system emphasized in this article are labelled.

(ii) the <u>pattern optics</u> (along with the image plane diaphragm) by which the spatial intensity pattern is formed on the sample (and the observation region defined);

(iii) the sample itself, which may or may not be adversely affec-ted by some sort of <u>photodamage</u>.

Each of these components is considered in detail in the following.

LIGHT INTENSITY MODULATION

Two fundamentally different light modulation protocols are used. In the most common (flash photobleaching), the laser beam is first kept fairly dim so as to excite fluorescence while avoiding significant photobleaching, then flashed at high intensity (at perhaps a 5000 to

TABLE 1

	Advantages	Disadvantages	References
Flash	Results obtained from one flash Easily interpreted	Relatively low photon count	Axelrod *et al.*, 1976
Continuous	High photon count rate	Decay shape very sensitive to bleaching kinetics. Several runs at different intensities needed. Interpretation not easy.	Peters *et al.*, 1981

20,000-fold increased intensity) for a time interval much shorter than the characteristic time for motion of a molecule across the illuminated region, so as to induce an initial inhomogeneous distribution of unbleached fluorophore. It is then returned to its low intensity level to observe the redistribution of unbleached fluorophore monitored as a recovery of fluorescence into the illuminated region. The rate of fluorescence recovery can be converted directly to a diffusion coefficient or flow rate. In the other extreme (continuous photobleaching), the laser is kept at constant and moderately high intensity so that photobleaching and redistribution occur simultaneously with similar characteristic times. The resulting decay of the fluorescence intensity is thereby slowed by lateral motion of unbleached fluorophore into the illuminated and observed region. Table 1 compares the flash and continuous protocols.

In the flash method, light modulation has been achieved in three different ways by using:

(i) a solenoid- or rotating disk-mounted neutral density filter by which a glass filter attenuating the beam intensity by about 10^4 is removed and returned back into the beam;

(ii) dual beamsplitters whereby a completely blockable bright beam and a continuous dim beam are superimposed;

(iii) an acousto-optic modulator by which the intensity of a first order diffracted laser beam is modulated by the amplitude of a radio-frequency voltage applied to the device.

TABLE 2

	Advantages	Disadvantages	References
Glass filter	High ratio of contrast between dim and bright intensities possible	Possible distortion of intensity profile by filter heating Slow (flashes longer than $\sim$5 ms) Possible beam shift from filter wedge	Barisas, 1980
Dual beam-splitter	High contrast ratio possible Insignificant beam shift	Moderately slow (flashes longer than $\sim$1 ms)	Koppel, 1979; Smith, Weis and McConnell, 1981)
Acousto-optic modulator	Fast (flashes longer than $\sim$1 μs)	Some intensity profile distortion Low contrast ratio (1:5000 at best) Some DC intensity drift	Garland, 1981

These systems are compared in Table 2.

PATTERN OPTICS

Several types of bleaching patterns have been used, all of which have both advantages and disadvantages.

Gaussian spot

A Gaussian spot is produced by placing a simple converging lens in the incident laser beam so that its focal point falls at the field diaphragm plane at the entrance to the epi-illumination system of the microscope. The spot on the sample can be enlarged by defocusing the converging lens. The spot has an intensity profile given by:

$$I(r) = I_0 e^{-2r^2/w^2} \qquad (1)$$

where r is the radial distance and w the characteristic e^{-2} radius which decreases with increasing angle of convergence of rays focussing on the sample. The radius w can be as small as 0.3 μm (Yoshida and Asakura, 1974). Analysis of the recovery curves for flash photobleaching is straightforward in principle (Axelrod et al., 1976). For diffusion, the diffusion coefficient is:

$$D = \gamma \ w^2/4t_{\frac{1}{2}} \tag{2}$$

where $t_{\frac{1}{2}}$ is the time for half of the recovery to be complete and γ is
a numerical factor near unity that increases slowly with the fraction
of fluorophores that are bleached (the bleaching "depth").

The chief advantages of the spot method are that it is easy to set
up, the illumination is very localized allowing study of small regions
on heterogeneous samples (such as small parts of cells), and it is
fast. (Diffusion coefficients ranging from 10^{-6} to 10^{-12} cm^2/sec can
be measured in a reasonably short experimental time). In addition,
closing the image-plane diaphragm around the image of the spot allows
one to discriminate against out-of-focus fluorescence sources in the
sample (Koppel et al., 1976).

The chief disadvantage of the spot method is that w is difficult
to measure precisely and the resulting uncertainty can lead to absolute errors of ±25% in D, although relative measurements of D from
one sample to the next are still reliable. Various methods for
measuring w have been described (Thompson and Axelrod, 1980; Schneider
and Webb, 1981; Sorscher and Klein, 1980).

Distinguishing linear flow from diffusion is possible by perform-
ing several runs with different w: the characteristic recovery time
is proportional to w for flow and to w^2 for diffusion. For systems
where a combination of flow and diffusion takes place, such as random
flows that might arise from motile cells, it is best to measure the
initial slope of the recovery immediately after bleaching, since this
is proportional to the average diffusion rate even in the presence of
linear flow.

The fluorescence may not return to its prebleach levels if some of
the fluorophore is immobile. The fraction (f) of fluorophores that
are mobile can be calculated (Axelrod et al., 1976). Equation (2)
above gives D for just the mobile fluorophores. The average diffusion
coefficient is $\overline{D} = f\ D$, and the initial slope of the recovery is
always proportional to $\overline{D}$, not to D.

Circular disk

A uniformly illuminated, sharp-edged circular disk has been employed by Kapitza and Sackmann (1980). The advantage here is that the shape and time scale of the recovery is completely insensitive to the bleaching depth. The disadvantages are that only large disks of at least ~5 μm diameter can be used so that the edge is not blurred by optical diffraction, and that the initial slope is always infinite and therefore contains no information about $\overline{D}$. Circular disk illumination can be attained by stopping down a uniformly illuminated field diaphragm or by imaging a bright circular object (e.g. the output end of an optical fibre bundle) onto the field diaphragm plane. For a uniformly illuminated circular disk:

$$D = 0.88 \ w^2/4t_{\frac{1}{2}} \tag{3}$$

Gaussian stripe

A Gaussian stripe pattern can be created by replacing the spherical converging lens just before the field diaphragm with a cylindrical converging lens. The stripe is useful in three applications:

(i) to examine one dimensional diffusion along elongated cells such as a bundle of fine neurite fascicles (Feldman *et al.*, 1981) with the stripe oriented transverse to the neurites;

(ii) to examine lateral diffusion at the edge of cells whose fluorescence labelling is so weak that it only appears to exceed the background at the edge due to tangential viewing;

(iii) to detect anisotropic diffusion by comparing results with the stripe oriented in two perpendicular directions on successive runs.

For a Gaussian stripe:

$$D = 3\gamma' \ w^2/4t_{\frac{1}{2}} \tag{4}$$

where γ' is a numerical factor near unity that increases slowly with bleaching depth.

Periodic stripe pattern

This pattern consists of parallel dark and light stripes of narrow inter-stripe spacing created by placing a Ronchi ruling in the laser beam at or before the field diaphragm plane (Smith and McConnell, 1978). The recovery curve is approximately an exponential with a characteristic time τ for $1/e$ recovery such that:

$$D = P^2/4\pi^2\tau \tag{5}$$

where P is the distance between adjacent stripes on the sample.

The periodic stripe pattern has the following advantages:

(i) the distance P is well defined by the Ronchi-ruling and geometrical optics so the absolute magnitude of D can be calculated reliably;

(ii) a large region is viewed so more photons can be collected, thereby reducing statistical uncertainty arising from shot noise;

(iii) the recovery rate and shape is only weakly dependent on the depth of bleaching or the exact profile of each stripe;

(iv) anisotropic motion can be detected in a fashion analogous to that using a single Gaussian stripe (Smith et al., 1979).

The disadvantages of the periodic stripe are mainly the following:

(i) the technique is non-local and therefore is best suited to spatially homogeneous and flat samples;

(ii) out-of-focus fluorescence cannot be excluded as well as with a focussed spot;

(iii) the minimum interstripe distance P as formed from a Ronchi ruling is generally larger than a typical w for a Gaussian spot, so experimental times for the periodic pattern can be considerably longer than for the spot method.

Smith (1982) has developed a method of imposing a submicron periodic pattern mask on a polymer film sample by chromium vapour deposition, thereby extending the applicability of the periodic pattern to measurement of very low diffusion coefficients.

Lateral diffusion of a membrane molecule can be retarded by reversible binding to some membrane structure that acts as an anchor. Koppel (1981) has shown theoretically that a periodic bleach pattern leads to a recovery that can be decomposed into a double exponential whose amplitudes and rates depend both on lateral diffusion coefficients and on the chemical binding kinetic rates.

Scanning methods

By laterally scanning a focussed spot or stripe in a sinusoidal
fashion after a flash bleach (Koppel, 1979), the relative contribution
of flow vs. diffusion can be determined. In addition, the scanning
technique is self-calibrating in the sense that D can be determined
without a separate (and possibly inaccurate) measurement of w. This
is a semi-local technique easily applicable to small regions on cell
surfaces.

A particularly elegant scanning system involves a linear rather
than sinusoidal motion of a periodic stripe pattern. The recovery
here also directly distinguishes between diffusion and flow and is
thereby well-suited for unambiguous measurement of D on samples that
may move during post-bleach observation. As a non-local illumination,
the scanning periodic pattern method is best suited to homogeneous
flat samples. The scanning periodic pattern has been produced by a
translating Ronchi ruling (Wang et al., 1982; Lanni and Ware, 1982) or
by a moving interference pattern generated from crossed laser beams
(Davoust et al., 1982).

Irregular large areas

One can selectively bleach large irregularly shaped areas simply by
moving a slightly defocussed spot around under manual control of a
joystick translator-mounted focussing lens (Stya and Axelrod, 1983).
In one application of this approach, possible exchange of fluorescently labelled acetylcholine receptors (AChR) between clustered and
non-clustered regions on the surface of a developing muscle cell in
culture was examined (Fig. 2). All of the clustered AChR on all of
the cells were totally bleached, leaving only the fluorescence of
the non-clustered AChR intact. Six hours later, without any
relabelling, new fluorescent clusters became visible, proving that
non-clustered AChR can spontaneously cluster.

Spherical samples

The special geometry of a round cell lends itself to bleaches of azimuthal symmetry. In fact, the very first fluorescence photobleaching
experiment (Peters et al., 1974) employed an azimuthally bleached
half of an erythrocyte ghost with a sharp dividing edge between the

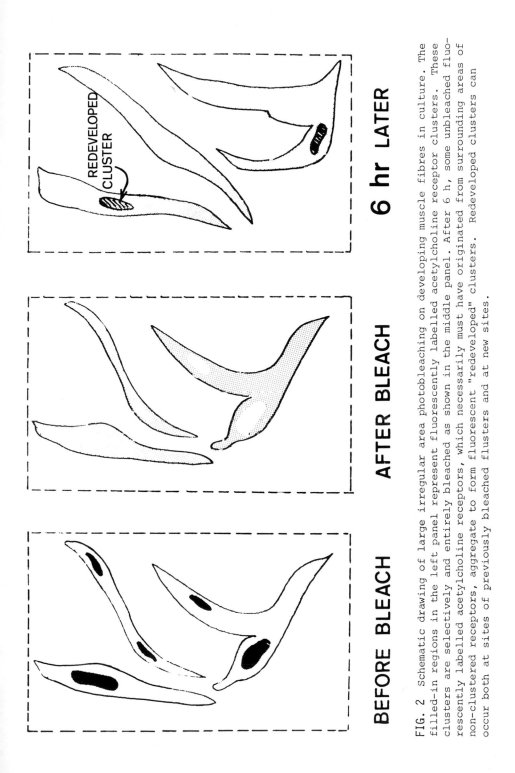

BEFORE BLEACH   AFTER BLEACH   6 hr LATER

FIG. 2 Schematic drawing of large irregular area photobleaching on developing muscle fibres in culture. The filled-in regions in the left panel represent fluorescently labelled acetylcholine receptor clusters. These clusters are selectively and entirely bleached as shown in the middle panel. After 6 h, some unbleached fluorescently labelled acetylcholine receptors, which necessarily must have originated from surrounding areas of non-clustered receptors, aggregate to form fluorescent "redeveloped" clusters. Redeveloped clusters can occur both at sites of previously bleached clusters and at new sites.

bleached and unbleached halves. However, under an appropriate obser-
vational scheme, it is not important to establish what fraction of
the cellular area is azimuthally bleached, nor even what the exact
bleaching intensity profile is (Koppel and Sheetz, 1980). In this
scheme, one scans the post-bleach fluorescence intensity $F(x,t)$ from
one pole of the cell to the other along a diameter line (defined as
the x-axis) and forms the "first moment":

$$U(t) \equiv \int_{-R}^{R} xF(x,t)\,dx \qquad (6)$$

where t is the time after bleaching and R is the spherical cell
radius. It turns out that $U(t)$ decays exponentially with a characteristic time τ' related to D by:

$$D = R^2/2\tau' \qquad (7)$$

Another elegant use of spherical sample geometry (Smith, McConnell
et al., 1981) employs a polarized bleaching and observation laser
beam defocussed to illuminate the entire cell uniformly. If the
dipole moments of the fluorophores are preferentially oriented in an
orderly manner (i.e. either radially or tangentially), then bleaching
occurs only at those ranges of angles around the cell where the dipole
moments and the light polarization are substantially parallel. This
effect leads to an azimuthally symmetric pattern of bleached fluorophore which then relaxes back to isotropic uniformity. The fluorescence recovery as observed by the same, but attenuated, defocussed
polarized beam is exponential with a characteristic time τ'' such that:

$$D = R^2/6\tau'' \qquad (8)$$

Because the bleach pattern depends on the polarization of the bleaching beam rather than focussing or imaging of any kind, this technique
should be applicable to spherical systems such as liposomes or vesicles too small to be resolved well in an optical microscope.

Surface illumination

The adsorption/desorption kinetics and possible surface diffusion of soluble fluorescent molecules reversibly adsorbed at a solid/liquid interface can be studied in chemical equilibrium by combining fluorescence photobleaching recovery with illumination and bleaching of just the surface region (Thompson et al., 1981). Such selective surface illumination can be achieved with the "evanescent wave" of a totally internally reflecting laser beam incident upon the interface at a sufficiently oblique angle. After flash bleaching the surface adsorbed molecules, the desorption rate can be deduced from the fluorescence recovery curve as unbleached molecules adsorb from bulk solution and bleached molecules desorb from the surface. By focussing the totally internally reflecting laser beam at the interface, surface diffusion of the adsorbate can be detected and distinguished from the adsorption/desorption process. Such lateral surface diffusion during reversible nonspecific adsorption, if it exists, may greatly speed the rate of interaction between agonists and cell surface receptors (Adam and Delbruck, 1968; Berg and Purcell, 1977). The total internal reflection/fluorescence photobleaching recovery system has been used to measure the molecular dynamics of bovine serum albumin adsorbed to a glass surface (Burghardt and Axelrod, 1981).

PHOTODAMAGE

Of crucial relevance in all variants of the fluorescence photobleaching technique is the possibility of damage to the sample resulting from exposure high light levels (up to ~ 100 mW/μm^2) during the bleaching flash. Although no absolute conclusions can be drawn as yet, a number of studies bear on whether photodamage is a problem.

Evidence FOR photodamage effect

(a) Prolonged illumination of fluorescently labelled erythrocyte or kidney cells leads to substantial crosslinking of membrane proteins. This can, however, be inhibited by certain scavengers or free radical quenchers (Sheetz and Koppel, 1979; Leplock et al., 1978).

(b) The polymerization kinetics of fluorescently labelled actin
are altered by excitation of the fluorophore (Lanni *et al.*, 1981).

Evidence AGAINST photodamage effect

(a) The diffusion coefficients of rhodopsin in rod outer disc
segments as measured by intrinsic absorption photobleaching and by
extrinsic fluorescence photobleaching are equal to each other
(Wey *et al.*, 1981).

(b) Total prebleaching of a fluorescein labelled cell surface does
not affect the fluorescence recovery observed after focussed-spot
bleaching of a rhodamine label on another cell surface receptor
(Wolf *et al.*, 1980).

(c) The rate of fluorescence redistribution after artificially
induced fusion of one fluorescently labelled erythrocyte with
another unlabelled one indicates the same diffusion coefficient
as is obtained on the same fused system by photobleaching techniques (Schindler *et al.*, 1980).

(d) Localized fluorescence photobleaching induces no morphological
changes in cells that can be seen by scanning electron microscopy,
and the cells remain impermeable to trypan blue (Jacobson *et al.*,
1978).

(e) The local temperature increase during typical spot photobleaching experiments on fluorescently labelled cell surface receptors
can be calculated to be less than 0.1°C (Axelrod, 1977).

(f) Fluorescence photobleaching data directly report the motion
of only those fluorophores which have not been bleached.

In general, photobleaching recovery experiments involve as little
(and usually far less) exposure to excitation light as conventional
experiments involving fluorescence microphotography of living cells.
The latter experiments should be subjected to the same critical
scrutiny with regard to possible photodamage.

SUMMARY

The photobleaching technique is now well-established for the measurement of lateral diffusion, and a variety of pattern geometries have
been used successfully in investigations of isotropic samples, arti-
ficial membranes and membrane vesicles and of cell surfaces.

Considerations of the biological implications have been presented by Cherry (1979) and Peters (1981). Examples of the application to lateral diffusion in cells are presented elsewhere in the present volume by Ware and by McConnell, and in model monolayer systems by Beck and Peters.

ACKNOWLEDGEMENTS

This work was supported by USPHS NIH grants NS 14565 and 17017.

REFERENCES

Adam, G. and Delbruck, M. (1968). *In* "Structural Chemistry and Molecular Biology" (Eds A. Rich and N. Davidson), pp. 198-215. W.H. Freeman, San Francisco.

Axelrod, D. (1977). *Biophys. J.* 18, 129-131.

Axelrod, D., Koppel, D.E., Schlessinger, J., Elson, E. and Webb, W.W. (1976). *Biophys. J.* 16, 1055-1069.

Barisas, B.G. (1980). *Biophys. J.* 29, 545-548.

Berg, H. and Purcell, E.M. (1977). *Biophys. J.* 20, 193-219.

Burghardt, T.P. and Axelrod, D. (1981). *Biophys. J.* 33, 455-468.

Cherry, R.J. (1979). *Biochim. Biophys. Acta* 559, 289-327.

Davoust, J., Devaux, P.F. and Leger, L. (1982). *EMBO J.* 1, 1233-1238.

Feldman, E.L., Axelrod, D., Schwartz, M., Heacock, A.M. and Agranoff, B.W. (1981). *J. Neurobiol.* 12, 591-598.

Garland, P.B. (1981). *Biophys. J.* 33, 481-482.

Jacobson, K., Hou, Y. and Wojcieszyn, J. (1978). *Exp. Cell Res.* 116, 179-189.

Kapitza, H.-G. and Sackmann, E. (1980). *Biochim. Biophys. Acta* 595, 56-64.

Koppel, D.E. (1979). *Biophys. J.* 28, 281-292.

Koppel, D.E. (1981). *J. Supramol. Struct.* 17, 61-67.

Koppel, D.E., Axelrod, D., Schlessinger, J., Elson, E.L. and Webb, W.W. (1976). *Biophys. J.* 16, 1315-1329.

Koppel, D.E. and Sheetz, M.P. (1980). *Biophys. J.* 30, 187-192.

Lanni, F., Taylor, D.L. and Ware, B.R. (1981). *Biophys. J.* 35, 351-364.

Lanni, F. and Ware, B.R. (1982). *Rev. Sci. Instrum.* 53, 905-908.

Leplock, J.R., Thompson, J.E. and Kruuv, J. (1978). *Biochem. Biophys. Res. Commun.* 85, 344-350.

Peters, R. (1981). *Cell Biol. Int. Reports* 5, 733-760.

Peters, R., Brunger, A. and Schulten, K. (1981). *Proc. Natl. Acad. Sci. USA* 78, 962-966.

Peters, R., Peters, J., Tews, K. and Bahr, W. (1974). *Biochim. Biophys. Acta* 367, 282-294.

Schindler, M., Koppel, D.E. and Sheetz, M.P. (1980). *Proc. Natl. Acad. Sci. USA* 77, 1457-1461.

Schneider, M.B. and Webb, W.W. (1981). *Appl. Optics* 20, 1382-1388.

Sheetz, M.P. and Koppel, D.E. (1979). *Proc. Natl. Acad. Sci. USA* 76, 3314-3317.

Smith, B.A. (1982). *Macromolecules* 15, 469-472.

Smith, B.A., Clark, W.R. and McConnell, H.M. (1979). *Proc. Natl. Acad. Sci. USA* 76, 5641-5644.

Smith, B.A. and McConnell, H.M. (1978). *Proc. Natl. Acad. Sci. USA* 75, 2759-2763.

Smith, L.M., McConnell, H.M., Smith, B.A. and Parce, J.W. (1981). *Biophys. J.* 33, 139-146.

Smith, L.M., Weis, R.M. and McConnell, H.M. (1981). *Biophys. J.* 36, 73-91.

Sorscher, S.M. and Klein, M.P. (1980). *Rev. Sci. Instrum.* 51, 98-102.

Stya, M. and Axelrod, D. (1983). *Proc. Natl. Acad. Sci. USA* 80, 449-453.

Thompson, N.L. and Axelrod, D. (1980). *Biochim. Biophys. Acta* 597, 115-165.

Thompson, N.L., Burghardt, T.P. and Axelrod, D. (1981). *Biophys. J.* 33, 435-454.

Wang, Y.-L., Lanni, F., McNeil, P.L., Ware, B.R. and Taylor, D.L. (1982). *Proc. Natl. Acad. Sci. USA* 79, 4660-4664.

Wey, C.-I., Cone, R.A. and Edidin, M.A. (1981). *Biophys. J.* 33, 225-232.

Wolf, D.E., Edidin, M.A. and Dragsten, P.R. (1980). *Proc. Natl. Acad. Sci. USA* 77, 2043-2045.

Yoshida, A. and Asakura, T. (1974). *Optik* 41, 281-292.

Translational Diffusion and Phase Separation in Phospholipid Monolayers: A Fluorescence Microphotolysis Study

K. BECK and R. PETERS

THE COMPLEXITY OF CELL MEMBRANE ARCHITECTURE

Cell membranes are involved in many dynamic processes such as endo-
and exo-cytosis, redistribution of surface antigens, insertion of
newly synthesized membrane proteins, cell division and cell motility.
Therefore models of membrane organization emphasizing dynamic aspects,
for instance the fluid mosaic model of Singer and Nicolson (1972),
have attracted much attention in the last decade. These models were
primarily based on quantitative studies of molecular motion in lipid
model membranes (Träuble and Sackmann, 1972; Devaux and McConnell,
1972) and on qualitative observations on living cells (Frye and
Edidin, 1970; Taylor et al., 1971; Pinto da Silva, 1972). Subsequently, spectroscopic techniques have been developed which permit
measurement of molecular motion in living cells. Among these
approaches, fluorescence microphotolysis (Peters et al., 1974;
Edidin et al., 1976; Jacobson et al., 1976; Axelrod et al., 1976;
Smith and McConnell, 1978) has been widely employed. Data obtained
by fluorescence microphotolysis (for reviews see Cherry, 1979; Peters,
1981; Edidin, 1981) and by various other techniques have helped to
clarify the relations between molecular motion and membrane function,
e.g. in the cell cycle (de Laat et al., 1977, 1980), mitogenesis
(Ferber et al., 1975), fertilization (Peters and Richter, 1981;
Wolf et al., 1981; Scandella et al., 1982), early development (Johnson et al., 1982), visual transduction (Liebman and Pugh, 1979),

receptor mediated processes (Pastan and Willingham, 1981) and electron transfer in mitochondria (Hackenbrock, 1981).

However, the techniques which were thought initially to quantify
the postulated fluid-mosaic character of cell membranes have shown,
and continue to show with increasing clarity, that cell membranes
are also characterized by properties in common with solids. In most
cases the translational and rotational diffusion coefficients of
integral membrane proteins are too small to be accounted for by a
fluid bilayer. Immobile fractions seem to be a common occurrence
among integral membrane proteins. Furthermore it is believed (Geiger,
1982; Baines, 1983) that strong interactions exist between cell membranes and sub-membranous cytoskeletal elements, although molecular
details have been elucidated so far only in the case of the erythrocyte membrane (for reviews see Lux, 1983; Goodman and Shiffer, 1983).
In addition, it is likely that extracellular coatings such as the
glycocalyx (Aplin and Hughes, 1982; Hay, 1981) also restrict and
modulate mobility in cell surface membranes (Jacobson and Wojcieszyn,
1981).

Cell membranes are highly anisotropic. It is likely, for instance,
that lipid phase segregation gives rise to a solid-fluid domain
structure in certain membranes (Galla *et al.*, 1979; Jain and White,
1977; Karnovsky *et al.*, 1982). Most cells show permanent specialized
surface regions such as microvilli, tight junctions or other crystalline arrays (e.g. the purple membrane of halobacteria).

Thus, future models of membrane structure will have to account for
the fluid-solid composite character of biological membranes in general
and for a large variety of structural and functional specializations
in particular. Such models (e.g. Loor, 1981) are still in an initial,
qualitative state.

TRANSLATIONAL DIFFUSION IN BILAYER MEMBRANES

The complexity of cell membrane architecture makes it difficult to
study the structure-function relationships of membrane components.
Parallel to the direct investigation of cell membranes, model membrane systems have been developed which allow one to vary several
parameters in a well-defined manner (for reviews see Fendler, 1982;
Szabo and Waldbillig, 1982), and thus permit the study of isolated

aspects of cell membrane function more easily. The molecular mobility
of lipids and proteins has been investigated in various types of
spherical bilayer membranes (for review see Vaz *et al.*, 1982) and
also in planar (black) lipid membranes (Fahey *et al.*, 1977; Wolf *et
al.*, 1977; Fahey and Webb, 1978).

From the vast body of experimental data we mention only a few facts
which are relevant to further discussion here. In bilayers prepared
from a single phospholipid species, the translational diffusion
coefficient of lipid probes ranges from 1-15 μm^2/s in the fluid phase,
i.e. at a temperature well above T_c, the gel-liquid crystalline main
phase transition temperature. Lowering the temperature below T_c
reduces the translational diffusion coefficient, within an interval
of 1-2°C around T_c, by 2-3 orders of magnitude. The mobility of
integral proteins in reconstituted fluid bilayers can be quantitatively described (Peters and Cherry, 1982) by the hydrodynamic continuum theory of Saffman and Delbrück (Saffman and Delbrück, 1975;
Saffman, 1976; Hughes *et al.*, 1981). In order to account for the
mobility of lipid probes and other small molecules, however, it
seems necessary (Vaz and Hallmann, 1983) to apply models which take
the discontinuous, molecular character of the bilayer into consideration. One of these models, the "free volume" model (Cohen and Turnbull, 1959) for diffusion in solutions, has been adapted previously
(Träuble and Sackmann, 1972; Galla *et al.*, 1979) to bilayer membranes
and will be applied in this study to monolayers (see also Peters and
Beck, 1983).

PREVIOUS MEASUREMENTS OF TRANSLATIONAL DIFFUSION IN MONOLAYERS

Conceptually, monomolecular films spread at the air-water interface
are an excellent model system for studies of lipid diffusion because
many relevant parameters can be measured and manipulated, e.g.

(i) molecular packing density, surface pressure Ⅱ, surface viscosity and surface potential are dependent variables which can
be controlled and adjusted;

(ii) experimental conditions can be adjusted over a wide range and
their influence on monolayer behaviour investigated (e.g. temperature of subphase and environment, composition of subphase, ionic

strength, ion composition, gaseous atmosphere;

(iii) composition of the monolayer (pure and mixed) and in particular the probe concentration can be fully controlled.

These vast possibilities, however, have been little explored and only
relatively few experiments have been reported. In most of the previous diffusion measurements, radiotracer methods have been employed
(Sakata and Berg, 1964; Good and Schechter, 1972; Stroeve and Miller,
1975; Wirz and Neumann, 1978; Vollhardt et al., 1980a,b, 1981). In
addition, Teissié et al. (1978) have described a special variant of
fluorescence microphotolysis ("dimerization photobleaching") which is
based on the photo-dimerization of 12(9-anthroyloxy)-stearic acid
(12-AS) (McGrath et al., 1976). Representative results from the
above are collated in Table 1.

It is obvious that most of the reported diffusion coefficients are
very large, even if a direct comparison with bilayer values is not
possible in each case. Compounds which form monolayers do not always
also form bilayers (e.g. fatty acids and pure cholesterol). Molecular
densities can be varied over a large range in the case of monolayers
whereas molecular packing in bilayers is frequently not exactly known.
Nevertheless, the data collected in Table 1 suggest profound differences between monolayers and bilayers. For phospholipid monolayers,
for instance, diffusion coefficients have been reported to be 1000-
2000 $\mu m^2/s$ in the fluid phase (corresponding to the liquid crystal
phase in bilayers) and 10-400 $\mu m^2/s$ in the crystalline phase (corresponding to the gel phase in bilayers), respectively (Teissié et al.,
1978) (D $\sim$ 10 $\mu m^2/s$ is thought to be the lower detection limit of the
method applied). Smaller values which approach bilayer values, and
are also compatible with values we have measured for different probes
and lipids, have been reported by only one group, namely Loughran et
al. (1980). The authors have measured translational diffusion in
monolayers prepared from mixtures of 12-(1-pyrene)-dodecanoic acid
(PDA) and oleic acid. Diffusion coefficients were derived from PDA
excimer fluorescence, basing data evaluation on an interesting Monte-
Carlo simulation of random walk in a lattice containing traps. Diffusion coefficients obtained in this manner are in the range of 100
$\mu m^2/s$.

TABLE 1   Previous measurements of translational diffusion in monolayers at the air-water interface

System	Method	T (°C)	$\bar{a}$ (nm²)	Π (mN/m)	D (μm²/s)	Reference
Myristic acid on water	radioactive tracer	22	("liquid expanded phase") ("intermediate phase")		30,000 3,000	Sakata and Berg (1964)
Palmitic acid on 10 mM HCl	radioactive tracer	22	0.75 0.50 0.33 0.23		4,100 4,410 3,300 2,300	Vollhardt et al. (1980b, 1981) *
Stearic acid on water	radioactive tracer	28	0.45 0.27		200–300 ∼100	Good and Schechter (1972)
Stearic acid on 10 mM HCl	radioactive tracer	22	0.75 0.50		2,800 2,100	Vollhardt et al. (1980b) *
12-(9-anthroyloxy-)stearic acid (12-AS) on 100 mM NaCl pH 6	dimerization photobleaching	20	0.40 0.375 0.35	21.8 25.8 26.4	5,000 1,000 950	Teissié et al. (1978)
12-(1-pyrene)dodecanoic acid in oleic acid on water	monomer-excimer dynamics		0.47		90–170	Loughran et al. (1980) **
12-AS in DPPC molar conc. 0.01 on 100 mM NaCl	dimerization photobleaching	20	0.95 0.56 0.48 0.42	0.2 5.5 8.1 15.7	1,300 1,000 550 400	Teissié et al. (1978)
NBD-egg-PE in DPPC molar conc. 0.01 on water	fluorescence microphotolysis	23	(solid condensed phase)		0.01	von Tscharner and McConnell (1981a) ***
Cholesterol on water	radioactive tracer	22	0.40 0.38	<1 15–40	200 40	Stroeve and Miller (1975)

* authors assume a simple linear relationship between D and $\bar{a}$; ** authors give more confidence to the lower limit of D; *** D estimated, not measured.

Some authors have reported that concentration profiles of radiotracer measurements sometimes indicated surface convection (Good and
Schechter, 1972; Wirz and Neumann, 1978). Vollhardt et al. (1980a,b,
1981) explicitly discussed the possibility of temperature-induced
instabilities in the subphase and tried to account for such effects.
It was, however, van Tscharner and McConnell (1981a) who first visualized surface convection in fluorescently-labelled phospholipid monolayers by light microscopy. Following this, McConnell and coworkers
have transferred monolayers onto alkylated glass slides using this
system as a specific target for various cells (see e.g. von Tscharner
and McConnell, 1981b; Hafeman et al., 1981, 1982; Huang and McConnell,
1983; Seul et al., 1983; Weis and McConnell, 1983; and the article
by McConnell elsewhere in this volume).

A NEW METHOD FOR MEASUREMENT OF TRANSLATIONAL DIFFUSION IN MONOLAYERS

Confirming the discovery of von Tscharner and McConnell (1981a) that
monolayers at the air-water interface can be subject to vigorous
streaming which interferes with diffusion measurements, we set out to
solve this problem by experimental means (Peters and Beck, 1983).
The rationale underlying our construction is essentially that of a
breakwater scaled down to the millimeter range. A small monolayer
area is contained in a circular compartment and protected against air
convection by a glass cover. The circular geometry of the compartment may be important in yielding a maximum area/circumference ratio.
The essential details of the device are shown in Fig. 1. A small
trough of 80 cm^2 surface area has been milled from a piece of poly-
tetrafluoroethylene (PTFE), cleaned, placed on the stage of a fluorescence microscope and filled to the brim with a suitable subphase
(double distilled water or Millipore Milli-Q water). A PTFE plate,
about 2 mm thick, containing circular holes and canals is placed
across the air-water interface at one end of the trough. The canals
in the PTFE plate connect the circular compartments with the free
interface. Thus, monolayer material spread at the free part of the
interface has access to the shielded circular compartment and surface
pressure can be adjusted. The canals are as narrow as compatible
with monolayer viscosity; in the case of the phospholipid species

FIG. 1 A trough for measurement of translational diffusion in monolayers spread at the air-water interface. A small trough is placed
on the stage of a fluorescence microscope. A surface balance of the
Wilhelmy-type (accuracy 0.2 mN/m) and a barrier provide for control
of surface pressure and monolayer area. In order to prevent surface
flow, a 2 mm thick plate is placed across the trough at one end. The
plate contains holes for microscopic and surface balance measurements,
respectively. Holes and free area are connected by a canal system.
The trough itself is thermostated and is situated in a thermostated
compartment built around the stage of the microscope.

investigated in the study described here, a canal system with a width
of 1 or 2 mm was found to be optimal. Different compartments of the
PTFE plate are used for diffusion and surface pressure measurements,
respectively. The small compartments are covered with an alkylated
glass slide and used for diffusion measurements by fluorescence
microphotolysis. A larger, uncovered compartment is used for surface
pressure measurements employing a Wilhelmy-type balance (Kuhn et al.,
1972) with a cylindrical plate. The trough is thermostated and placed
in a closed box built around the stage of the microscope.

   The concepts and procedures applied in fluorescence microphotolysis
have been summarized previously (Peters, 1981; Koppel, 1983; see also
the article by Axelrod in this volume). Details of the instrumentation are given in Peters and Richter (1981). In the present experiments, long distance objective lenses (16 X and 40 X) have been used
and the illuminating light beam had a Gaussian intensity profile with
an e^{-2}-radius of 18.5 μm or 7.0 μm, respectively. Data have been
evaluated by two procedures. The half-time of fluorescence recovery
has been determined, and used to calculate diffusion coefficients
according to Axelrod et al. (1976). Examples of original measuring
traces are given in Fig. 2a. In addition reciprocal plots have been
produced (Fig. 2b) according to Yguerabide et al. (1982). The
effectiveness of this device in eliminating surface convection is
best assessed by microscopic observation of fluorescently labelled
monolayers. Micrographs (see later, Fig. 6) clearly demonstrate
absence of flow because sharp images are obtained even with long
exposure times (5-10 s). It is true, however, that even with all of
the precautions described, careful controls are necessary. After
changing surface pressure it is necessary to wait for some minutes to
obtain a calm surface and to ensure that the surface pressure is in
equilibrium in the compartments as well as on the free surface.
Furthermore, effective temperature control is essential.

LIPID PROBE DIFFUSION IN FLUID MONOLAYERS AND THE "FREE VOLUME MODEL"

Similar to bulk material, monolayers can exist in various phases which
have been referred to by analogy as gaseous, liquid, and solid (for a
review see Gaines, 1966). The actual state of a monolayer depends on
parameters such as packing density, surface pressure, temperature and
subphase composition. The conventional manner to study monolayer
phases and phase transitions is by surface-pressure/area isotherms
and temperature/area isobars (e.g. Albrecht et al., 1978, 1981). Let
us first consider lipid probe diffusion in monolayers of the phospholipid L-α-dilauroylphosphatidylcholine (DLPC) which is in a fluid
state at 22°C throughout the whole surface pressure range (Phillips
and Chapman, 1968).

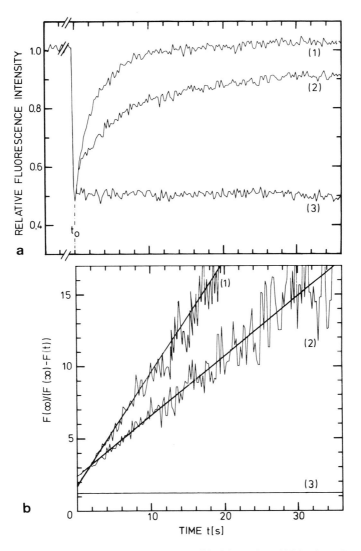

FIG. 2 Effect of surface pressure on lipid probe diffusion in DPPC monolayers: (a) original measuring traces; (b) same data as in (a) replotted in a reciprocal manner according to Yguerabide *et al.* (1982). NBD-egg-PE was used as the probe and added to DPPC in a molar ratio of 0.01. Measurements were taken with an illuminated area of Gaussian intensity profile and an e^{-2}-radius of 18.5 μm. The fluorescence intensity was measured before bleaching, immediately after a 17 ms high intensity flash at t_0, and during the course of fluorescence recovery F(t). Surface pressure, half-time of fluorescence recovery and diffusion coefficients are: (1) 1.2 mN/m, 2.2 s, 44 $μm^2/s$; (2) 7.8 mN/m, 5.5 s, 17 $μm^2/s$; (3) 10 mN/m, D < 0.3 $μm^2/s$. In panel (b) the linearity is a sensitive indicator for absence of flow.

TABLE 2 Translational diffusion in monolayers at the air-water interface as measured by fluorescence microphotolysis (Peters and Beck, 1983).

DLPC			DPPC			Mixtures of DLPC and DPPC		
Π (mN/m)	$\bar{a}$ (nm^2)	D (μm^2/s)	Π (mN/m)	$\bar{a}$ (nm^2)	D (μm^2/s)	X_{DPPC}	$\bar{a}$ (nm^2)	D (μm^2/s)
1.0	0.98	110	1.0	0.83	47	0.2	0.64	34
5.0	0.85	68	5.0	0.71	31	0.4	0.63	30
10.0	0.76	59	7.5	0.51	24	0.6	0.59	25
15	0.69	47	8.0	0.50	14	0.7	0.57	12
20	0.65	41	9.0	0.49	2.7	0.8	0.52	0.1
30	0.59	25	10.0	0.48	0.29			
38	0.55	15	20	0.45	0.02	(Π = 20 mN/m)		

Fluorescent probe: NBD-egg-PE (Avanti Polar Lipids, Birmingham, Alabama, USA)

Probe concentration ratio: 0.01

Lipids from Sigma (Taufkirchen).

Subphase: double distilled water or Millipore Milli Q-water

T = 21-22°C.

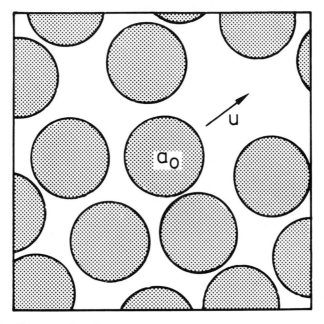

FIG. 3 The "free volume" model of translational diffusion in fluid
lipid monolayers. Lipid molecules are modelled by hard rods of a
minimum molecular area a_0. Free area is created at random by density
fluctuations. If the free area exceeds a critical value a neighbouring molecule can undergo a diffusional displacement.

Diffusion coefficients of the fluorescent probe N-4-nitro-benzo-
2-oxa-1,3-diazolyl egg phosphatidylethanolamine (NBD-egg-PE) in DLPC
monolayers measured by fluorescence micropnotolysis (Peters and
Beck, 1983) are presented in Table 2.   The diffusion coefficient
decreases from 110 μm^2/s at a surface pressure of $\Pi = 1$ mN/m to
15 μm^2/s at 38 mN/m. We have shown (Peters and Beck, 1983) that
the dependence of D on Π or on mean molecular area $\bar a$, respectively,
can be quantitatively accounted for by the "free volume" model
of diffusion (Cohen and Turnbull, 1959; Träuble and Sackmann,
1972; Galla et al., 1979). As indicated in Fig. 3, this approach
models lipid molecules as hard rods moving in a plane with constant
velocity u. Between these constantly moving particles, free area is
created at random. Sometimes, free area exceeds a critical value and
thus enables a diffusional displacement of a neighbouring molecule.
Applied to monolayers the "free volume" model leads to the following

simple relationship (Peters and Beck, 1983):

$$\ln D = \ln D_{max} - [k / (\bar{a} - a_0)] \tag{1}$$

where D_{max} is an upper limit of the diffusion coefficient in the fluid
monolayer, k is a constant, $\bar{a}$ is the mean molecular area and a_0 is the
minimum molecular area. Plotting lipid probe diffusion coefficients
according to Equation (1) as $\ln D$ vs $1/(\bar{a} - a_0)$ yields a linear rela-
tionaship (see also Peters and Beck, 1983). In this procedure the
minimum molecular area a_0 is not known initially. However if a_0 is
varied systematically, the linearity of the plot is optimal (correla-
tion coefficient $|r|$ maximal) for $a_0 = 0.425$ nm^2 (Fig. 4). This a_0-
value is in excellent agreement with the area occupied by diacylphos-
phatidylcholine molecules in crystalline monolayers as measured by
both X-ray diffraction and surface balance methods (Hui *et al.*, 1975),
and supports the notion that the "free volume" model correctly describes lipid probe diffusion in fluid monolayers.

THE MONOLAYER PHASE TRANSITION

L-α-dipalmitoylphosphatidylcholine (DPPC) is known to undergo a phase
transition from a fluid to a crystalline state upon raising the surface pressure at constant temperature (Phillips and Chapman, 1968;
Nagle, 1976; Albrecht *et al.*, 1978). In the surface-pressure/area
diagram the transition is apparent as a "shoulder" which, at 22°C,
extends from about 4 to 14 mN/m (see later: Fig. 7). The fact that,
in the transition region, the Π versus $\bar{a}$ isotherm is not completely
horizontal (i.e. Π not constant) has given rise to much speculation
(Phillips and Chapman, 1968; Nagle, 1975; Albrecht *et al.*, 1978;
Baret, 1981). Some authors even assumed that the transition is second
order. Albrecht *et al.* (1978) have argued that cooperativity in monolayers may be restricted to units of finite size thus making the
transition region a two-phase region in which the DPPC monolayer is
composed of fluid and crystalline arrays. It appears that this old
question can now be answered by diffusion measurements and fluorescence microscopy.

Diffusion coefficients of NBD-egg-PE in DPPC monolayers measured
by fluorescence microphotolysis are also present in Table 2. D-values

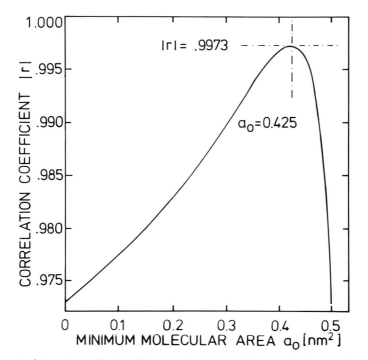

FIG. 4 Application of the "free volume" model to fluid monolayers. The data for NBD-egg-PE diffusion in DLPC monolayers (Table 2) have been plotted according to Equation (1), yielding a linear relationship between $\ln$ D and reciprocal free area (see text). The graph shows that the linearity of the plot as indicated by maximization of the correlation coefficient of linear regression $|r|$ is optimal for a minimum molecular area a_0 = 0.425 nm^2 which agrees with X-ray data on crystalline monolayers.

of a different fluorescent probe, dioctadecyloxacarbocyanine [diO-C18(3)], are shown in Fig. 5. In both cases the prominent feature is a steep decline of the diffusion coefficient between 8 and 10 mN/m. The appearance at various surface pressure values of a DPPC monolayer which has been doped with NBD-egg-PE is shown in Fig. 6. These photomicrographs reveal that monolayer fluorescence is homogeneous in the fluid (i.e., Π < ∿4 mN/m at 22°C) and in the crystalline phase (i.e., Π >14 mN/m). However, in the intermediate region (i.e., ∿4 mN/m ≤ Π ≤ 14 mN/m) the monolayer is composed of fluorescent and non-fluorescent areas. NBD-egg-PE, a derivative of a "fluid" phospholipid with a heterogeneous population of acyl chains, is expected to have a partition preference for fluid phases, and we therefore assume

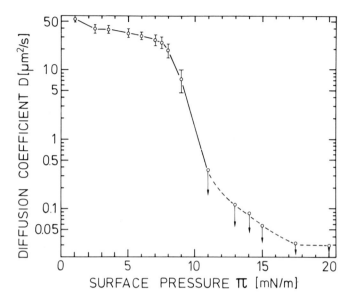

FIG. 5 Translational diffusion of the oxacarbocyanine dye diO-C18(3) in a DPPC monolayer (molar probe/DPPC concentration ratio: 0.01). The diffusion behaviour of diO-C18(3) is very similar to that of NBD-egg-PE (Table 2). Translational diffusion decreases with decreasing mean molecular area $\bar{a}$ (for the dependence of Π on $\bar{a}$ see Fig. 7). Corresponding to the transition of the monolayer from a fluid to a crystalline state which is also apparent in the $\Pi/\bar{a}$ - isotherm (Fig. 7), the diffusion coefficient decreases by more than an order of magnitude between 8 and 10 mN/m. The values at higher surface pressure which are marked by arrows represent upper limits (T $\cong$ 22°C).

FIG. 6 Direct visualization of phase separation during the fluid-to-crystalline phase transition of DPPC monolayers by fluorescence microscopy. The lipid probe was NBD-egg-PE (molar concentration ratio: 0.01, T = 22°C).
(a) Π = 2.5 mN/m: fluorescence is homogeneous;
(b) Π = 4.0 mN/m: first indications of a phase separation become apparent. Small dark patches surrounded by large brightly fluorescent areas are to be seen;
(c) Π = 7.5 mN/m and (d) Π = 8.5 mN/m: with increasing surface pressure the dark areas become more prominent;
(e) Π = 10.0 mN/m: the geometry of the dark patches changes from more or less regular to sometimes bizarre forms;
(f) Π = 15.0 mN/m: at high surface pressure, fluorescence becomes again homogeneous. Because of the low diffusion coefficient (crystalline phase) fluorescence has been partially bleached during the exposure time.
Photomicrographs were taken on Kodak Tri-X film with exposure times of 5-10 s. The illuminated area had a Gaussian intensity profile with an e^{-2}-radius of 18.5 µm (bar: 10 µm).

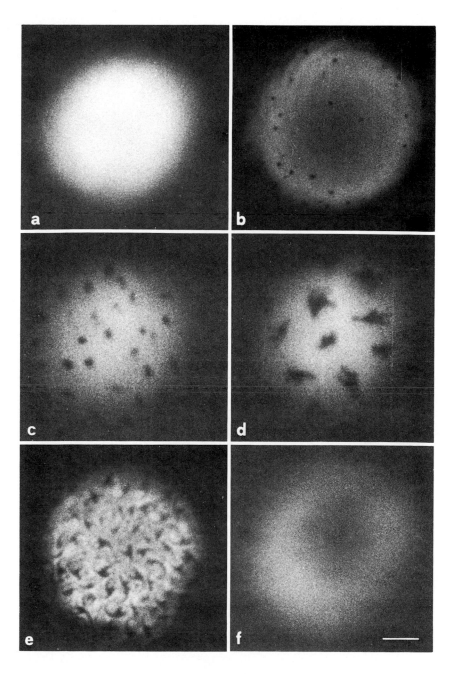

FIG. 6

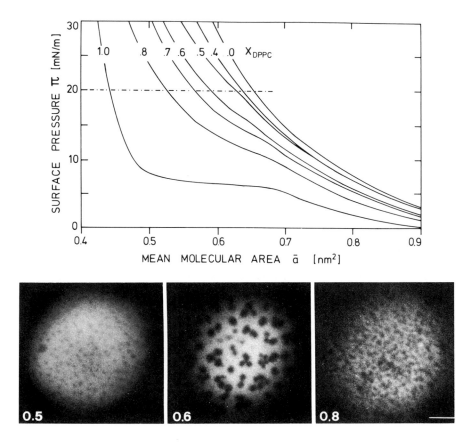

FIG. 7 Surface pressure/area-isotherms for DLPC, DPPC and various mixtures of DLPC and DPPC (all with NBD-egg-PE at a molar concentration ratio of 0.01; T = 21.5°C). Isotherms are smooth for monolayers with a molar fraction of DPPC (X) <0.5. At X = 0.5 first indications of a shoulder appear which become increasingly prominent with increasing X_{DPPC} indicating a phase separation. Whereas monolayer fluorescence looks perfectly homogeneous in pure DLPC and in pure DPPC at 20 mN/m, in mixed monolayers dark patches first appear at X_{DPPC} = 0.5 and become larger with increasing X_{DPPC}. Photomicrographs show a monolayer appearance at 20 mN/m for X_{DPPC}: 0.5, 0.6, and 0.8 (from left to right). Diffusion coefficients are given in Table 2 (bar: 10 μm).

that the fluorescent and non-fluorescent areas seen in Fig. 6 are fluid and solid domains, respectively.

The interpretation of the DPPC data is supported by studies of
monolayers prepared from mixtures of DLPC and DPPC whose surface pres-
sure/area isotherms (Fig. 7) clearly show that these monolayers are
subject to phase separation phenomena which depend on both the molar
fraction of DPPC and surface pressure. For a molar fraction of DPPC
less than or equal to 0.4, the Π versus $\overline{a}$ isotherms are smooth. At
X_{DPPC} = 0.5, the first indications of a shoulder, which becomes more
and more prominent with increasing X_{DPPC}, appear. Both diffusion
coefficients (Table 2) and fluorescence microscopic appearance provide further evidence for phase separation. The diffusion coefficient
shows a steep decline if X_{DPPC} is increased above $\sim$0.5. The fluorescence of DLPC/DPPC monolayers is homogeneous for X_{DPPC} <0.5, but for
$0.5 \leqq X_{DPPC} \leqq 0.8$ it becomes patchy in much the same manner as that
of pure DPPC monolayers in the transition region.

Taken together these measurements and observations strongly suggest
that the fluid to crystalline phase transition of DPPC monolayers pro-
ceeds via a two-phase region in which fluid and solid domains coexist.
Thus the notion is substantiated that the fluid-to-crystalline phase
transition of DPPC monolayers is first order. For reasons which are
still unknown the cooperativity is restricted to units of finite size.

CONCLUSION

In this chapter a new experimental approach to the measurement of
translational diffusion in monolayers at the air-water interface has
been described. If surface convection is eliminated, lipid probe
diffusion coefficients are similar in monolayers and bilayers. The
dependence of the diffusion coefficient on surface pressure and molecular area, in fluid monolayers, is well described by the "free volume" model. The isothermal phase transition of DPPC monolayers from
a fluid to a crystalline state has been shown to proceed via a two-
phase region in which the monolayer is made up of fluid and solid
domains. The study thus demonstrates that, despite earlier misgivings,
monolayers at the air-water interface do provide a useful model system for studies of two dimensional diffusion in membranes, provided
the appropriate experimental precautions are observed.

ACKNOWLEDGEMENTS

We should like to thank Dr O. Albrecht of the University of Mainz
for permission to use his surface balance. We are also grateful to
P. Meller for providing the computer program for the reciprocal plots
shown in Fig. 2.

REFERENCES

Albrecht, O., Gruler, H. and Sackmann, E. (1978). J. Phys. (Paris)
    39, 301-313.
Albrecht, O., Gruler, H. and Sackmann, E. (1981). J. Colloid Inter-
    face Sci. 79, 319-338.
Aplin, J.D. and Hughes, R.C. (1982). Biochim. Biophys. Acta 694, 375-
    418.
Axelrod, D., Koppel, D.E., Schlessinger, J., Elson, E. and Webb, W.W.
    (1976). Biophys. J. 16, 1055-1069.
Baines, A.J. (1983). Nature 301, 377-378.
Baret, J.F. (1981). Progr. Surf. Membr. Sci. 14, 291-351.
Beck, K. and Peters, R. (1982). Hoppe Seyler's Z. Physiol. Chem.
    363, 894.
Beck, K. and Peters, R. (1983). Eur. J. Cell Biol. Suppl. 2, 5.
Cherry, R.J. (1979). Biochim. Biophys. Acta 559, 289-327.
Cohen, M.H. and Turnbull, D. (1959). J. Chem. Phys. 31, 1164-1169.
Devaux, Ph. and McConnell, H.M. (1972). J. Amer. Chem. Soc. 94,
    4475-4481.
Edidin, M. (1981). In "Membrane Structure" (Eds J.B. Finean and R.H.
    Michell), pp. 37-82. Elsevier/North-Holland, New York.
Edidin, M., Zagyanski, Y. and Lardner, T.J. (1976). Science 191,
    466-468.
Fahey, P.F., Koppel, D.E., Barak, L.S., Wolf, D.E., Elson, E.L. and
    Webb, W.W. (1977). Science 195, 305-306.
Fahey, P.F. and Webb, W.W. (1978). Biochemistry 17, 3046-3053.
Fendler, J.H. (1982). "Membrane Mimetic Chemistry". Wiley, New York.
Ferber, E., de Pasquale, G.G. and Resch, K. (1975). Biochim. Bio-
    phys. Acta 398, 364-376.
Frye, L.D. and Edidin, M. (1970). J. Cell Sci. 7, 319-335.
Gaines, G.L. Jr. (1966). In "Insoluble Monolayers at Liquid-Gas Inter-
    faces", pp. 156-288. Wiley, New York.
Galla, H.J., Hartmann, W., Theilen, U. and Sackmann, E. (1979). J.
    Membr. Biol. 48, 215-236.
Geiger, B. (1982). Trends Biochem. Sci. 7, 389-390.
Good, P.A. and Schechter, R.S. (1972). J. Colloid Interface Sci. 40,
    99-106.
Goodman, S.R. and Shiffer, K. (1983). Amer. J. Physiol. 244, C121-
    C141.
Hackenbrock, C.R. (1981). Trends Biochem. Sci. 6, 151-154.
Hafeman, D.G., von Tscharner, V. and McConnell, H.M. (1981). Proc.
    Natl. Acad. Sci. USA 78, 4552-4556.
Hafeman, D.G., Smith, L.M., Fearon, D.T. and McConnell, H.M. (1982).
    J. Cell Biol. 94, 224-227.
Hay, E.E. (1981). J. Cell Biol. 91, 205s-223s.
Huang, L. and McConnell, H.M. (1983). Biophys. J. 41, 115a.

Hughes, B.D., Pailthorpe, B.A. and White, L.R. (1981). *J. Fluid Mech.* 110, 349-372.

Hui, S.W., Cowden, M., Papahadjopoulos, D. and Parsons, D.F. (1975). *Biochim. Biophys. Acta* 382, 265-275.

Jacobson, K., Wu, E. and Poste, G. (1976). *Biochim. Biophys. Acta* 433, 215-222.

Jacobson, K. and Wojcieszyn, J. (1981). *Comments Mol. Cell. Biophys.* 1, 189-199.

Jain, M.K. and White, H.B. (1977). *In* "Advances in Lipid Research" (Eds R. Paoletti and D. Kritchevsky), vol. 15, pp. 1-60. Academic Press, New York.

Johnson, P., Garland, P.B., Campbell, P. and Kusel, J.R. (1982). *FEBS Lett.* 141, 132-135.

Karnovsky, M.J., Kleinfeld, A.M., Hoover, R.L. and Klausner, R.D. (1982). *J. Cell Biol.* 94, 1-6.

Koppel, D.E. (1983). *In* "Fast Methods in Physical Biochemistry and Cell Biology" (Eds R.I. Sha'afi and S.M. Fernandez), pp. 339-367. Elsevier/North-Holland, Amsterdam.

Kuhn, H., Möbius, D. and Bücher, H. (1972). *In* "Physical Methods of Chemistry" (Eds A. Weissberger and B. Rossiter), Vol. 1, part 3B, pp. 651-653. Wiley, New York.

De Laat, S.W., van der Saag, P.T. and Shinitzky, M. (1977). *Proc. Natl. Acad. Sci. USA* 74, 4458-4461.

De Laat, S.W., van der Saag, P.T., Elson, E.L. and Schlessinger, J. (1980). *Proc. Natl. Acad. Sci. USA* 77, 1526-1528.

Liebman, P.A. and Pugh, E.N. Jr. (1979). *Vision Res.* 19, 375-380.

Loor, F. (1981). *In* "Cell Surface Reviews" (Eds G. Poste and G.L. Nicolson), pp. 253-335. Elsevier/North-Holland, Amsterdam.

Loughran, T., Hatlee, M.D., Patterson, L.K. and Kozak, J.J. (1980). *J. Chem. Phys.* 72, 5791-5797.

Lux, S.E. (1983). *In* "The Metabolic Basis of Inherited Disease", 5th Edn. (Eds J.B. Stanbury, J.B. Wyngaarden, J.L. Goldstein and M.S. Brown), pp. 1573-1605, McGraw-Hill, New York.

McGrath, A.E., Morgan, C.G. and Radda, G.K. (1976). *Biochim. Biophys. Acta* 426, 173-185.

Nagle, J.F. (1975). *J. Chem. Phys.* 63, 1255-1261.

Nagle, J.F. (1976). *J. Membr. Biol.* 27, 233-250.

Pastan, I.H. and Willingham, M.C. (1981). *Science* 214, 504-509.

Peters, R. (1981). *Cell. Biol. Int. Reports* 5, 733-760.

Peters, R., Peters, J., Tews, K.H. and Bähr, W. (1974). *Biochim. Biophys. Acta* 367, 282-294.

Peters, R. and Richter, H.-P. (1981). *Dev. Biol.* 86, 285-293.

Peters, R. and Cherry, R.J. (1982). *Proc. Natl. Acad. Sci. USA* 79, 4317-4321.

Peters, R. and Beck, K. (1983). *Proc. Natl. Acad. Sci. USA* 80, 7183-7187.

Phillips, M.C. and Chapman, D. (1968). *Biochim. Biophys. Acta* 163, 301-313.

Pinto da Silva, P. (1972). *J. Cell Biol.* 53, 777-787.

Saffman, P.G. (1976). *J. Fluid Mech.* 73, 563-602.

Saffman, P.G. and Delbrück, M. (1975). *Proc. Natl. Acad. Sci. USA* 72, 3111-3113.

Sakata, E.K. and Berg, J.C. (1964). *Ind. Eng. Chem. Fundamentals* 3, 132-139.

Scandella, C., Campisi, J., Elhai, J. and Selak, M. (1982). *Biophys. J.* 37, 16-17.

Seul, M., Weis, R.M. and McConnell, H.M. (1983). *Biophys. J.* 41, 212a.

Singer, S.J. and Nicolson, G.L. (1972). *Science* 175, 720-731.

Smith, B.A. and McConnell, H.M. (1978). *Proc. Natl. Acad. Sci. USA* 75, 2759-2763.

Stroeve, P. and Miller, J. (1975). *Biochim. Biophys. Acta* 401, 157-167.

Szabo, G. and Waldbillig, R.C. (1982). *In* "Methods of Experimental Physics, Biophysics" (Eds G. Ehrenstein and H. Lecar), Vol. 20, pp. 513-543. Academic Press, New York.

Taylor, R.B., Duffus, W.P.H., Raff, M.C. and de Petris, S. (1971). *Nature New Biol.* 233, 225-229.

Teissié, J., Tocanne, J.F. and Bundras, A. (1978). *Eur. J. Biochem.* 83, 77-85.

Trauble, H. and Sackmann, E. (1972). *J. Amer. Chem. Soc.* 94, 4499-4510.

Vaz, W.L.C., Derzko, Z.I. and Jacobson, K.A. (1982). *In* "Membrane Reconstitution" (Eds G. Poste and G.L. Nicholson), Vol. 8, pp. 83-136. Elsevier/North-Holland, Amsterdam.

Vaz, W.L.C. and Hallmann, D. (1983). *FEBS Lett.* 152, 287-290.

Vollhardt, D., Zastrow, L. and Schwartz, P. (1980a). *Colloid Polymer Sci.* 258, 1176-1182.

Vollhardt, D., Zastrow, L., Heybey, J. and Schwartz, P. (1980b). *Colloid Polymer Sci.* 258, 1289-1295.

Vollhardt, D., Zastrow, L. and Schwartz, P. (1981). *J. Colloid Interface Sci.* 83, 643-644.

von Tscharner, V. and McConnell, H.M. (1981a). *Biophys. J.* 36, 409-419.

von Tscharner, V. and McConnell, H.M. (1981b). *Biophys. J.* 36, 421-427.

Weis, R.M. and McConnell, H.M. (1983). *In* "Progress in Clinical and Biological Research, Cell Function and Differentiation" (Eds G. Akoyunoglou, A.E. Evangelopoulos, J. Georgatsos, G. Palaiologas, A. Trakatellis and C.P. Tsiganos), Vol. 102A, pp. 331-336. Alan R. Liss Inc., New York.

Wirz, J.H. and Neumann, R.D. (1978). *J. Colloid Interface Sci.* 63, 583-589.

Wolf, D.E., Schlessinger, J., Elson, E.L., Webb, W.W., Blumenthal, R. and Henkart, P. (1977). *Biochemistry* 16, 3476-3483.

Wolf, D.E., Edidin, M. and Handyside, A.H. (1981). *Dev. Biol.* 85, 195-198.

Yguerabide, J., Schmidt, J.A. and Yguerabide, E.E. (1982). *Biophys. J.* 39, 69-75.

Biophysical Studies of Cell-Cell Recognition

H.M. McCONNELL

INTRODUCTION

In spite of its importance, remarkably little is known about the
molecular basis of specific cell-cell adhesion. This is doubtless
related in part to the technical difficulties involved. The cell
surface biochemistry of eukaryotic cells is still in a relatively
primative state, and putative complementary ligands and receptors on
most cell types have not been identified. Even if ligand-receptor
pairs were identified and isolated, the study of their interaction
might be a difficult biophysical problem, since many specific cell-
cell interactions probably involve a relatively large number of weak,
but specific bonds.

In order to approach this general problem of cell-cell recognition
we have adopted two strategies. First, we replace one plasma membrane of a cell-cell pair by a reconstituted membrane. Second, we
focus on the cellular components of the immune system. It is almost
certain that some immune responses do require cell-cell recognition
at the level of plasma membrane-plasma membrane contact. Moreover,
in some (but not all) cases, the molecules involved in cellular immune
responses have been thoroughly studied. This is also true for antibody-dependent cellular responses. Such antibody-dependent responses
provide us with relatively simple experimental models of cell-cell
recognition, but it remains to be seen whether or not specific anti-

body-dependent recognition events are generally similar to other cell-cell recognition events.

The difficulty of investigating the biochemistry and biophysics of cell-cell interactions is nicely illustrated in Fig. 1. Here one sees a scanning electron microscope micrograph of a natural killer (NK) cell, and an NK cell conjugate with a tumour target cell (Frey et al., 1982). In this case little is known about putative conjugate ligand-receptor pairs on the two cells or the mechanism(s) of cyto-toxicity. The photograp graphically illustrates the fine structure of the contacts, and the probable role of cytoskeletal components in the binding and triggering process.

RECONSTITUTED MEMBRANES: LIPOSOMES CONTAINING A VIRAL PROTEIN AND A HISTOCOMPATIBILITY ANTIGEN ARE RECOGNIZED BY PRECYTOTOXIC T-LYMPHOCYTES

Substantial progress in understanding the molecular basis of immune cell-target cell interaction has been achieved by replacing the target cell membrane by reconstituted membranes - that is, liposomes or lipid vesicles containing specific haptens or antigens in the outer membrane (Tom and Six, 1980). Such studies make it possible to delineate the chemical compositions and concentrations of molecules in the target membrane that are required for effector cell binding and triggering. Notable examples are reconstituted membranes con-taining class I histocompatibility antigens (H-2K^k) and viral pro-teins (vesicular stomatitis virus G protein (VSV-G)), which can elicit virus-specific H-2 restricted T-cell cytotoxic responses. The pres-ence of both molecules in the same target lipid membrane has been shown to be necessary (Finberg et al., 1978; Loh et al., 1979; Cartwright et al., 1982).

An additional motivation for using reconstituted membranes in such studies has been the hope that structural information can be obtained at the molecular level. For example, in the case of H-2 restricted, virus-specific T-cell responses, there is the much discussed question as to whether H-2 molecules and viral protein molecules are bound by physically separate receptors on the T-cell surface ("dual recogni-tion"), or whether there is a single receptor that binds H-2 and viral protein in close association with one another ("modified self").

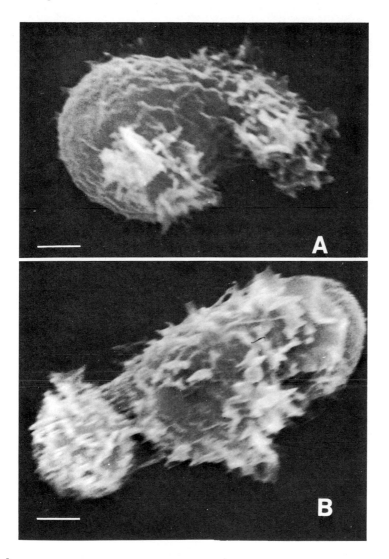

FIG. 1 A. Scanning electron micrograph of a murine NK cell. The asymmetric morphology is typical of these cells. B. Scanning electron micrograph of an NK cell-Moloney virus-transformed mouse lymphoma (YAC) cell conjugate. There are numerous microvilli (filopodia) in the region between the two cells, presumably originating from the NK cell. Bars: 2 μm. Taken from Frey et al. (1982).

In an attempt to approach this problem from a biophysical point of view, we first prepared fluoresceinated $H-2K^k$ ($FH-2K^k$) and fluoresceinated VSV-G protein (FG), and showed that liposomes containing

FH-2K^k and G, or H-2K^k and FG, elicit a G-specific, H-2K^k restricted secondary T-cell response (Cartwright *et al.*, 1982). Fluorescence photobleaching studies showed no evidence for an effect of G on the diffusion of FH-2K^k, or for an effect of H-2K^k on the diffusion of FG. Such experiments definitely rule out an extensive co-oligomerization of H-2K^k and G in the liposome, but do not answer a number of other important questions (Cartwright *et al.*, 1982). Perhaps the most intriguing question is whether the H-2K^k and G molecules become significantly associated with one another only during their mutual interaction with the T-cell receptor. This type of structural question is quite general. One may pose similar, related questions for a number of other cellular responses involving membrane-membrane interactions.

Unfortunately, it is virtually impossible to use conventional epifluorescence microscopy to obtain information on the motion and distribution of putative linker molecules such as FH-2K^k and FG that are jointly bound to liposomes and effector cells. In order to approach this probelm we have used lipid monolayers on solid substrates as targets for effector cells of the immune system, especially those cells that are linked to lipid haptens in the monolayer by means of specific antibodies.

THE BINDING AND TRIGGERING OF CELLS BY SENSITIZED LIPID MONOLAYERS ON SOLID SUBSTRATES

We have employed lipid monolayers on glass or quartz slides in order to observe the fluorescence of "linker" molecules held jointly by an effector cell membrane and a monolayer target membrane. The preparation of these supported monolayers is described in detail elsewhere (von Tscharner and McConnell, 1981). Briefly, a microscope slide is rendered hydrophobic by alkylation with octadecyltrichlorosilane. The alkylated slide is then pushed through a lipid monolayer on an air-water interface. When fluorescent lipid probes are included in these monolayers, lateral diffusion coefficients can be determined, e.g. by measuring fluorescence recovery after pattern photobleaching. The lateral diffusion coefficients so obtained are quite similar to those observed for phospholipid bilayers, if the lipid monolayers are coated at a lateral pressure of about 40 dynes/cm. As judged by

abrupt changes in lateral diffusion coefficients, the Chapman chain
melting transitions of such monolayer phosphatidylcholines occur at
very nearly the same temperatures as do the corresponding transitions
in bilayers and multilayers of phosphatidylcholines. Most remarkably,
the monolayers also appear to exhibit the lower temperature phase
transitions found in phosphatidylcholine bilayers and multibilayers
(M. Seul *et al.*, 1983). On the basis of these results we have confidence that the lipid monolayers have a high degree of perfection and
are equivalent to lipid bilayers in many respects.

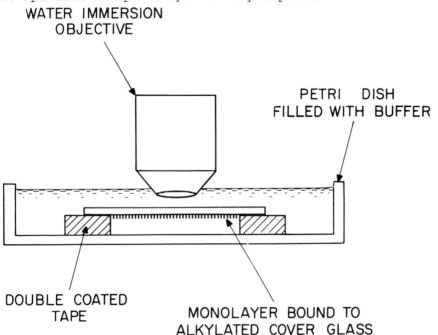

FIG. 2 Lipid monolayer membrane on glass (or quartz) substrate
mounted so as to facilitate photobleaching measurements of lateral
diffusion and microscopic observation of adherent cells. The configuration shown is "monolayer down", and is particularly convenient for
observing specific binding. Taken from Hafeman *et al.* (1981).

   The experimental system used to observe the epifluorescence of
lipid monolayers containing fluorescent lipid probes is sketched in
Fig. 2 (Hafeman *et al.*, 1981). A laser beam passes through a Ronchi
ruling, down through the microscope objective lens and is focussed
onto the monolayer containing a fluorescent lipid probe. A short

burst of laser radiation bleaches a pattern of parallel rectangular
fluorescent stripes. This pattern disappears with a time constant
that provides a quantitative measure of the lateral diffusion coefficient. Similar measurements have been made on the lateral diffusion
of fluoresceinated antibodies bound to lipid haptens in monolayers.
Just as was found for lipid bilayers, the fluoresceinated antibodies
diffuse with very nearly the same lateral diffusion coefficients as
do the lipids. When the lateral diffusion of the lipids is fast
(10^{-7}-10^{-8} cm^2/s, so is the lateral diffusion of the bound fluoresceinated antibody; when the lateral diffusion of the lipids is slow
(10^{-10}-$10^{-12} cm^2/s$, the lateral diffusion of the bound antibodies is
also found to be slow (Hafeman et al., 1981).

Cells having appropriate Fc receptors bind to and are triggered by
laterally mobile (fluoresceinated) antibodies specifically bound to
lipid haptens in supported monolayers. For example, in experiments
carried out by Hafeman et al. (1981), a nitroxide (spin label) lipid
hapten was first incorporated into a supported lipid monolayer, and
fluoresceinated rabbit antinitroxide antibodies allowed to bind to
these lipid haptens. Guinea pig peritoneal macrophages were then
added to this "opsinized" monolayer. After a few minutes, the slide
was turned upside down. Virtually all of the macrophages remained
bound. If antibody and/or lipid hapten was omitted from the monolayer, virtually all the macrophages fell off. In these experiments
the monolayer fluorescence due to the fluoresceinated antibodies is
uniform. When the macrophages are added to the monolayer, the intensity of fluorescence in the contact area increases dramatically, as
illustrated in Fig. 3. The laterally diffusing antibodies bind to
and are trapped by the Fc receptors of the macrophage. When the
monolayer membrane is "fluid" (lateral diffusion coefficient,
10^{-8} cm^2/s) this diffusion to the macrophage traps is quite rapid,
taking on the order of one hour (see Fig. 3). When the monolayer
membrane is "solid" (lateral diffusion coefficient, $\lesssim 10^{-10}$ cm^2/s),
diffusion to the trap is much slower. Experiments have not yet
established if any of the macrophage-associated fluorescence is due
to antibody internalization.

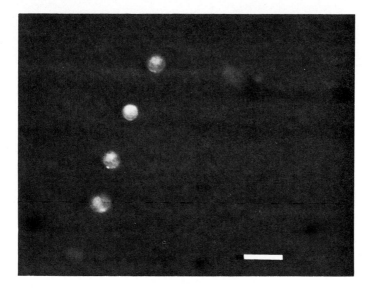

FIG. 3 An epifluorescence photomicrograph showing increased intensity
of monolayer-bound fluorescent IgG in the region of macrophage contact after 37°C incubation for one hour on a fluid monolayer. Spread
(triggered) macrophages on the same monolayer are not observed due to
photobleaching facilitated by superoxide anion, O_2^- (absence of superoxide dismutase in the cell buffer). Note that the fluorescence
intensity under the macrophages is not uniform. Bar: 20 µm. Taken
from Hafeman et al. (1981).

Supported lipid monolayers also provide an ideal system to observe
cellular triggering. In the presence of specific lipid hapten bound
antibody, the macrophages described above are triggered when the
temperature is increased from room temperature to 37°C (Hafeman et al.
1981). When triggered, the cells release superoxide anion, O_2^-,
which leads to enhanced photobleaching of fluorescent groups near the
cells. This effect can be prevented by including superoxide dismutase in the cell buffer. The release of enzymes such as cathepsin B
can be detected by their reaction with synthetic substrates that
yield fluorescent products.

EVANESCENT WAVE EXCITATION OF THE FLUORESCENCE OF LINKER
MOLECULES BOUND BOTH TO MONOLAYERS AND TO CELLS

The technique of evanescent wave excitation of fluorescence provides
an ideal way to study the motion and distribution of specific linker
molecules involved in the binding and triggering of cells by lipid

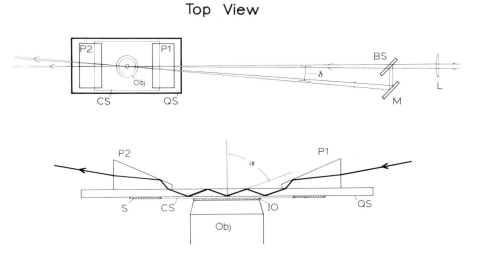

FIG. 4 System used to observe the stimulation of fluorescence by molecules near a lipid monolayer ($\lesssim 700$ Å), using evanescent radiation. See text for details. Taken from Weis *et al.* (1982).

monolayers. Figure 4 shows the experimental system employed in our laboratory (Weis *et al.*, 1982). The underside of the microscope slide is coated with a lipid monolayer. Cell buffer solution is held in position under the microscope slide by means of a cover slip (CS) and thin (24 μ) spacers (S). A focussed beam from an argon ion laser is introduced into the quartz microscope slide (QS) by means of prism P1, undergoes total internal reflection, and exits the slide by means of the prism P2. In most experiments a 63X objective lens (Obj) is used with immersion oil (IO). For some experiments, it is useful to use two focussed laser beams, produced by a beam splitter (BS) and a mirror (M), as shown in Fig. 4. The interference pattern produced at the monolayer can be used for pattern photobleaching to measure the lateral diffusion of fluorescent molecules bound to the monolayer. The intensity of the evanescent wave field decreases exponentially with distance as one moves away from the monolayer and quartz slide. The 1/e distance is approximately 700 Å for our present system.

Our first experiments with this technique involved studies of the binding and triggering of rat basophil leukaemia cells. These cells

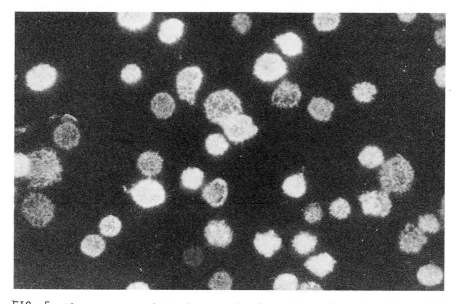

FIG. 5 Fluorescence photomicrograph of monolayer-bound rat basophil leukaemia cells. The fluorescence arises from fluoresceinated mono- colonal IgE molecules bound to lipid haptens in the monolayer. Fluorescence is excited by evanescent radiation with the apparatus depicted in Fig. 5. Taken from Weis *et al.* (1982).

are known to bind to and be triggered by liposomes containing lipid hapten in the presence of hapten specific IgE antibodies. This sys- tem was selected for study because it has been the subject of exten- sive previous investigations of the molecular basis of triggering (Metzger, 1979; Ishizaka and Ishizaka, 1978).

Figure 5 shows an evanescent wave epifluorescence microscope view of RBL cells specifically bound to lipid haptens in a monolayer by means of fluoresceinated IgE molecules (FIgE) (Weis *et al.*, 1982). An ordinary epifluorescence microscope view of these cells shows a uni- form fluorescence. The apparent non-uniform fluorescence in Fig. 5 arises from the non-uniform three-dimensional attachment of these cells to the monolayer. Evidently, filopodia having FIgE molecules at their tips form limited numbers of specific links to the monolayer surface. The binding of these cells to the monolayer surface depends strongly on lipid hapten density, time, temperature and the presence of Ca^{++} and Mg^{++} in the cell buffer. The RBL cells are also triggered

to release [^{3}H] serotonin. A detailed description of these experiments is given elsewhere (Weis *et al.*, 1982).

SYNOPSIS AND PROGNOSTICATION

It is fascinating to note the overall similarity of the effector cell
morphologies in Figs. 1 and 5. In both cases, the filopodia reach
from the effector cell to the target, a tumour cell in the first case
and a lipid hapten-containing monolayer in the second case. It is
tempting to speculate that this is a general cellular morphology
characteristic of specific cell-cell recognition. What are the biochemical correlates of this morphology? One reasonable guess is that
specific receptors are concentrated in the regions of membrane-membrane contact. The result shown in Fig. 5 for fluorescent antibodies
bound to monolayers certainly supports this view as far as the monolayer membrane is concerned. Unpublished experiments by R. Weis and
the present author, using mixtures of fluoresceinated and nonfluores-
ceinated, specific and non-specific IgE on RBL cells also support this
view for IgE-Fc complexes on the RBL cell membrane.

   The reader will doubtless appreciate that we have merely "scratched
the surface" in our studies of membrane-membrane interactions. The
planar supported monolayers provide an ideal geometry for physical
and chemical studies of these interactions. Soon we hope to be able
to support lipid bilayers in a similar way, which should facilitate
studies of transmembrane proteins and the transmembrane potentials
and conductances that are relevant to cell-cell recognition.

ACKNOWLEDGEMENTS

This work was supported by NIH grant 5R01 AI13587 and NSF grant PCM
801993.

REFERENCES

Cartwright, G.A., Smith, L.M., Henizelmann, E.W., Ruebush, M.J.,
    Parce, J.W. and McConnell, H.M. (1982). *Proc. Natl. Acad. Sci.
    USA* 79, 1506-1510.
Finberg, R., Mescher, M. and Burakoff, S. (1978). *J. Exp. Med.* 148,
    1620-1627.
Frey, T., Petty, H.R. and McConnell, H.M. (1982). *Proc. Natl. Acad.
    Sci. USA* 79, 5317-5321.
Hafeman, D.G., von Tscharner, V. and McConnell, H.M. (1981). *Proc.
    Natl. Acad. Sci. USA* 78, 4552-4556.
Ishizaka, T. and Ishizaka, K. (1978). *J. Immunol.* 120, 800-805.

Loh, D., Ross, A.H., Hale, A.H., Baltimore, D. and Eisen, H.N. (1979).
    J. Exp. Med. 150, 1067-1074.
Metzger, H. (1979). *Annu. Rev. Pharmacol. Toxicol.* 19, 427-445.
Seul, M., Weis, R.M. and McConnell, H.M. (1983). *Biophys. J.* 41,
    212a.
Tom, B.H. and Six, H.R. (1980). *In* "Liposomes and Immunobiology".
    Elsevier/North-Holland, New York.
von Tscharner, V. and McConnell, H.M. (1981). *Biophys. J.* 36, 421-
    427.
Weis, R.M., Balakrishnan, K., Smith, B.A. and McConnell, H.M. (1982).
    J. Biol. Chem. 257, 6440-6445.

ESR Probes for Structure and Dynamics of Membranes

D. MARSH

INTRODUCTION

Spin-label ESR is a probe technique which uses a stable nitroxide
free radical to report on molecular structure and dynamics in biological systems. The principal ESR properties of nitroxide spin
labels, and the molecular information which may be obtained from
them, are listed in Table 1. In this chapter we shall be concerned
primarily with the first three sets of properties, which deal with
the orientational order and the rotational dynamics of the labelled
molecules. All three properties depend essentially on the angular
anisotropy of the spectrum with respect to the magnetic field direction.

The spectral anisotropy is described by the spin Hamiltonian for
the magnetic energy of the nitroxide unpaired electron spin, $\underline{S}$:

$$H = \beta \, \underline{H} \cdot \underline{\underline{g}} \cdot \underline{S} + \underline{I} \cdot \underline{\underline{A}} \cdot \underline{S} \qquad (1)$$

where β is the Bohr magneton, I is the ^{14}N nuclear spin and $\underline{H}$ is the
laboratory magnetic field vector. The g-value and the hyperfine
splitting, A, are both tensor quantities, having the principal ele-
ments: $(g_{xx}, g_{yy}, g_{zz}) = (2.0088, 2.0058, 2.0021)$ and (A_{xx}, A_{yy}, A_{zz})
$= (5.9, 5.4, 32.9)$ G, as determined from single crystal measurements
(Jost et $al.$, 1971). The orientation of the x, y, z principal axes

TABLE 1 ESR probe properties of nitroxide spin labels

ESR parameter	Molecular information
1. Anisotropy (A_{xx}, A_{yy}, A_{zz}) (hyperfine splittings; g-values)	Angular orientation and amplitudes of motion.
2. Spin-spin relaxation, T_2 (line-widths)	Rotational correlation times: $10^{-11}s \gtrsim \tau_R \gtrsim 10^{-8}s$
3. Spin-lattice relaxation, T_1 (saturation transfer ESR)	Slow rotational motions: $10^{-7}s < \tau_R \gtrsim 10^{-3}s$
4. Spin density, a_0 ($\propto \rho_N$) (isotropic hyperfine splitting)	Environmental polarity.
5. Spin-spin interactions:	
(exchange, J)	Diffusion, aggregation.
(dipolar, μ^2/r^3)	Distances.
6. Chemical reduction $>N\text{-}\overset{\bullet}{O} \rightarrow >N\text{-}OH$ (line-heights).	Location and translocation.
7. Intensity (double-integral/line-heights)	Binding titrations/ kinetics.

relative to the nitroxide group, found from the single crystal studies, is given in Fig. 1.

The anisotropy of the hyperfine splitting and g-values thus allows determination of the orientation of the molecule. Molecular motion which is rapid compared with the anisotropy ($A_{zz} - A_{xx} \sim 70$ MHz; $(g_{xx} - g_{zz})\beta H \sim 30$ MHz) gives rise to motional averaging of the aniso-tropy, and the degree of averaging is a measure of the angular ampli-tude of the molecular rotation. The line-widths in the spectrum are determined by the rate of modulation of the anisotropies by molecular rotation, and hence can be used to measure the rotational correlation times. This method is applicable to motions in the correlation time range $10^{-11}s \gtrsim \tau_R \gtrsim 10^{-8}$ s. The effects of rotational modulation of the anisotropy on the saturation characteristics of the spectrum are

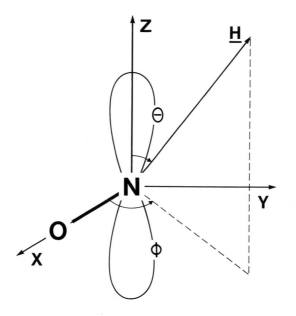

FIG. 1 Principal axis system for the nitroxide hyperfine and g-tensor. The z-axis is parallel to the free radical 2p-π orbital and the
x axis is directed along the N O bond direction. Polar coordinates
of the magnetic field direction, H(θ,φ), are given relative to this
system. The direction cosines of the magnetic field are: $\ell = \cos\phi\cdot\sin\theta$, $m = \sin\phi\,\sin\theta$, $n = \cos\theta$.

used in saturation transfer spectroscopy to extend the range of
motional sensitivity to correlation times in the regime 10^{-7} s $\overset{\sim}{<} \tau_R$
$\overset{\sim}{<} 10^{-3}$ s.

   Further discussion of these and experimental aspects of spin-label
ESR may be found in Marsh (1981, 1982) and in Knowles et al. (1976).
In the present contribution we consider three particular aspects in
detail, namely (i) the rapid rotation of small molecules in an isotropic medium, (ii) the ordering of the molecular motion in anisotropic systems, e.g. membranes, and (iii) the slow rotation of molecules in highly viscous systems. The last of these approaches the
limits of motional sensitivity of conventional nitroxide ESR spectroscopy. Examples of applications of these methods of analysis are
given by: (i) the effects of osmotic stress on the aqueous cytoplasm
of a marine alga studied via the rotational motion of the small spin

label, TEMPONE, (ii) the anisotropic molecular motion of spin-labelled ganglioside analogues incorporated into the cell membrane of a permanent mammalian cell line, and (iii) the motional restriction of lipid chains arising from lipid-protein interactions in membranes.

RAPID ROTATIONAL MOTION

Correlation times

Nitroxide spin labels undergoing rapid rotational motion in an isotropic medium have their spectral anisotropy completely averaged and give rise to a three-line spectrum, the relative line-broadenings of which are a measure of the rate of rotational averaging. The homogeneous line-broadening is given by the transverse or spin-spin relaxation time T_2, which depends upon the rotational correlation time for molecular motion.

Time-dependent perturbation theory may be used to calculate the spin-spin relaxation time in the fast motional narrowing regime, i.e. for correlation times: 10^{-11} s $\stackrel{\sim}{<} \tau_R \stackrel{\sim}{<} 3.10^{-9}$ s. The spin-spin relaxation rate is determined by the intensity of the magnetic field fluctuations $<H_f^2>$ arising from the motional modulation of the spectral anisotropies, and by the frequency distribution or spectral density, $J(\omega)$, of these fluctuations:

$$1/T_2 \sim <H_f^2> \cdot J(\omega) \tag{2}$$

The spectral density is given by the Fourier transform of the correlation function, $G(\tau)$, of the magnetic field fluctuations:

$$J(\omega) = \int_{-\infty}^{\infty} G(\tau) \cdot e^{i\omega\tau} \cdot d\tau \tag{3}$$

The correlation function specifies the characteristic persistence time of the molecular rotation and is defined by:

$$G(\tau) = <F_f^*(t + \tau) \cdot F_f(t)> \tag{4}$$

where the angular brackets represent a time (or rotational) average, and $F_f(t)$ is the time-dependent part of the magnetic field fluctuations.

Assuming a Brownian diffusion model, the probability density of the time-dependent molecular orientation is given by the solution of the rotational diffusion equation. The correlation function then has an exponential τ-dependence and is of the form (Freed, 1964):

$$G_L(\tau) \sim \sum_{m=-L}^{+L} g_{Lm} \, e^{-\tau/\tau_{Lm}} \tag{5}$$

The rotational correlation times τ_{Lm} are given by the eigenvalues of the diffusion equation for a symmetric top:

$$\tau_{Lm}^{-1} = L(L + 1) \, D_\perp + (D_\parallel - D_\perp) m^2 \tag{6}$$

where $D_\parallel$, $D_\perp$ are the principal values of the presumed axial molecular diffusion tensor. For modulation of the hyperfine and Zeeman interactions (second order spherical harmonics) we require the $L = 2$ terms. For isotropic rotational diffusion: $D_\parallel = D_\perp = D_R$, and the rotational correlation time is given by:

$$\tau_R = \tau_{20} = 1/6D_R \tag{7}$$

For anisotropic rotational diffusion there are two independent rotational correlation times, $\tau_{20}^{-1} = 6D_\perp$ and $\tau_{2\pm2}^{-1} = 2D_\perp + 4D_\parallel$, of interest for ESR. The corresponding correlation times for rotation around the axis of symmetry, $\tau_\parallel = 1/6D_\parallel$, and for rotation perpendicular to the axis, $\tau_\perp = 1/6D_\perp$, are given by:

$$\tau_\parallel = \frac{2\tau_{20} \cdot \tau_{22}}{3\tau_{20} - \tau_{22}} \tag{8}$$

and

$$\tau_\perp = \tau_{20} \tag{9}$$

Since the fluctuating field $\langle H_f^2 \rangle$ in Equation (2) involves modulation of both the hyperfine and Zeeman anisotropies, the line-width will involve all possible quadratic cross-products between these two terms of the spin Hamiltonian (c.f. Equation (1)). The line-width therefore has the following dependence on the nitrogen nuclear spin quantum number:

$$\Delta H_{m_I} = A + B\ m_I + C\ m_I^2 \tag{10}$$

where $m_I = -1$, 0, +1 for the high-field, central and low-field hyperfine lines, respectively. The A-term contributes a constant broadening to all three lines and cannot be easily distinguished from other
(inhomogeneous) broadening mechanisms. The B-term gives a differential broadening of all three lines, and the C-term gives a symmetrical
broadening about the central line. Thus B and C may be determined
separately from the measured line-widths.

For _isotropic_ rotation the following expressions are obtained for
the line-width coefficients B and C (Fraenkel, 1967; Schreier _et al._,
1978):

$$B = 1.81 \times 10^6\ H\ [\Delta g\ \Delta A + 3\ \delta g \cdot \delta A]j(0) \tag{11}$$

$$C = 1.81 \times 10^6\ [(\Delta A)^2 + 3(\delta A)^2][j(0) - \frac{3}{8}\ j(\omega_A)] \tag{12}$$

where B, C and the magnetic field H are in gauss, and the hyperfine
anisotropies (also in gauss) are given by:

$$\Delta A = A_{zz} - \frac{1}{2}\ (A_{xx} + A_{yy}) \tag{13}$$

$$\delta A = \frac{1}{2}\ (A_{xx} - A_{yy}) \tag{14}$$

with exactly similar expressions for the g-value anisotropies, Δg
and δg. The reduced spectral densities, for isotropic liquids and
exponential correlation functions, are given by:

$$j(\omega) = \tau_R/(1 + \omega^2 \tau_R^2) \tag{15}$$

The frequency $\omega_A = \frac{1}{2}\ a_0 \approx 10^9\ s^{-1}$, corresponds to the hyperfine or
nuclear spin-flip frequency in radian·s^{-1}. Making the approximation
$j(\omega_A) \approx \tau_R$, which requires $\tau_R \stackrel{\scriptstyle \sim}{\scriptstyle <} 10^{-9}$ s, it is found that $|C/B| \sim 1$
for _isotropic rotation_, and using the tensor elements from Jost _et_
_al._ (1971) the rotational correlation time is given (Schreier _et al._,
1978) by:

$$\tau_B^{iso} \ (s) \ = \ -1.22 \cdot 10^{-9} \cdot B \, (G) \tag{16}$$

or

$$\tau_C^{iso} \ (s) \ = \ 1.19 \cdot 10^{-9} \ C \, (G) \tag{17}$$

where Equation (16) has been calculated for H = 3300 G.

For _anisotropic_ rotation which is axially symmetric about the
nitroxide z-axis, and with the same approximations as for Equations
(16) and (17), the correlation times can be calculated from the
following equations (Polnaszek, 1984; Goldman _et al._, 1972):

$$\tau_0 = \frac{1.11 \times 10^{-7}}{H \cdot \Delta A} \cdot \frac{5(\delta A) B - 8(\delta g) H \cdot C}{\Delta g \cdot \delta A - \delta g \cdot \Delta A} \tag{18}$$

and

$$\tau_{22} = \frac{3.69 \times 10^{-8}}{H \cdot \delta A} \ \frac{8 \ \Delta g \ H \ C - 5 \Delta A \ B}{\Delta g \ \delta A - \delta g \Delta A} \tag{19}$$

where again B, C, H, ΔA and δA are all in gauss, and the correlation
times τ_0 and τ_{22} (in seconds) are given by Equation (6) above. The
same expressions (Equations (18) and (19)) hold for axially symmetric
rotation about the x and y axes if the indices in Equations (13) and
(14), and the corresponding expressions for the g-value anisotropy,
are cyclically permuted.

The numerical expressions for the correlation times corresponding
to axial anisotropic rotation may be written in the form:

$$\tau_0 = c_1 [C + c_2 B] \tag{20}$$

and

$$\tau_{22} = b_1 [B + b_2 C] \tag{21}$$

Values for the numerical constants c_1, c_2, b_1 and b_2, corresponding
to correlation times in seconds and line-width coefficients in gauss,
are given in Table 2. Since B is negative and C is positive, the
requirement that the correlation times be positive leads to limits on
the possible relative values of the line-width coefficients. The
ranges of values for the ratio $|C/B|$ which are compatible with the
various different anisotropic rotations are given in Table 2. The

TABLE 2 Numerical constants relating line-width coefficients to correlation times; and range of validity of the ratio $|C/B|$, for axial rotation: $\tau_{20}=c_1(C+c_2B)$; $\tau_{22}=b_1(B+b_2C)$. (Adapted from Polnaszek, 1984).

Rotation axis	$|C/B|$	c_1 $(s \cdot G^{-1})$	c_2	b_1 $(s \cdot G^{-1})$	b_2	tensors*
isotropic	≈ 1	$1.19 \cdot 10^{-9}$	0	$-1.22 \cdot 10^{-9}$	0	a
z-axis	<1.01	$1.16 \cdot 10^{-9}$	-0.0316	$-0.437 \cdot 10^{-9}$	0.988	a
	<0.89	$1.28 \cdot 10^{-9}$	-0.0115	$-0.335 \cdot 10^{-9}$	1.130	b
y-axis	0.8-8.8	$4.95 \cdot 10^{-9}$	0.775	$-1.38 \cdot 10^{-9}$	0.113	a
x-axis	0.5-1.5	$-2.85 \cdot 10^{-9}$	1.447	$1.28 \cdot 10^{-9}$	1.912	b

*Spin Hamiltonian tensors taken for the following labels: (a) Doxyl (4,4-dimethyl-N-oxyl-oxazolidinyl) propane (Jost et al., 1971). (b) TEMPONE (Snipes et al., 1974). With the exception of z-axis rotation, the values are not very sensitive to the choice of single crystal tensor values.

$|C/B|$ ratio increases with increasing rotational anisotropy $(D_\parallel/D_\perp)$ for y-azis rotation, and for x-axis rotation it decreases (Goldman et al., 1972).

Rotational correlation times in the aqueous cytoplasm of
P. tricornutum.

The rotational correlation times of small nitroxide molecules can be used to determine the effective microviscosity of the aqueous cyto-plasm of cells. The method is illustrated in Fig. 2. A nitroxide such as TEMPONE (2,2,6,6-tetramethyl-4-piperidone-N-oxyl, which has intrinsically narrow hyperfine lines and which penetrates the cell membrane is used. The ESR signal from the external label is completely quenched by adding Ni^{2+} or some other inert paramagnetic ion which cannot enter the cell. The spin-spin interaction with the Ni^{2+} is so strong that the external spin-label signal is broadened out, while the Ni^{2+} itself does not give rise to an ESR signal at normal tempera-tures. The observed ESR spectrum thus arises solely from TEMPONE within the cells.

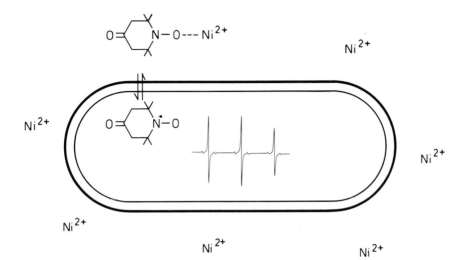

FIG. 2   Use of the TEMPONE spin label (2,2,6,6-tetramethyl-4-piperi-
done-N-oxyl) to study the cytoplasmic viscosity of cells. Externally
added 0.2 M Ni^{2+} cannot cross the cell membrane and thus broadens out
the ESR signal from extracellular TEMPONE. An ESR signal is observed
only from TEMPONE in the cytoplasm.

The spectra from TEMPONE in cells of *P. tricornutum* under various
conditions of osmotic stress are given in Fig. 3. The line-width in
the cytoplasm of the normal cells is seen to be considerably greater
than that in the external medium, and is increased further in the
cells subjected to osmotic stress. In the adapted cells the line-
width is almost, but not completely, restored to that of the normal
cells. The line-width coefficients are obtained from the relative
line-heights of the TEMPONE spectra:

$$B = (\Delta H_0/2) (\sqrt{h_0/h_{+1}} - \sqrt{h_0/h_{-1}}) \qquad (22)$$

$$C = (\Delta H_0/2) (\sqrt{h_0/h_{+1}} + \sqrt{h_0/h_{-1}} - 2) \qquad (23)$$

where ΔH_0 is the line-width of the central line and h_{+1}, h_0, h_{-1} are
the line-heights of the low-field, central and high-field lines, respectively. The ratios of the line-width coefficients and rotational
correlation times deduced from them are given in Table 3. The $|C/B|$
ratio is in all cases greater than one and the correlation times

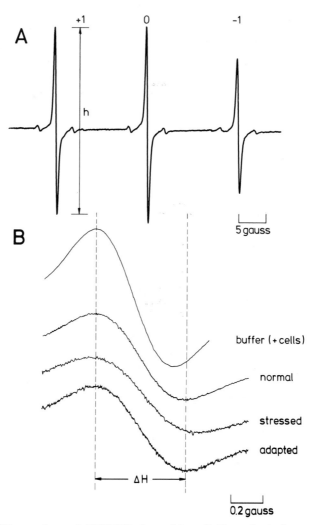

FIG. 3 ESR spectra of TEMPONE in cells of *Phaeodactylum tricornutum*
with 0.2 M $NiCl_2$ in the external medium, T = 2^oC. (A) Complete spectrum from cells in normal medium. (B) High-field line from cells
under different conditions of hyperosmotic stress. Normal: 0.8 osmol
/l; stressed: 1.9 osmol/l; adapted: 1.9 osmol/l + 0.05 M proline.
The spectrum: buffer(+ cells) is in the absence of Ni^{2+}. (Schobert
and Marsh, 1982).

corresponding to isotropic motion are listed. The only anisotropic
rotation which is consistent with the measured |C/B| ratio is y-axis
rotation, which seems less likely on the basis of the molecular geometry.

TABLE 3 Influence of hyperosmotic stress on the rotational correlation times of TEMPONE in the aqueous phase of *P. tricornutum*.
(Schobert and Marsh, 1982).

Condition	osmol/l	V_c^* (%)	C/B	τ_{20}^{iso} (10^{-11} s)
Buffer (+ cells, no NiCl$_2$)	0.8	–	1.27	2.9
Normal medium	0.8	100	1.58 ± 0.16	7.3 ± 0.7
Normal medium + NaCl (t = 10 min)	1.9	40	1.67 ± 0.07	12.1 ± 0.8
Normal medium + NaCl (t = 3 days)	1.9	80	1.88 ± 0.14	10.0 ± 0.1
Normal medium + NaCl + 0.05 M proline (t = 3 days)	1.9	92	1.93 ± 0.25	9.6 ± 0.4

*Cell volume measured using the TEMPO spin label, from the ratio of
the membrane-bound to the cytoplasmic ESR signal in the presence of
extracellular Ni^{2+}.

The correlation times in the cells are about 2 to 3 times greater
than in the external medium, and show a 1.5 to 2-fold increase under
conditions of high osmotic stress. This is only partially alleviated
on adaptation for three days, the recovery being greater in the presence of the amino acid proline but still being considerably less than
that observed in the cell volume. The correlation time may be related
to an effective microviscosity, η^{eff}, either via the Debye equation:

$$\tau^{iso} = 4\pi\eta^{eff}a^3/3kT \tag{24}$$

where a ≈ 3 Å is the molecular radius of TEMPONE, or via calibrations
from glycerol-water mixtures (Keith and Snipes, 1974). The effective
viscosity for the buffer is 1 cP and for the normal, stressed and
adapted cells is approx. 2.5 cP, 4.0 cP and 3.5 cP, respectively
using Equation (24) or 8 cP, 16 cP and 12 cP, respectively using the
glycerol-water calibrations. Clearly the effective cytoplasmic viscosity is considerably greater than that of the external medium, is

increased on osmotic stress, and after adaptation is still greater than for the non-stressed state.

ANISOTROPIC MOTION

Anisotropic motional averaging - ordering matrix (Saupe, 1964)

Molecular motion gives rise to an averaging of the anisotropy of the nitroxide ESR spectrum. For the averaging to take place, the motion must be rapid compared with the characteristic anisotropies in the spectrum, i.e. for rotational correlation times $\tau_R \ll h(A_{zz}-A_{xx})^{-1} \approx 10^{-8}$ s and $\tau_R \ll h[(g_{xx}-g_{zz})\beta H]^{-1} \approx 3.10^{-8}$ s. The extent of spectral averaging depends on the angular amplitude of molecular motion. The static spin Hamiltonian of Equation (1) can be replaced by one containing time-averaged tensors:

$$<H> \;=\; \beta \underline{H} \cdot \underline{\underline{<g'>}} \cdot \underline{S} \;+\; \underline{I} \cdot \underline{\underline{<A'>}} \cdot \underline{S} \tag{25}$$

where the angular brackets indicate a time (or spatial) average over the amplitude of the molecular motion. The motional averaging not only reduces the anisotropy of the tensor elements but also re-defines the direction of the principal axes and the symmetry of the tensors. For lipid molecules in bilayers or membranes the motion has axial symmetry about the ordering axis (or director) which usually coincides with the membrane normal. The time-average axial spin Hamiltonian can then be written:

$$<H> \;=\; g_\perp \beta (H_x S_x + H_y S_y) + g_\parallel \beta H_z S_z$$

$$+\, A_\perp (I_x S_x + I_y S_y) + A_\parallel I_z S_z \tag{26}$$

where the $\parallel$ suffix refers to the ordering axis and the $\perp$ suffix to the directions in the plane perpendicular to the ordering axis (see Fig. 4).

The time-averaged tensors may be expressed in terms of the nitroxide-fixed tensors by a transformation to the membrane-fixed axes, with rotational averaging about the ordering axis: $\underline{\underline{<g'>}} = <\underline{\underline{R}} \cdot \underline{\underline{g}} \cdot \underline{\underline{R}}^{-1}>$ and $\underline{\underline{<A'>}} = <\underline{\underline{R}} \cdot \underline{\underline{A}} \cdot \underline{\underline{R}}^{-1}>$ where $\underline{\underline{R}}$ is the matrix of direction cosines connecting

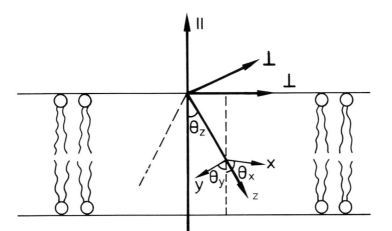

FIG. 4 Instantaneous orientation of the nitroxide x,y,z-axes relative to the symmetry axis, ‖, of the molecular ordering (here the
membrane normal). The molecular motion is such as to produce a time-
average axial symmetry relative to the ordering axis (‖). (From Marsh,
1981).

the two systems of axes. The transformation can be carried out
directly by matrix multiplication followed by integration over the
ϕ-rotation angle around the ordering axis. Alternatively, one can
take full advantage of the tensor transformation properties and the
symmetry of the motional averaging. The parallel component of the
motionally averaged hyperfine tensor is given by the tensor transformation:

$$A = \sum_{ij} A_{ij} <\cos\theta_i \cdot \cos\theta_j> \qquad (27)$$

where $i,j = x,y,z$ are the nitroxide axes and θ_i are the angles between
the x,y,z axes and the ordering axis (c.f. Fig. 4). An alternative
way to write Equation (27) is:

$$A_\| = \frac{1}{3} \sum_i A_{ii} + \frac{2}{3} \sum_{i,j} S_{ij} A_{ij} \qquad (28)$$

where S_{ij} are the elements of the <u>ordering tensor</u>, defined by:

$$S_{ij} = \frac{1}{2} (3 <\cos\theta_i \cdot \cos\theta_j> - \delta_{ij}) \qquad (29)$$

The perpendicular component of the time-averaged hyperfine tensor may be obtained from the well-known invariance of the trace under the similarity transformation $\underline{\underline{A}}' = \underline{\underline{R}} \cdot \underline{\underline{A}} \cdot \underline{\underline{R}}^{-1}$. Thus:

$$2A_\perp + A_\parallel = \sum_i A_{ii} \tag{30}$$

and from Equation (28)

$$A_\perp = \frac{1}{3} \sum_i A_{ii} - \frac{1}{3} \sum_{ij} S_{ij} A_{ij} \tag{31}$$

Clearly from the definition in Equation (29), the ordering tensor is symmetric: $S_{ij} = S_{ji}$, and from the orthogonality relation for direction cosines: $\sum_i \cos^2\theta_i = 1$ the trace is zero:

$$\sum_i S_{ii} = 0 \tag{32}$$

Thus only two of the diagonal elements are independent ($-S_{yy} = S_{xx} + S_{zz}$), and in general the ordering tensor will have five independent components. These may be further restricted by the symmetry of the molecule. A simple plane of symmetry reduces the number of independent elements to three, two orthogonal planes of symmetry reduce the number to two, and a three-fold or higher symmetry axis reduces the number to one (axial symmetry). In general, however, it is likely that the nitroxide group will destroy, at least partly, any inherent molecular symmetry.

Using Equation (32) together with Equations (28) and (31) and the nitroxide axis system given in Fig. 4, the time-averaged elements of the hyperfine tensor can be written:

$$A_\parallel = a_0 + \frac{2}{3}(A_{xx}-A_{yy})S_{xx} + \frac{2}{3}(A_{zz}-A_{yy})S_{zz} \tag{33}$$

$$A_\perp = a_0 - \frac{1}{3}(A_{xx}-A_{yy})S_{xx} - \frac{1}{3}(A_{zz}-A_{yy})S_{zz} \tag{34}$$

where $a_0 = \frac{1}{3}(A_{xx} + A_{yy} + A_{zz})$ is the isotropic hyperfine constant, corresponding to the hyperfine splitting which would be obtained for isotropic motion where $\langle\cos^2\theta_i\rangle = 1/3$ and $S_{ii} = 0$.

Exactly similar expressions to Equations (33) and (34) hold also for
the time-averaged elements of the g-tensor:

$$g_{\parallel} = g_0 + \frac{2}{3} (g_{xx} - g_{yy}) S_{xx} + \frac{2}{3} (g_{zz} - g_{yy}) S_{zz} \qquad (35)$$

$$g_{\perp} = g_0 - \frac{1}{3} (g_{xx} - g_{yy}) S_{xx} - \frac{1}{3} (g_{zz} - g_{yy}) S_{zz} \qquad (36)$$

Since the static hyperfine tensor is almost axial: $A_{xx} \approx A_{yy}$, the
following approximation is often made:

$$A_{\parallel} \simeq a_0 + \frac{2}{3} (A_{zz} - A_{yy}) S_{zz} \qquad (37)$$

$$A_{\perp} \simeq a_0 - \frac{1}{3} (A_{zz} - A_{yy}) S_{zz} \qquad (38)$$

    To within the accuracy of this approximation, the order parameter
S_{zz} can be determined from measurements of the hyperfine splitting
alone. The S_{xx} element, and hence $S_{yy} = -(S_{xx}+S_{zz})$, can then be
determined from the g-values. If the approximation of Equations (37)
and (38) is not used then measurements of both hyperfine splittings
and g-values are required to obtain the individual order tensor elements.

    The nitroxide axes do not necessarily have to coincide with the
molecular axes of motional symmetry. The ordering tensor elements
in the nitroxide system, S_{ij}, may be expressed in terms of ordering
tensor elements in a second molecule-fixed axis system, S'_{lm}, by the
usual orthogonal transformation:

$$S_{ij} = \sum_{l,m} \cos\alpha_{il} \cdot \cos\alpha_{jm} S'_{lm} \qquad (39)$$

where the $\cos\alpha_{il}$ are the direction cosines relating the axes of the
two systems. Two examples of commonly employed lipid spin labels are
given in Fig. 5. For phospholipids or fatty acids with the doxyl
(4,4-dimethyl-N-oxyl-oxazolidinyl) group attached to the hydrocarbon
chain, the nitroxide z-axis is perpendicular to the plane of the
doxyl group, which corresponds to the HCH plane of symmetry for

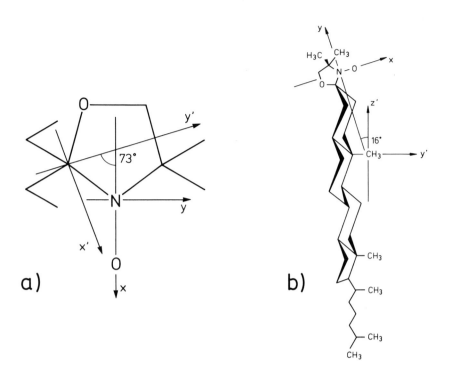

FIG. 5 Orientation of the nitroxide principal axes (x,y,z) relative
to the molecular symmetry axes (x',y',z'). (a) The doxyl (4,4-
dimethyl-N-oxyl-oxazolidinyl) group attached to a saturated hydrocarbon chain. The nitroxide z-axis is parallel to the z'-axis (which
is directed along the long axis of the all-trans chain). (b) The
doxyl group attached equatorially at the three-position of cholestane.
The nitroxide z-axis is nearly perpendicular to the long molecular
axis (z'). In both diagrams the doxyl ring lies in the plane of the
paper.

the unlabelled methylene group. This corresponds to the natural axis
z' for ordering of a CH_2 segment, since it is parallel to the long
axis of an all-<u>trans</u> chain. The nitroxide x,y-axes do not, however,
coincide with the symmetry axes of the unlabelled methylene group.
It is seen from Fig. 5a that the nitroxide x-axis is rotated by $17°$
from the x'-axis which is perpendicular to the CCC plane of symmetry
of the unlabelled methylene group. From Equation (39) the two systems are related by:

$$S_{xx} = S_{x'x'}\cos^2\alpha + S_{y'y'}\sin^2\alpha - 2 S_{x'y'} \sin\alpha \cos\alpha \quad (40)$$

$$S_{yy} = S_{x'x'} \sin^2\alpha + S_{y'y'} \cos^2\alpha + 2 S_{x'y'} \sin\alpha \cos\alpha \qquad (41)$$

where $\alpha = 17°$. Since only S_{xx} and S_{yy}, but not S_{xy} can be determined experimentally, the transformation can only be made if, for instance it is assumed that $S_{x'y'} \approx 0$ for symmetry reasons. Then $S_{x'x'} \simeq 1.1$ $S_{xx} - 0.1 S_{yy}$ and $S_{y'y'} \simeq 1.1 S_{yy} - 0.1 S_{xx}$. For the cholestane spin label, the nitroxide z-axis is nearly perpendicular to the long molecular z'-axis and the nitroxide y-axis is oriented at $\alpha \simeq 16°$ to the z'-axis. Thus in the molecular system: $S_{x'x'} \simeq S_{zz}$ and there are relations analogous to Equations (40) and (41) relating S_{yy} and S_{xx} to $S_{z'z'}$, $S_{y'y'}$ and $S_{y'z'}$. Again it is necessary to assume that the symmetry of the motion is such that $S_{y'z'} \approx 0$ in order to make the transformation. Then: $S_{z'z'} \simeq 1.1 S_{yy} - 0.1 S_{xx}$ and $S_{y'y'} \simeq 1.1 S_{xx} - 0.1 S_{yy}$.

Anisotropic lipid motion in cell membranes

The ESR spectra of spin labelled (doxyl) analogues of the ganglioside G_{M1}II incorporated in cell membranes of a permanent mouse fibroblast cell line are given in Fig. 6 (Schwarzmann et al., 1981, 1983). The extent of motional averaging of the spectral anisotropy is seen to increase for the series of doxyl-labelled gangliosides 5-G_{M1} to 13-G_{M1} as the spin label group is located further down the lipid chain towards the terminal methyl group. This increasing amplitude of angular rotation or "fluidity gradient" is a characteristic hallmark of the motion of lipid chains in fluid bilayer membranes, and confirms that the exogenously added ganglioside labels have been fully inte-grated into the fibroblast plasma membrane in a manner analogous to that of the endogenous lipids.

The axial symmetry of the powder-type membrane spectra is obvious, and the method of measuring the motionally-averaged hyperfine splitt-ings and g-values is shown in Fig. 6. The $A_\parallel$ hyperfine splitting is obtained directly from the separation of the outer hyperfine peaks and $g_\parallel$ from their mid-point as indicated. The $A_\perp$ hyperfine splitting is obtained from the apparent splitting $A_\perp'$ of the two inner hyperfine peaks and $g_\perp$ from their midpoint specified by $g_\perp'$. Simulations of the powder spectra (Griffith and Jost, 1976) indicate that the following empirical

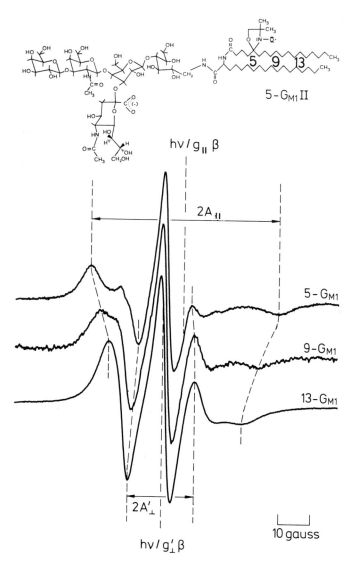

FIG. 6 ESR spectra of the positional isomers of the doxyl ganglioside analogue spin labels, 5-G_{M1}II, 9-G_{M1}II and 13-G_{M1}II, in permanent mouse fibroblast Cl-1D cell membranes at 35°C. The central peaks of the three spectra have been aligned, thus the relative field centres have no absolute significance (Schwarzmann *et al.*, 1981, 1983).

corrections must be made to $A_\perp'$:

$$A_\perp \simeq A_\perp' + 1.32G + 1.86G \times \log[1-(A_\parallel-A_\perp')/(A_{zz}-\overline{A_{xx}})]$$

$$\text{for} \quad (A_\parallel-A_\perp')/(A_{zz}-\overline{A_{xx}}) \geq 0.45 \qquad (42)$$

and

$$A_\perp \simeq A_\perp' + 0.85G \quad \text{for } (A_\parallel - A_\perp')/(A_{zz} - \overline{A_{xx}}) > 0.45 \tag{43}$$

For the g-values the corresponding corrections (for $\nu=9.5$ GHz) are:

$$g_\perp \simeq g_\perp' + 5.3 \cdot 10^{-4}\ (A_\parallel - A_\perp')/(A_{zz} - \overline{A_{xx}}) - 1.2 \cdot 10^{-4} \tag{44}$$

$$\text{for} \quad (A_\parallel - A_\perp')/(A_{zz} - \overline{A_{xx}}) \geq 0.33$$

and

$$g_\perp \simeq g_\perp' + 5.9 \cdot 10^{-5} \quad \text{for } (A_\parallel - A_\perp')/(A_{zz} - \overline{A_{xx}}) < 0.33 \tag{45}$$

where $\overline{A_{xx}} = \frac{1}{2}(A_{xx} + A_{yy})$, since the hyperfine tensor has approximate axial symmetry. The correction terms in Equations (42)-(45) have been obtained from the data tabulated by Griffith and Jost (1976). Equation (42) corresponds to the correction term quoted by them. Gaffney (1976) has proposed an alternative expression which is numerically closely equivalent. Neither method explicitly accounts for the change in differential broadening which frequently accompanies the changes in line splittings. At low anisotropies such factors become particularly important.

The effective order parameters and isotropic hyperfine splitting constants for the ganglioside spin label analogues in the mouse fibroblast cell membranes are given in Table 4. The order parameters give quantitative expression to the flexibility gradient in the fluid membrane. The isotropic hyperfine constants, which are sensitive to the environmental polarity, indicate that the spin label groups positioned closer to the methyl end of the chain are situated deeper in the hydrophobic interior of the membrane, further demonstrating that the ganglioside analogues are incorporated into the membrane in a similar manner to the endogenous lipids.

Although there is clear definition of the partial motional averaging in the spectra of Fig. 6, the line-widths are rather broad. Thus it cannot be excluded that the spectra are approaching the slow motional regime (see next section), in which not only the amplitude but also the rate of motion contributes to the observed line splittings. For this reason the order parameters and hyperfine splitting factors of Table 4 must be considered as effective values.

TABLE 4 Effective order parameters and isotropic hyperfine splitting constants of ganglioside G_{M1}II analogue nitroxide spin labels incorporated into Cl-1D mouse fibroblasts, $T = 35^{\circ}C$ (Schwarzmann *et al.*, 1983).

	a_o^{eff} (G)	S_{zz}^{eff}
$5-G_{M1}II$	15.4	0.73
$9-G_{M1}II$	15.3	0.51
$13-G_{M1}II$	14.3	0.36

SLOW MOTIONS

Slow motion of spin labels - Correlation times

For rotational correlation times longer than 3×10^{-9} s, perturbation calculations can no longer be applied and motional-narrowing theory breaks down. Spectral line-shapes in this slow motion regime can be calculated either from the stochastic Liouville equation (Freed, 1976) or from the diffusion-coupled Bloch equations (McCalley *et al.*, 1972). For spectra without extensive motional narrowing, correlation time calibrations exist based on the decrease in separation and increase in line-width of the outer hyperfine extrema, compared with their rigid limit values (c.f. Fig. 7).

The outer extrema in the slow motion spectra correspond to those nitroxides with their z-axes oriented along the magnetic field direction. Molecular rotation causes these nitroxides to exchange with those of different orientation in other parts of the spectrum. By analogy with the case of two-site exchange in magnetic resonance, it is expected that as the motional rate of the spin label increases from the rigid limit, the outer hyperfine extrema will first broaden and then shift closer together. For two-site exchange, the increase in line-width (in angular frequency units) is given simply by $1/\tau_1$, where τ_1 is the lifetime of the exchanging state.

Thus in this simple model the rotational correlation time is given by:

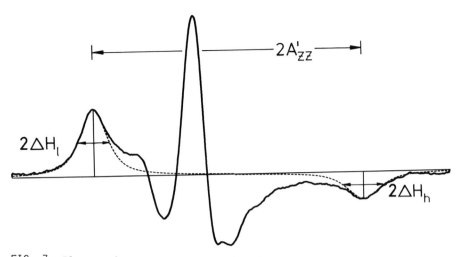

FIG. 7 Slow motion ESR spectrum of a doxyl phosphatidic acid spin label, 14-PASL, in association with the myelin proteolipid apoprotein, T = 30 °C. The outer extrema are fitted by absorption peaks with a line-shape which is a linear combination of 75% Lorentzian + 25% Gaussian. The method of measuring the line-widths ΔH_1, ΔH_h and hyperfine splitting, A'_{ZZ} is indicated (Brophy *et al.*, 1984; Figure by courtesy of L.I. Horvath).

$$\tau_R = \left(\frac{\hbar}{g\beta\Delta H_m^R} \right) \left(\frac{\Delta H_m}{\Delta H_m^R} - 1 \right)^{-1} \tag{46}$$

where ΔH_m^R is the rigid limit value of ΔH_m, the line-width at half-height of the outer extrema, with m = ℓ or h for the low-field or high-field extrema, respectively (see Fig. 7).

In the two-site exchange model the peak separation, 2A (in angular frequency units), is decreased by the factor $[1-2(\tau A)^{-2}]^{1/2} \simeq [1-(\tau A)^{-2}]$, where $\tau/2$ is the lifetime at one of the sites. Within this model the rotational correlation time can therefore be approximated by the change in splitting of the outer extrema according to:

$$\tau_R = \left(\frac{\hbar}{g\beta A_{ZZ}^R} \right) \left(1 - \frac{A'_{ZZ}}{A_{ZZ}^R} \right)^{-1/2} \tag{47}$$

where A_{ZZ}^R is the rigid limit value for the outer hyperfine splitting, A'_{ZZ}, defined for the slow motion spectrum in Fig. 7.

Slow motion simulations have been performed for isotropic rotation using the stochastic Liouville approach (Freed, 1976). Empirical correlation time calibrations have been deduced from the simulations using the following generalized analogues of Equations (46) and (47):

$$\tau_R = a'_m \left(\frac{\Delta H^m_m}{\Delta H^R_m} - 1 \right)^{b'_m} \tag{48}$$

$$\tau_R = a \left(1 - \frac{A'_{zz}}{A^R_{zz}} \right)^b \tag{49}$$

The calibration constants a'_m, b'_m of Equation (48) and a, b of Equation (49) are given in Tables 5 and 6 respectively, for various diffusion models and for various values of the peak-to-peak derivative Lorentzian line-width, δ.

For the simple two-site exchange model, the corresponding values of the line-width calibration constants from Equation (46) are: $b'_m = -1$, with $a'_m = 2.9 \cdot 10^{-8}$ s and $6.57 \cdot 10^{-8}$ s for $\delta = 3.0$G and $\delta = 1.0$G, respectively. These are in quite good agreement with the values of Table 5, supporting the concept of lifetime broadening. In fact, the increase in line-width of the outer extrema is, to within a factor of 2 (or 1/2), given by $1/\tau_R$ for the entire range of simulations (Freed, 1976). The two-site exchange values of the calibration constants for the splitting are given from Equation (47) by: $b = -1/2$ and $a = 1.78 \cdot 10^{-9}$ s (for $A^R_{zz} = 32$G). These are similar to the values of Table 6 for strong jump diffusion, for which the exchange analogy is most likely to hold.

The successful application of the calibrations of Equations (48) and (49) depends on the correct choice of the rigid limit parameters, especially A^R_{zz} since this is also sensitive to the polarity of the environment. The rigid limit line-widths are determined principally by unresolved proton hyperfine structure, and are less likely to be affected by polarity. The outer extrema in the first derivative spectra accurately reproduce the absorption line-shape (Hubbell and McConnell, 1971).

TABLE 5   Parameters for fitting the correlation time calibration: $\tau_R = a'_m (\Delta H_m/\Delta H_m^R - 1)^{b'_m}$ (Freed, 1976)

Diffusion model	δ (G)	m	a'_m $(10^{-8}s)$	b'_m
Free diffusion	3.0	l	1.29	-1.033
	3.0	h	1.96	-1.062
	1.0	l	5.32	-1.076
	1.0	h	7.97	-1.125
Brownian diffusion	3.0	l	1.15	-0.943
	3.0	h	2.12	-0.778
	1.0	l	5.45	-0.999
	1.0	h	9.95	-1.014

More precisely, it is found from simulations that the half-widths of the outer extrema at half-height, ΔH_m^R, are given (Freed, 1976) by:

$$2\Delta H_l^R = 1.59\ \delta \qquad (50)$$

$$2\Delta H_h^R = 1.81\ \delta \qquad (51)$$

where δ is the peak-to-peak derivative Lorentzian line-width. This is close to the predicted value of $\Delta H^R = (\sqrt{3}/2)\delta$. Various experimental studies suggest that $\delta \approx 3.0G$ for doxyl nitroxides, which is the value used for the majority of the simulations. Thus in the absence of more detailed experimental values it is suitable to take $\Delta H_l^R = 2.39G$, $\Delta H_h^R = 2.72G$, corresponding to $\delta = 3.0G$.

The calibrations of Equations (48) and (49) are obtained from simulations for isotropic motion. However, since they are based on the outer extrema of the spectra, they reflect essentially only the motion of the nitroxide z-axis. For rapid axial rotation about the z-axis the formalism is unchanged and the calibrations give the correlation time for the slow rotation of the z-axis, i.e. $\tau_R = \tau_\perp =$ $1/6D_\perp$. For preferentially more rapid rotation about the x or y axis, with relatively small anisotropy ($D_\parallel \lesssim 3D_\perp$), the calibrations approximately yield the compound correlation time $\tau_R = 1/6\sqrt{D\ D_\perp}$ (Freed, 1976).

For anisotropic rotation about the nitroxide y-axis, and very much slower motion perpendicular to this axis, correlation time calibrations have been given by Polnaszek et al. (1981). In the slow motional regime (2.10^{-9} s $\lesssim \tau_\parallel \lesssim 7.5 \cdot 10^{-8}$ s), the spectra resemble the rigid limit spectra and the calibration in terms of the outer splitting (Equation (49)) may be used with the calibration constants for y-axis rotation given in Table 6. Values are given for both Brownian diffusion and including the effects of time-dependent fluctuations in the order parameter, such as are found for the cholestane spin label in gel-phase lipid bilayers.

Slow motion of spin-labelled lipids in membranes

As discussed in an earlier section, spin-labelled lipid molecules in fluid bilayer membranes undergo anisotropic molecular rotations in the nanosecond time range, which give rise to partial motional averaging of the spectrum. These characteristic spectra are also observed in biological membranes with relatively high lipid/protein ratio, confirming that the majority of the lipids exist in a bilayer-like arrangement in the membrane (c.f. for instance Fig. 6). In membranes with a high protein content, or in reconstituted lipid-protein systems, a second spectral component is often observed corresponding to lipids which are more motionally restricted than the fluid bilayer lipids. This second component, which frequently lies within the slow motion regime, is attributed to lipids interacting directly with the hydrophobic surface of integral membrane proteins (for a review, see Marsh and Watts, 1982).

An example is given in Fig. 8 of the spectrum of a doxyl phosphatidylcholine spin label, 14-PCSL (Brophy et al., 1984) in a myelin proteolipid apoprotein recombinant with dimyristoyl phosphatidylcholine at a lipid/protein mole ratio of 12:1. The spectrum (Fig. 8a) is dominated by a sharp component from the fluid lipids, which is closely similar to the spectrum from bilayers of the lipids alone

TABLE 6 Parameters for fitting the correlation time calibration:
$\tau_R = a \left(1 - A'_{zz}/A^R_{zz}\right)^b$ (Freed, 1976; Polnaszek et al., 1981).

Diffusion model	δ (G)	a (10^{-9} s)	b
Isotropic rotation			
Brownian diffusion	0.3	0.257	-1.78
	3.0	0.54	-1.36
	5.0	0.852	-1.16
	8.0	1.09	-1.05
Free diffusion	0.3	0.699	-1.20
	3.0	1.10	-1.01
Strong jump	0.3	2.46	-0.589
	3.0	2.55	-0.615
y-axis rotation			
Brownian diffusion	---	0.2596	-1.396
Fluctuating torques	---	0.2616	-1.301

(Fig. 8c). A second, broader component is, however, clearly discerned
in the outer wings of the spectrum, which is not present in the spectrum of the lipids alone. The broader component is better visualized
on subtracting out the fluid component. The difference spectrum in
Fig. 8b clearly lies in the slow motion regime and, apart from the
sharp discontinuities arising from an imperfect match of the fluid
components, very closely resembles the spectrum from the delipidated
apoprotein.

Similar slow-motion lipid components to those of Fig. 8b are
observed in natural membranes. Table 7 gives the effective correlation times obtained for spin-labelled fatty acids and a spin-labelled
steroid in acetylcholine receptor-rich membranes from *T. marmorata*.
Equations (48) and (49) were used to analyse the slow-motion components
yielding effective rotational correlation times $\tau_R \sim 50$ ns (Table 7)
compared with $\tau_R \sim 1$-5 ns for the fluid lipid component. Effective

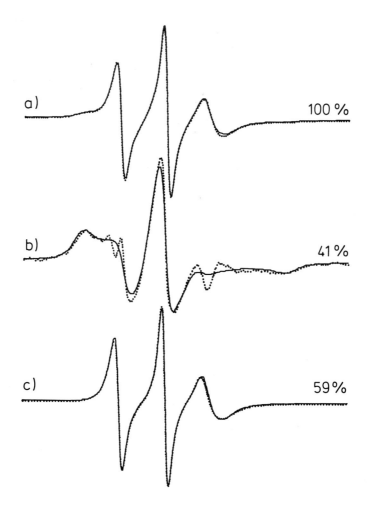

FIG. 8 Spectral subtraction and addition with the phosphatidylcholine
nitroxide spin label 14-PCSL. Full lines-original spectra (a) myelin
proteolipid apoprotein/dimyristoyl phosphatidylcholine recombinant of
lipid/protein ratio 29:1, (b) proteolipid apoprotein alone, (c)
dimyristoyl phosphatidylcholine alone. Dotted lines-summed spectra
and difference spectra: (a) 59% lipid-alone spectrum plus 41% protein
alone spectrum, (b) recombinant minus 59% lipid-alone spectrum, (c)
recombinant minus 41% protein-alone spectrum. Percentages refer to
double-integrated spectral intensity. Total scan width = 100 gauss
(Brophy *et al.*, 1984; Figure by courtesy of L.I. Horvath).

correlation times of the same order of magnitude have been obtained
for the motionally restricted lipids in cytochrome oxidase-dimyristoyl
phosphatidylcholine complexes (Knowles *et al.*, 1979). In rod outer

TABLE 7 Rotational correlation times of lipid spin labels in the
motionally restricted lipid regions of acetylcholine receptor-rich
membranes from *T. marmorata* (Marsh and Barrantes, 1978).

Spin label	T ($^{\circ}$C)	τ_R (ns)*	τ_R (ns)[†]	τ_R (ns)[§]
12-doxyl-stearic acid	34	50	46	12
16-doxyl-stearic acid	-4	34	41	45
doxyl androstanol	14	45	73	47

*Deduced from the low-field line-width, ΔH_l.
[†]Deduced from the high-field linewidth, ΔH_h.
[§]Deduced from the outer splitting, A'_{zz}.

segment disc membranes the effective correlation time for the motionally restricted lipids has been found to decrease from $\tau_R \sim 50$ ns at
3°C to $\tau_R \sim 20$ ns at 24°C (Watts *et al.*, 1982), indicating the
expected temperature sensitivity.

There are two motions which can contribute to the effective correlation time of the lipids interacting directly with the protein.
These are motions of the lipid chains in contact with the protein and
the exchange of the lipids on and off the protein surface. Thus a
lower limit for the lifetime of the lipids on the surface of the protein is given by $\tau_{ex} \sim \tau_R \sim 50$ ns. This is somewhat shorter than the
average lipid-lipid exchange lifetime in fluid lipid bilayers result-
ing from lateral diffusion: $\tau_{ex} = \langle x^2 \rangle / 4D_{latt} \sim 100\text{-}200$ ns (Träuble
and Sackmann, 1972; Devaux *et al.*, 1973). In addition, simulations
of exchange between the two spectral components have suggested lipid
lifetimes at the protein interface in the range $\tau_{ex} \sim 100\text{-}200$ ns
(Davoust and Devaux, 1982). Thus it seems likely that there is some
independent lipid motion on the surface of the protein (protein rotational correlation times are in the range 20-100 μs), in addition to
exchange.

Spectral subtractions such as those of Fig. 8 also allow quantitation of the relative proportions of the two spin label components,
giving the stoichiometry and (with different spin-labelled phospholipids) the specificity of the lipid-protein interaction (see, e.g.,

Marsh and Watts, 1982). The lipid-protein association equilibrium can
be described by the following equation (Brotherus et $al.$, 1981):

$$n_f^*/n_b^* = n_t/(N_1 K_r) - 1/K_r \tag{52}$$

where n_f^*/n_b^* is the ratio of fluid to motionally restricted components
in the spectrum and n_t is the total lipid/protein ratio in the sample.
N_1 is the effective number of lipid molecules associated with each
protein molecule, and K_r is the association constant of the spin-
labelled lipid relative to that of the host lipid. For the myelin
proteolipid recombinants with dimyristoyl phosphatidylcholine and the
phosphatidylcholine spin label it is found that $N_1 = 10$ and $K_r = 1$
(Brophy et $al.$, 1983,1984) and for a range of different integral membrane proteins it is found that the values of N_1 correlate both with
the protein molecular weight and with the estimated number of lipids
which can be accommodated around the intramembranous perimeter of the
protein - see Table 8. The selectivity of the protein for different
lipids is revealed by K_r values different from unity. For myelin
proteolipid apoprotein-dimyristoyl phosphatidylcholine recombinants
the relative association constants are in the order: stearic acid >
phosphatidic acid > diphosphatidylglycerol > phosphatidylserine >
phosphatidylglycerol $\approx$ phosphatidylcholine > phosphatidylethanolamine >
cholestane (Brophy et $al.$, 1983,1984). For the diphosphatidylglycerol
spin label and different membrane proteins, the relative association
constants are in the order Na^+,K^+-ATPase>~cytochrome oxidase > rhodopsin (Marsh et $al.$, 1982). These results indicate the way in which
the difference in motional rates between the fluid bilayer lipids and
the lipids at the protein-lipid interface may be used in a very direct
way to study lipid-protein interactions in biological membranes (for
more discussion see Marsh and Watts, 1982).

ACKNOWLEDGEMENTS
I would like to acknowledge my colleagues whose collaborative work is
reviewed briefly here, in particular Drs B. Schobert, G. Schwarzmann,
P. Hoffmann-Bleihauer, K. Sandhoff, L.I. Horvath and P.J. Brophy. I
would also like to thank Dr C.F. Polnaszek for correspondence and for
a preprint of his review article.

TABLE 8 Stoichiometries of the motionally-restricted lipid spin
label component in various lipid-protein systems (Marsh *et al.*, 1982)

Protein/membrane	N_1^{exp} (mole/mole)	$N_1^{exp}/\sqrt{MW}$	N_1^{calc} (mole/mole)
Cytochrome oxidase – DMPC	55 ± 5	0.123 ± 0.011	50
Bovine rod outer segment disc/rhodopsin	24 ± 3	0.125 ± 0.016	24
Frog rod outer segment disc/rhodopsin	22 ± 2	0.114 ± 0.010	(24)
Na^+/K^+-ATPase shark rectal gland	58 ± 4	0.112 ± 0.008	($\sim$60)
Acetylcholine receptor-rich membrane/*T. marmorata*	45%*	--	52-55

N_1^{exp} is the effective number of motionally restricted lipids per protein deduced from the spin label experiments. N_1^{calc} is the estimated
number of lipids which can be accommodated around the intramembranous
perimeter of the protein. MW is the protein molecular weight.

*Percentage of total lipid motionally restricted, proteins other than
acetylcholine receptor also being present in significant amounts and
the labels used also exhibiting a specificity for the protein.

REFERENCES

Brophy, P.J., Horvath, L.I. and Marsh, D. (1983). *Biochem. Soc.
    Trans.* 11, 159-160.
Brophy, P.J., Horvath, L.I. and Marsh, D. (1984). *Biochemistry* 23,
    860-865.
Brotherus, J.R., Griffith, O.H., Brotherus, M.O., Jost, P.C., Silvius,
    J.R. and Hokin, L.E. (1981). *Biochemistry* 20, 5261-5267.
Davoust, J. and Devaux, P.F. (1982). *J. Magn. Reson.* 48, 475-494.
Devaux, P.F., Scandella, C.J. and McConnell, H.M. (1973). *J. Magn.
    Reson.* 9, 474-485.
Fraenkel, G.K. (1967). *J. Phys. Chem.* 71, 139-171.
Freed, J.H. (1964). *J. Chem. Phys.* 41, 2077-2083.
Freed, J.H. (1976). *In* "Spin Labelling Theory and Applications".
    (Ed. L.J. Berliner), Vol. 1, pp. 53-132, Academic Press, New York.
Gaffney, B.J. (1976). *In* "Spin Labelling. Theory and Applications".
    (Ed. L.J. Berliner), Vol. 1, pp. 567-571, Academic Press, New York.
Goldman, S.A., Bruno, G.V., Polnaszek, C.F. and Freed, J.H. (1972).
    J. Chem. Phys. 56, 716-735.
Griffith, O.H. and Jost, P.C. (1976). *In* "Spin Labelling. Theory and
    Applications". (Ed. L.J. Berliner), Vol. 1, pp. 453-523, Academic
    Press, New York.

Hubbell, W.L. and McConnell, H.M. (1971). *J. Amer. Chem. Soc.* 93, 314-326.
Jost, P.C., Libertini, L.J., Hebert, V.C. and Griffith, O.H. (1971). *J. Molec. Biol.* 59, 77-98.
Keith, A.D. and Snipes, W. (1974). *Science* 183, 666-668.
Knowles, P.F., Marsh, D. and Rattle, H.W.E. (1976). "Magnetic Resonance of Biomolecules". Wiley, London, New York.
Knowles, P.F., Watts, A. and Marsh, D. (1979). *Biochemistry* 18, 4480-4487.
Marsh, D. (1981). *In* "Membrane Spectroscopy". (Ed. E. Grell), pp.51-142. Springer-Verlag, Berlin, Heidelberg, New York.
Marsh, D. (1982). *In* "Techniques in Lipid and Membrane Biochemistry". (Eds J.C. Metcalfe and T.R. Hesketh), Vol. B4/II, pp.B426/1-B426/44. Elsevier Biomedical Press, Ireland.
Marsh, D. and Barrantes, F.J. (1978). *Proc. Natl. Acad. Sci. USA* 75, 4329-4333.
Marsh, D. and Watts, A. (1982). *In* "Lipid-Protein Interactions". (Eds P.C. Jost and O.H. Griffith), Vol. 2, pp. 53-126. Wiley, New York.
Marsh, D., Watts, A., Pates, R.D., Uhl, R., Knowles, P.F. and Esmann, M. (1982). *Biophys. J.* 37, 265-274.
McCalley, R.C., Shimshick, E.J. and McConnell, H.M. (1972). *Chem. Phys. Lett.* 13, 115-119.
Polnaszek, C.F. (1984). *In* "Spin Labelling in Pharmacology". (Ed. J.L. Holtzmann), Academic Press.
Polnaszek, C.F., Marsh, D. and Smith, I.C.P. (1981). *J. Magn. Reson.* 43, 54-64.
Saupe, A. (1964). *Z. Naturforsch.* 19a, 161-171.
Schobert, B. and Marsh, D. (1982). *Biochim. Biophys. Acta* 720, 87-95.
Schreier, S., Polnaszek, C.F. and Smith, I.C.P. (1978). *Biochim. Biophys. Acta* 515, 375-436.
Schwarzmann, G., Schubert, J., Hoffmann-Bleihauer, P., Marsh, D. and Sandhoff, K. (1981). *In* "Glyco-conjugates. Proceedings of the Sixth International Symposium on Glycoconjugates". (Eds. T. Yamakawa, T. Osawa and S. Handa), pp. 333-334. Scientific Societies Press, Tokyo.
Schwarzmann, G., Hoffmann-Bleihauer, P., Schubert, J., Sandhoff, K. and Marsh, D. (1983). *Biochemistry* 22, 5041-5048.
Snipes, W., Cupp, J., Cohn, G. and Keith, A.D. (1974). *Biophys. J.* 14, 20-32.
Trauble, H. and Sackmann, E. (1972). *J. Amer. Chem. Soc.* 94, 4499-4510.
Watts, A., Volotovski, I.D., Pates, R.D. and Marsh, D. (1982). *Biophys. J.* 37, 94-95.

Saturation Transfer EPR and Triplet Anisotropy: Complementary Techniques for the Study of Microsecond Rotational Dynamics

D.D. THOMAS, T.M. EADS, V.A. BARNETT, K.M. LINDAHL, D.A. MOMONT and T.C. SQUIER

INTRODUCTION

Saturation transfer electron paramagnetic resonance (ST-EPR) spectroscopy is the term associated with a family of EPR techniques used to
measure rotational motions of nitroxide spin-labels in the microsecond
to millisecond time range, thus complementing the conventional EPR
technique, which is only sensitive to submicrosecond rotational motions
of nitroxides. As a result of its sensitivity to slower motions,
ST-EPR has found its widest applications in probing the large-scale
rotational dynamics of biological macromolecules, particularly in
organized assemblies, such as muscle protein filaments, membranes,
and chromosomes. In this regard, the relationship in EPR between conventional and saturation transfer methods is quite similar to the
relationship in optical spectroscopy between singlet methods (e.g.
fluorescence depolarization) and triplet methods (e.g. transient
dichroism, phosphorescence depolarization). The primary purposes of
the present contribution are to introduce the principles of ST-EPR and
to illustrate these principles with selected results from our study
of spin-labelled muscle proteins (Thomas et al., 1980; Barnett and
Thomas, 1984). Because of the development of complementary methods
in optical spectroscopy, particularly triplet state spectroscopy, we
will compare both the principles and results of these ST-EPR studies
with those of triplet anisotropy decay (Eads et al., 1984), illustrating the common principles and contrasting nature of the two techniques.

SPECTROSCOPY AND THE DYNAMICS
OF MOLECULAR BIOLOGICAL SYSTEMS

239

THEORETICAL PRINCIPLES

Conventional EPR

Sensitivity to orientation The principles and biophysical applications of conventional EPR are discussed in detail in the previous chapter, so we will only outline them here. For both conventional and saturation transfer EPR, the sensitivity of the experiment to rotational motion arises from the high degree of orientational resolution in the nitroxide EPR spectrum: due to anisotropic magnetic interactions, each position (resonance magnetic field value H_{res}) in the spectrum corresponds to a narrow range of orientations of the spin-label relative to the applied DC magnetic field. At some positions in the spectrum, this orientational resolution, measured as the full width at half maximum of orientations excited, is as small as 8° (Thomas *et al.*, 1976). In contrast, the full width at half maximum for excitation (photoselection) by polarized light in optical spectroscopy is 90°, more than ten times worse than in nitroxide EPR. As shown in Fig. 1, the orientational sensitivity of nitroxide EPR manifests itself mainly as a variation in the splitting among the three hyperfine lines in the spectrum (corresponding to nitrogen nuclear spin quantum numbers of +1, 0, and -1). This splitting depends almost exclusively on $\cos^2\theta$, where θ is the angle between the nitroxide's principal axis and the applied field. Thus, the top three spectra in Fig. 1 could arise from a uniformly oriented population of probes in a single crystal, or in a uniformly oriented collection of supramolecular assemblies having helical or cylindrical symmetry, such as a bundle of muscle fibres or a stack of planar membranes. Figure 2 illustrates that, even in the absence of perfect orientational order, the spectrum provides a direct and unambiguous readout of the distribution of orientations, providing relatively independent information on the average orientation (θ_0 in Fig. 2), the orientational disorder ($\Delta\theta$ in Fig. 2), and the number of discreet preferred orientations in the population. Although polarized optical spectroscopy is also sensitive to the orientational distribution to chromophores, this high orientational resolution in the frequency domain of the spectrum is not available. As a result, orientational information from optical

Θ_0

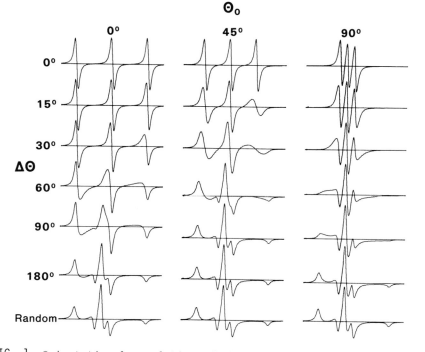

FIG. 1 Orientational resolution of the conventional EPR spectrum of nitroxide spin-labels, obtained by computer simulation. The derivative of absorption is plotted against the resonance magnetic field strength (H_{res}). The top three spectra correspond to uniformly oriented populations of spin-labels, each having a single value of $\cos^2\theta$, where θ is the angle between the applied magnetic field and the principal axis of the nitroxide spin-label (approximately perpendicular to the N-O bond). The slight dependence of the spectrum on rotation about the principal axis is neglected. The bottom spectrum corresponds to a randomly oriented population of spin-labels, immobile in the nanosecond time range (a "powder" spectrum). Lines at 0° and 90° illustrate that, although there is no preferred orientation in this population, the orientational resolution (anisotropy) remains in the spectrum: each position in the spectrum corresponds essentially to a particular value of $\cos^2\theta$.

spectroscopy is more ambiguous than from EPR. This difference provides EPR with its main advantage over optical techniques, and helps make the two complementary.

Isotropic rotational motion In the absence of orientational order of the supramolecular assemblies being studied (e.g., in a solution or in a randomly-oriented dispersion of muscle fibres or membranes), no orientational information can be derived directly from the EPR spectrum

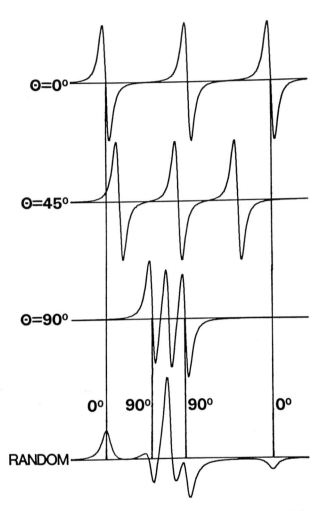

FIG. 2 Effect of varying amounts of orientational disorder on the nitroxide EPR spectrum. Each column corresponds to a different value of θ_0, the mean angle between the probe's principal axis and the applied field. Each row corresponds to a different value of $\Delta\theta$, the full width at half maximum of an assumed Gaussian distribution of probe orientations (from Barnett and Thomas, 1984).

but the orientational <u>resolution</u> remains in the spectrum (see Fig. 1, bottom), so that a very narrow distribution of orientations (as small as $8°$) is excited at one spectral position. This high-resolution photoselection, based on spectral position, provides EPR (both conventional and saturation transfer) with its sensitivity to rotational

motion. Since a change in the orientation of a probe by about 8° can
change its position in the spectrum significantly, rotational motion
transfers spins from one part of the spectrum to another. If the time
required for an 8° rotation (roughly equal to 0.1 τ_R, where τ_R is the
rotational correlation time) is comparable to or less than the inverse
of the frequency-resolution in the spectrum (roughly equal to T_2, the
transverse relaxation time), the spectrum will be narrowed (averaged)
by the rotational motion. For a typical nitroxide radical, T_2 is about
20 ns, so motional narrowing can be detected only when the rotational
correlation time is comparable to or less than 200 ns. Conventional
EPR is performed under conditions where the shape of the nitroxide
spectrum, in the absence of preferred orientation, is determined
almost exclusively by motional narrowing, and is thus sensitive only
to submicrosecond rotational motions. This principle is illustrated,
in the case of isotropic rotational diffusion, on the left side of
Fig. 3.

Anisotropic rotational motion In the case of more complex motions
(e.g. motion that is restricted in angular amplitude), the details of
the motion (i.e. the rate as well as the amplitude) can often be
determined from the EPR spectrum. Motion that is restricted in angular amplitude will cause motional narrowing in only a part of the
spectrum, resulting in characteristic line-shapes that are often
distinguishable from those corresponding to isotropic (unrestricted)
rotational motion. Examples of EPR spectra corresponding to restricted motion are shown in Fig. 4. Similarly, the presence of two populations of spin-labels having very different rotational correlation
times will often give rise to two clearly resolved spectral components.
Thus, even though EPR is a steady-state (as opposed to time-resolved)
technique, its orientational resolution permits the study of anisotropic motion and motional heterogeneity in the time scale of conventional EPR (submicrosecond). In this sense, when comparing conventional EPR to fluorescence depolarization, the ability of EPR to
reflect motional anisotropy or heterogeneity is more comparable to
that of time-resolved than of steady-state fluorescence methods.

V₁   **V₂'**

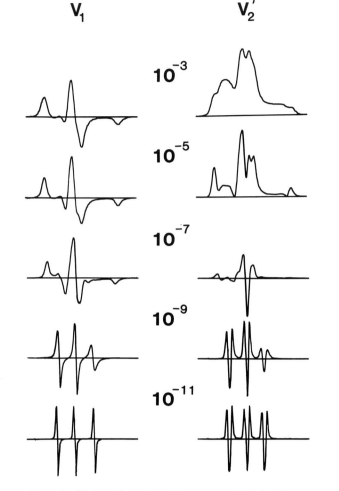

FIG. 3 Conventional (V₁) and saturation transfer (V₂') EPR spectra
corresponding to isotropic Brownian rotational diffusion at known
rotational correlation times (Squier and Thomas, 1984). The bottom
two rows of spectra were obtained from solutions of small spin-labels,
and the rest were obtained from spin-labelled haemoglobin. The viscosity was varied by varying the glycerol concentration and temperature, and the rotational correlation time was calculated from the
Stokes-Einstein-Debye equation for isotropic rotational diffusion:
$\tau_R = V\eta/(kT)$, where V is the volume of the equivalent sphere.

Saturation transfer EPR

General principles The term "conventional EPR" indicates an experiment that probes the linear response of a spin system to the exciting
microwave radiation, implying that the spin system is never signifi-

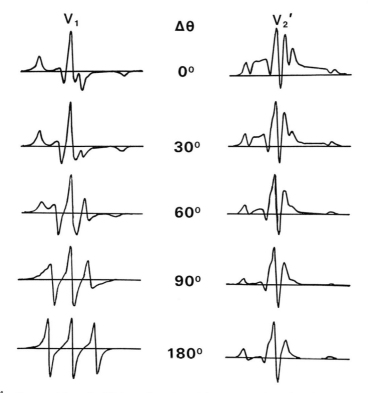

FIG. 4 Conventional (V_1) and saturation transfer (V_2') EPR spectra corresponding to restricted rotational motion, with a variable angular range of $\Delta\theta$. For V_1 and V_2' spectra, the correlation time is fixed in each column, corresponding essentially to values of 0.1 ns and 0.1 μs, for V_1 and V_2', respectively, i.e. near the isotropic motion limits for maximal motional narrowing and saturation transfer, respectively. Note that an increase in the angular amplitude $\Delta\theta$ of subnanosecond motions causes an increase in motional narrowing in V_1 spectra (left), while an increase in $\Delta\theta$ for the slower motions only changes the relative intensities at different spectral positions in V_2' spectra (right).

cantly perturbed from equilibrium, having a constant excess population in the ground state, as determined by the Boltzmann distribution of electron spin states. If the microwave field strength H_1 is intense enough, the spin system becomes saturated, i.e. the ground state becomes significantly depleted, and the net steady-state absorption is no longer proportional to H_1. The amount of saturation depends on the competition between the rate of excitation and the rate constant for relaxation back to the ground spin state. In the absence of rotational motion, this rate constant for recovery from saturation is $1/T_1$, where T_1 is the longitudinal relaxation time. T_1 is analogous to the excited-state lifetime in optical spectroscopy.

However, rotational motion can increase this rate of recovery by causing saturation transfer, according to the following argument (Thomas *et al.*, 1976). At any one time during a spectral scan in a steady-state EPR experiment, only a small fraction of the probes, corresponding to a narrow angular range (as little as 8°), are at resonance and are therefore subjected to saturating radiation. There-fore, rotational diffusion that transfers these probes out of this saturated angular range (and transfers other probes into it) during the excited-state lifetime T_1 will transfer saturation to other spectral regions, thus decreasing the saturation (ground-state deple-tion) at the resonance position. Just as in conventional EPR, the minimum time required for a transfer of spins out of the narrow reson-ance range (i.e. an 8° rotation) is about 0.1 τ_R. Since T_1 is about 10 μs for slowly rotating nitroxides, saturation transfer occurs when-ever the rotational correlation time (τ_R) is less than or comparable to 100 μs, optimum sensitivity to τ_R occurs in the range of 10-100 μs, and saturation transfer becomes maximal when τ_R is less than 1 μs. Since T_1 is much greater than T_2, any EPR experiment becomes sensitive to slower rotational motion when saturation is imposed. The develop-ment of ST-EPR instrumentation and methodology has involved a search for an EPR experiment that is optimally sensitive to saturation, hence optimally sensitive to saturation transfer, and hence optimally sensitive to microsecond rotational motion.

Comparison with the principles of optical spectroscopy   Both ST-EPR and the motion-sensitive optical spectroscopies (anisotropy of emis-sion or ground-state depletion) depend on relaxation processes. Both depend on competition between (1) relaxation processes that are inde-pendent of rotational motion and determine the excited-state lifetime (designated T_1 in EPR and τ in optical spectroscopy) and (2) apparent relaxation processes that depend on rotational motion, and are observ-able only by using orientation-dependent photoselection (obtained by varying the applied magnetic field in EPR or the direction of polar-ization in optical spectroscopy). In this sense, the principles of polarized optical spectroscopy are much more analogous to those of ST-EPR than to those of conventional EPR, and the optical technique most analogous to ST-EPR is polarized ground-state depletion, usually observed by detecting the transient absorption dichroism of triplet probes. In fact, time-resolved ST-EPR could be performed similarly to these time-resolved optical ground-state depletion experiments. The absorbance could be monitored as a function of time after a saturating pulse of excitation, and sensitivity to rotation would be obtained by varying the spectral position (orientation) of the saturating (photo-selecting) and observing radiation fields. The technology for such a time-resolved saturation transfer (saturation recovery) experiment has been developed in recent years, and signal/noise levels sufficient for biophysical applications should be possible in the near future. However, in the meantime, steady-state methods must be used in ST-EPR as established over the past decade (Hyde and Dalton, 1972; Thomas *et al.*, 1976; reviewed by Hyde and Thomas, 1980, and by Thomas, 1984).

Specific methodology: modulation spectroscopy Although the use of steady-state, as opposed to time-resolved, methods in ST-EPR limits the resolution of this technique, two factors give ST-EPR advantages over the analogous steady-state ground-state depletion in optical spectroscopy. The first factor, already mentioned above, is the superior orientational resolution of EPR. The second is that the modulation techniques and phase-sensitive detection used in EPR permit a more direct look at saturation, and hence at saturation transfer. That is, ST-EPR experiments are usually not performed by using constant excitation and simply observing the steady-state absorption, which would be decreased due to saturation (ground-state depletion), and hence increased due to rotational motion (saturation transfer). Instead, the excitation condition is modulated by a small-amplitude field modulation (usually at 50 kHz, a frequency near the inverse of the relaxation time T_1), and phase-sensitive detection is used to monitor selectively the signal component that lags $90°$ behind the modulation. This out-of-phase signal increases with saturation and, therefore, decreases with saturation transfer. A motion that would cause a small fractional change in the total steady-state absorption spectrum causes a large fractional change in this more selective out-of-phase signal. In many ways, the modulation and phase-sensitive detection methods used in ST-EPR are analogous to those introduced recently in fluorescence for a similar purpose (Weber, 1977). However, the orientational resolution of EPR provides an important advantage over fluorescence: because the orientational resolution varies with the position in the spectrum, different regions of the spectrum have different sensitivities to rotational motion. Therefore, the out-of-phase EPR signal responds to rotational motion not only by decreasing its overall intensity, but also by changing the shape of the spectrum, due to greater intensity decreases in the more sensitive regions of the spectrum (for more details, see Thomas *et al.*, 1976). Hyde and Thomas (1973) found that the shape of the spectrum is most sensitive to rotational motion if the absorption signal is detected at the second harmonic (100 kHz) of the 50 kHz modulation, and this detection scheme (absorption, second harmonic, out-of-phase, designated V_2', as opposed to the conventional absorption, first harmonic, in-phase, designated V_1) has been the ST-EPR method of choice ever since.

Isotropic rotational motion Figure 3 shows the sensitivity of ST-EPR spectral line-shapes to isotropic rotational motion, and shows clearly the superior sensitivity of V_2' spectra in the microsecond time range. For both conventional and ST-EPR, the quantitative analysis of data is performed by comparing spectra with reference spectra, either obtained from experimental model systems, as in Fig. 3, or from theoretical computer simulations (Thomas and McConnell, 1974; Thomas *et al.*, 1976). A spectrum is analysed either by fitting the complete spectrum to a reference spectrum, or, more commonly, by comparing spectral parameters that are especially sensitive to rotational motion. Since conventional EPR spectra change by motional narrowing, a splitting between peaks is most often used there. Since microsecond motion changes only the intensities of various spectral positions of ST-EPR spectra, but not the positions themselves, peak height ratios are most often used (Thomas *et al.*, 1976).

Anisotropic motion In the case of complex motions, it is virtually impossible to produce an experimental model system for which the

details of the motions are known, independently of the EPR spectra,
so theoretical computer simulations are essential for producing reliable reference spectra. Robinson and Dalton (1980, 1981) have simulated ST-EPR spectra corresponding to the anisotropic motion of
freely tumbling ellipsoids. Another type of motion, which is of
particular interest in studying segmental flexibility within organized
systems, is rotational motion restricted in angular amplitude.
Figure 4 shows examples of spectra corresponding to this type of
motion, and compares the line-shape changes that occur in conventional
(V_1) and ST-EPR (V_2') spectra (Lindahl and Thomas, 1982, 1984). The
variable in these spectra is the angular amplitude $\Delta\theta$, not the rotational correlation time. In contrast to Fig. 3, adjacent V_1 and V_2'
spectra do not correspond to the same rate of motion in Fig. 4. All
of the V_1 spectra correspond to a rotational correlation time of 0.1
ns, close to the limit for maximal motional narrowing, and the V_2'
spectra correspond to a time of 0.1 μs, close to the limit for maximal saturation transfer without motional narrowing. Qualitatively,
the changes in spectral line-shapes as $\Delta\theta$ increases (Fig. 4) are
similar to the changes as τ_R decreases (Fig. 3). Motional narrowing
shifts spectral positions inward in V_1 spectra, while saturation
transfer decreases the intensities at some spectral positions more
than at others in V_2' spectra. These line-shape changes are not
exactly the same for the two types of motions, so information about
the anisotropy of motion can, in principle, be derived from the spectrum. However, the dependence of V_2' line-shapes on the type of
microsecond motion is more subtle than that of V_1 line-shapes on the
type of nanosecond motion (Lindahl and Thomas, 1982, 1984). It is
often quite difficult to distinguish a decrease in the rate of motion
from a decrease in its amplitude. Similarly, it is more difficult to
detect the presence of two or more different motions in the microsecond range than in the nanosecond range. An approach that has
proved successful in reducing the ambiguity of interpreting ST-EPR
spectra is to supplement the ST-EPR data with conventional EPR experiments on oriented systems (e.g. muscle fibres or membranes) with their
symmetry axes oriented parallel to the applied magnetic field. These
data can be analysed with reference spectra like those in Figs. 1 and

2, yielding direct information about the distribution of orientations seen in a 1-microsecond "snap-shot", thus providing an estimate of the angular amplitude of slower rotational motions that affect the ST-EPR spectrum.

Time-resolved optical polarization spectroscopy, using triplet probes, can be used to obtain information about anisotropic rotational motion in the microsecond time range, as described in detail in other articles in this volume. This information is obtained by comparing experimental results (the time-dependence of anisotropy after an excitation pulse) with those predicted by theory (e.g. Kinosita *et al.*, 1977). Amplitudes and rates of motion are related to amplitudes and rates of anisotropy decay. As in ST-EPR, detailed analysis usually requires some assumptions about the particular type (geometry) of motion taking place. In contrast to ST-EPR, actual (rather than effective) correlation times can be determined directly from the data, and certain model-independent conclusions can be drawn. For example, the residual anisotropy (that does not decay to zero in the time range of interest) indicates that the rotational motion is restricted in amplitude, with a certainty that compared favourably with that of conventional EPR and greatly exceeds that of ST-EPR. However, while the time-resolution of optical spectroscopy is clearly superior to that of EPR, its orientational resolution is not. For example, while conventional EPR on an oriented sample is quite powerful in estimating the orientational distribution of the microsecond motions observable by ST-EPR, steady-state fluorescence depolarization is much less useful in estimating the orientational distribution of microsecond motions observed by triplet anisotropy decay. Thus, a combination of EPR (conventional EPR on oriented samples, as well as ST-EPR) and triplet anisotropy decay techniques can be an effective approach to resolving questions about the details of slow anisotropic rotational motions. The following section illustrates these principles with examples from the study of rotational dynamics in myosin.

APPLICATION TO MUSCLE PROTEINS

Introduction to muscle contraction

The contractile machinery of skeletal muscle consists essentially of
a system of interdigitating protein filaments, the thick filaments,
each containing several hundred myosin molecules, and the thin fila-
mments, each containing several hundred actin molecules. The filaments interact through "crossbridges", parts of myosin containing a
"neck" (designated S-2) and two "heads" (designated S-1), that pro-
ject from the thick to the thin filaments. Each head contains an
ATPase active site and an actin-binding site. Force is produced by
a cyclical interaction among myosin, actin, and ATP, resulting in a
sliding of the filaments to increase their overlap. A widely discussed model for the mechanism of this force-production postulates a
flexible myosin molecule that rotates during the ATPase cycle while
attached to actin, thus "rowing" the filaments past each other
(Huxley, 1969). In order to test this and other models for the role
of myosin motions in muscle contraction, it is essential to obtain
direct and quantitative information about the rotational dynamics of
myosin, using spectroscopic probes. Here, we will focus on EPR
studies; for a more comprehensive review of molecular studies in
muscle, see Morales *et al.* (1982).

EPR Studies

ST-EPR A series of ST-EPR experiments have been carried out on spin-
labels (both maleimide and iodoacetamide derivatives of TEMPO)
attached specifically and covalently to myosin heads, in order to ask
directly whether these rotational motions of myosin heads occur, in
purified myosin as well as in muscle fibres (reviewed by Thomas, 1982).
Experiments on the isolated myosin head (S-1) showed that the spin-
labels are rigidly fixed to the protein framework, so that the spectra
report large-scale head motions (Thomas *et al.*, 1975, 1980). The
results of ST-EPR experiments on isolated myosin heads (S-1), myosin
monomers (which exist only at high ionic strength), and myosin filaments (which exist at physiological ionic strength) are summarized in
Table 1. The spectral line-shapes were all similar to reference
spectra corresponding to isotropic rotational motion, so effective

TABLE 1 Effective correlation times (τ_R) determined from ST-EPR spectra of iodoacetamide-spin-labelled myosin heads at 20°C.

	τ_R (μs)
Subfragment-1 (S-1, the isolated myosin head)*	0.18
Myosin monomers (0.5 M KCl)*	0.30
Myosin filaments (0.1 M KCl)[+]	10
Actomyosin ([actin]/[myosin] = 4)[+]	$\geqq$400

Effective correlation times were determined by comparing line-shapes with those of reference spectra corresponding to isotropic motion (Thomas *et al.*, 1976). *From Thomas *et al.* (1975). Values for S-1 and myosin are averages of results for conventional and ST-EPR, since both techniques were applicable in this time range. [+]From Thomas *et al.* (1980).

rotational correlation times (τ_R) were assigned by comparison with reference spectra corresponding to isotropic motion. As shown in Table 1, the τ_R value for myosin monomers was less than twice that of isolated S-1, and that of myosin filaments was only about ten times that in myosin monomers. These ratios are much less than the ratios of molecular weights. Careful analysis of these values, including information about the known shapes of these systems, results in the conclusion that there must be considerable flexibility within myosin, permitting myosin heads to rotate relatively independently of the rest of myosin (Thomas *et al.*, 1975). This mobility is abolished when actin binds to myosin (Thomas *et al.*, 1975, 1980). The mobility was clearly less in myosin than in S-1, less in filaments than in monomers, and much less in the presence of actin, but the spectral line-shapes did not contain enough information to conclude whether the decreased mobility was due to a decrease in the rate or the amplitude of the motion, or to changes in both. For example, it remained possible that the motion in myosin filaments was just as fast as in myosin monomers, corresponding to an actual correlation time much less than the effective one (10 μs - see Table 1), with the main change coming from a severe restriction in the range of the motion.

Conventional EPR Conventional EPR experiments on oriented fibres were performed to help reduce the ambiguity of interpretation in the

TABLE 2 Orientational distributions of iodoacetamide spin-labelled myosin heads in muscle fibres, determined from conventional EPR.

	θ_0	$\Delta\theta$
Full overlap (heads bound to actin)*	68° ($\pm1^\circ$)	17° ($\pm1^\circ$)
No overlap (heads free)[+]		$>90^\circ$

Conventional (V_1) EPR spectra were fitted to spectra like those in Fig. 2. θ_0 is the mean angle and $\Delta\theta$ is the full width at half maximum of the Gaussian distribution of orientations θ. *From Thomas and Cooke (1980). [+]From Barnett and Thomas (1984).

ST-EPR spectra. Selected results are summarized in Table 2. A procedure was developed for labelling myosin heads specifically (at a single cysteine residue) in functional, glycerinated (demembranated) muscle fibres (Thomas et al., 1980; Thomas and Cooke, 1980). When the fibres were oriented parallel to the magnetic field, under conditions where myosin heads were bound to actin, conventional EPR spectra indicated that the probes, and therefore the myosin heads, were almost uniformly oriented with respect to the fibre axis. The full width at half maximum of the orientational distribution, determined with the use of reference spectra like those in Fig. 2, was only 17°, providing an estimate for the angular range of the motion. Using this value to analyse the ST-EPR spectra of myosin bound to actin, it is possible to conclude that both the rate and amplitude of myosin head motion is greatly restricted by binding to actin. In order to determine the angular range of myosin head orientations in myosin filaments in the absence of actin, fibres were stretched to eliminate contact between the two proteins (Thomas and Cooke, 1980; Barnett and Thomas, 1984). The result was a large change in the spectrum, indicating that the angular range of the labelled heads was greater than 90° (Barnett and Thomas, 1984). This tends to rule out the possibility that the myosin heads undergo small amplitude motions with an actual correlation time much shorter than the effective one obtained from ST-EPR spectra (10 µs). The picture that emerges is one of myosin heads rotating relatively freely through large angles in the microsecond time range.

Time-resolved triplet-state spectroscopic studies

This picture of myosin flexibility has been confirmed and further
clarified by recent studies of myosin labelled with eosin-iodoacetamide,
a probe with a stable triplet state, which makes possible time-
resolved studies of either emission (phosphorescence) or absorption
(ground-state depletion, or triplet-triplet absorption) in the microsecond time range (Eads *et al.*, 1984). Table 3 summarizes some of

TABLE 3 Rotational amplitudes (A) and correlation times (ϕ) of eosin-
labelled myosin heads, determined by absorption anisotropy, at 4°C.

	A_1	ϕ_1 (μs)	A_2	ϕ_2 (μs)	A_3
Subfragment-1	0.275	0.200	----	----	0.002
Myosin monomers	0.14	0.40	.084	2.6	0.015
Myosin filaments	0.080	0.73	.058	4.9	0.049

These values were determined by least-squares fits of the anisotropy
decays to the function $A_1\exp(-t/\phi_1) + A_2\exp(-t/\phi_2) + A_3$ (Eads *et al.*,
1984). Except for the temperature, conditions were essentially the
same as in the EPR experiments (Tables 1 and 2).

TABLE 4 Comparison of EPR and triplet anisotropy results for myosin
filaments.

	Correlation time (μs)	angular amplitude ($^{\circ}$)
EPR	10	$>90^{\circ}$
Triplet	0.73 and 4.9	102°

These results are from Tables 1, 2, and 3. The correlation time from
ST-EPR is an effective one (τ_R, Table 1), assuming isotropic motion
($\Delta\theta=180^{\circ}$), whereas those from triplet anisotropy are directly observed
correlation times (ϕ, Table 3), independent of motional models. The
angular amplitude from EPR is the full width ($\Delta\theta$) of orientations
observed directly in oriented stretched muscle fibres (Table 2), while
that from triplet anisotropy is determined from the residual anisotropy (A_3 in Table 3), assuming motion within a cone of full width
$\Delta\theta$ (Kinosita *et al.*, 1977).

the results of time-resolved absorption (ground-state depletion) anisotropy studies and Table 4 shows a comparison of the correlation times and angular amplitudes of motion obtained by EPR and triplet anisotropy for myosin filaments. The probe on isolated S-1 in aqueous solution rotates with a single correlation time around 200 ns, agreeing with EPR results and with previous fluorescence experiments (Mendelson et al., 1973), and indicating that the probe reports the large-scale motion of the myosin head. Experiments on labelled S-1 in glycerol solutions confirms this (data not shown). Also consistent with EPR, heads in intact myosin monomers rotate only a little more slowly than in the isolated S-1, indicating considerable flexibility within myosin. However, a small-amplitude decay component is observed corresponding to slower motion, indicating that the rapid motions are moderately restricted in amplitude. Analysis by the theory of Kinosita et al. (1977), which assumes rotational motion in a cone of angular width $\Delta\theta$, indicates a value of 88° for $\Delta\theta$. In myosin filaments, two decay components are clearly evident, having comparable amplitudes (and, therefore, comparable values of $\Delta\theta$), but correlation times of 0.73 and 4.9 μs, both smaller than the effective correlation time of 10 μs obtained from ST-EPR. An explanation for this longer effective ST-EPR correlation time is suggested by the observation that the absorption anisotropy does not decay completely to zero in the microsecond time range, but has a residual anisotropy value of 0.049, 26% of the total initial anisotropy. Thus the motion is restricted in angular amplitude. Analysis by the theory of Kinosita et al. (1977), indicates that the two motions add up to a $\Delta\theta$ value of 102°. Although this is quite consistent with the result of conventional EPR ($\Delta\theta > 90^{\circ}$; a value of 102° would probably be too great to be detected by conventional EPR), the restriction in amplitude is sufficient to cause significant changes in the ST-EPR spectrum (Fig. 4). The effective correlation time, obtained by assuming isotropic motion (using reference spectra like those in Fig. 3), is longer than that obtained by assuming restricted motion (using reference spectra like those in Fig. 4, details to be published elsewhere). These results help illustrate the point that the effective correlation time obtained from ST-EPR, assuming isotropic motion, should usually be considered an upper bound for the actual correlation time.

CONCLUSION

The EPR and triplet anisotropy decay results are consistent with each
other and, considering other information about the structure of myosin, lead to the following simple and plausible model for the rotational flexibility of myosin in the thick filaments of muscle. As
long as the myosin heads do not contact actin, they are free to rotate
through large angles ($>90°$) in the sub-millisecond time range. At
least two types of motion are present. One motion, having a correlation time about 1 μs, probably corresponds to the motion of the heads
(S-1) with respect to each other and with respect to the neck (S-2).
A slower motion, having a correlation time of several microseconds,
probably corresponds to rotation of the neck with respect to the rest
of myosin (anchored in the thick filament). This picture provides a
well-defined starting point for considering how myosin motions might
be coupled to force-generation.

The two techniques provide similar kinds of information, but are
also clearly complementary, as discussed above. The orientational
resolution of conventional EPR, coupled with the microsecond motional
sensitivity of ST-EPR, describes clearly the large amplitude and
approximate time range of the motion. The time-resolution of triplet
anisotropy decay demonstrates clearly the presence of two motional
components, and the non-zero residual anisotropy clearly demonstrates
that motion is restricted.

There are other important ways in which these two methods are
complementary. Triplet probes can be studied at somewhat lower concentrations. Since spin labels are much smaller than triplet probes,
they are expected to perturb the system less. In addition, spin
labels tend to bind more rigidly than triplet probes, probably also
due to their small size, which might allow them to fit in small binding pockets. Spin labels can be studied in very turbid samples (e.g.
muscles), which offer serious experimental difficulties for optical
techniques. Because of these advantages of spin labels, and because
of EPR's orientational resolution, one of the most promising areas of
future instrumental development is that of time-resolved ST-EPR,
which would combine many of the advantages of both EPR and optical
techniques, and should make possible, in a single experiment, a more

detailed and less ambiguous analysis of complex rotational motions than is now possible with any present technique.

REFERENCES

Barnett, V.A. and Thomas, D.D. (1984). *J. Mol. Biol.*, in press.
Eads, T.E., Austin, R.A. and Thomas, D.D. (1984). *J. Mol. Biol.*, in press.
Huxley, H.E. (1969). *Science* 164, 1356-1366.
Hyde, J.S. and Dalton, L.R. (1972). *Chem. Phys. Lett.* 16, 568-572.
Hyde, J.S. and Thomas, D.D. (1973). *Ann. N.Y. Acad. Sci.* 222, 680-692.
Hyde, J.S. and Thomas, D.D. (1980). *Annu. Rev. Phys. Chem.* 31, 293-317.
Kinosita, K. Jr., Kawato, S. and Ikegami, A. (1977). *Biophys. J.* 20, 289-305.
Lindahl, K.M. and Thomas, D.D. (1982). *Biophys. J.* 37, 71a.
Lindahl, K.M. and Thomas, D.D. (1984). Submitted for publication.
Mendelson, R.A., Morales, M.F. and Botts, J. (1973). *Biochemistry* 12, 2250-2255.
Morales, M.F., Borejdo, J., Botts, J., Cooke, R., Mendelson, R.A. and Takashi, R. (1982). *Annu. Rev. Phys. Chem.* 33, 319-351.
Robinson, B.R. and Dalton, L.R. (1980). *J. Chem. Phys.* 72, 1312-1324.
Robinson, B.R. and Dalton, L.R. (1981). *Chem. Phys.* 54, 253-259.
Squier, T.C. and Thomas, D.D. (1984). Submitted for publication.
Thomas, D.D. (1978). *Biophys. J.* 24, 439-462.
Thomas, D.D. (1982). *In* "Mobility and Function in Proteins and Nucleic Acids" (Ciba Foundation Symposium 93), (Ed. R. Porter), pp. 165-185. Pitman, London.
Thomas, D.D. (1984). *In* "The Enzymes of Biological Membranes". (Ed. A. Martonosi), Plenum Press, in press.
Thomas, D.D. and McConnell, H.M. (1974). *Chem. Phys. Lett.* 25, 470-475.
Thomas, D.D., Seidel, J.C. and Gergely, J. (1975). *Proc. Natl. Acad. Sci. USA* 72, 1729-1733.
Thomas, D.D., Dalton, L.R. and Hyde, J.S. (1976). *J. Chem. Phys.* 65, 3006-3024.
Thomas, D.D. and Cooke, R. (1980). *Biophys. J.* 32, 891-906.
Thomas, D.D., Ishiwata, S.I., Seidel, J.C. and Gergely, J. (1980). *Biophys. J.* 32, 873-890.
Weber, G. (1977). *J. Chem. Phys.* 66, 4081-4091.

Time-resolved Fluorescence Spectroscopy: Some Applications of Associative Behaviour to Studies of Proteins and Membranes

L. BRAND, J.R. KNUTSON, L. DAVENPORT, J.M. BEECHEM, R.E. DALE, D.G. WALBRIDGE and A.A. KOWALCZYK

INTRODUCTION

Fluorescence of organic molecules often exhibits dramatic sensitivity to the microenvironment of the fluorophore. This is due in part to the different physical and chemical characteristics of the excited state and the ground state. Excited-state processes such as solvation, proton transfer, electron ejection, exciplex and excimer formation, energy transfer and conformational rearrangements can each lead to environment-dependent changes in the emission properties.

Electronic transitions may be characterized in terms of their energy, transition probability, and direction. Fluorescence excitation and emission spectra provide information about energies of transitions. Fluorescence depolarization is determined by the orientations of the electronic transitions in absorption and emission relative both to each other and to the principal axes of rotation of the molecules in which they are embedded when these have some rotational freedom. Decay times and quantum yields provide relative information regarding the probability of transitions. The decay of the fluorescence intensity provides crucial information about the competition between various pathways for the return of an excited molecule to the ground state. In contrast, the decay of the emission anisotropy is capable of providing detailed information on the <u>rotational</u> behaviour of molecules during the excited-state lifetime.

Decay kinetics and heterogeneity

A pure fluorophore dissolved in a non-interacting solvent may show

SPECTROSCOPY AND THE DYNAMICS
OF MOLECULAR BIOLOGICAL SYSTEMS

259

monoexponential fluorescence decay kinetics:

$$I(\lambda,t) = \alpha(\lambda)\exp[-(t/\tau)] \tag{1}$$

Multiexponential decay kinetics may arise from a mixture of two or more fluorophores:

$$I(\lambda,t) = \sum_i \alpha_i(\lambda)\exp[-(t/\tau_i)] \tag{2}$$

In this context a single chromophore in different microenvironments may be considered as a mixture of chromophores. An example of interest in biophysics would be two or more tryptophan or tyrosine residues at different locations in a protein molecule. A protein that has only a single tryptophan residue may still exhibit complex fluorescence decay, if the protein exists in more than one conformation. It is of interest that even the free amino acid, tryptophan, can exist in several conformational states leading to deviations from monoexponential fluorescence decay kinetics (Rayner and Szabo, 1978; Robbins *et al.*, 1980; Szabo and Rayner, 1980; Ross *et al.*, 1981a; Jameson, 1983; Chang *et al.*, 1983; Petrich *et al.*, 1983). Another example of micro-heterogeneity in biophysics is the situation of a fluorescence probe at different locations in a bilayer membrane.

Excited-state reactions

Multi- or non-exponential decay may also arise from a pure fluorophore that undergoes excited-state reactions of the type mentioned above. The fluorescence decay kinetics for some excited-state processes are well understood. For example, excited-state reactions such as excited-state proton transfer, excitation energy transfer and exciplex or excimer formation may, under suitable conditions, fall into the class of two-state reversible excited-state reactions. The kinetics of emission for the initially excited species A and for the species B formed in the excited state may be represented (Birks, 1970; Porter, 1972; DeToma and Brand, 1977; Gafni and Brand, 1978; Laws and Brand, 1979; Lakowicz and Balter, 1982) by:

$$I_A(\lambda,t) = I_A(\lambda)\{\alpha_1\exp[-(t/\tau_1)] + \alpha_2\exp[-(t/\tau_2)]\}$$

$$I_B(\lambda,t) = I_B(\lambda)\beta\{\exp[-(t/\tau_1)] - \exp[-(t/\tau_2)]\} \tag{3}$$

under most circumstances.

In these expressions, the lifetimes τ_1 and τ_2 are respectively identical for both the emission of the species excited directly and of that formed during the excited state. The emission originating from the species generated in the excited-state shows biexponential kinetics with pre-exponential terms which are equal in magnitude (β) and opposite in sign. If, for unit excitation of A, $I_A(\lambda)$ and $I_B(\lambda)$ are the emission spectral distributions of A and B normalized by their respective radiative rates, i.e. by the inverse of their radiative ("natural") lifetimes, the pre-exponential terms appearing in Equation 3 may be expressed as $\alpha_1 = (\gamma_2 - X_A)/(\gamma_2 - \gamma_1)$, $\alpha_2 = (X_A - \gamma_1)/(\gamma_2 - \gamma_1)$ and $\beta = k_{AB}/(\gamma_2 - \gamma_1)$. In these expressions, $X_A = k_{AB} + k_A$, k_{AB} is the rate constant for formation of B from A, and k_A is the decay rate of A (including radiative and non-radiative processes, but excluding that for formation of B), while $\gamma_{1,2} = \tau_{1,2}^{-1} = (1/2)\{(X_A + X_B) \mp [(X_A - X_B)^2 + 4k_{AB}k_{BA}]^{1/2}\}$ where X_B and k_{BA} are defined analogously to X_A and k_{AB} respectively.

In the irreversible case, e.g. at high proton concentrations in an excited-state protonation reaction or, often, for excitation energy transfer, $k_{BA} = 0$ leading to $\gamma_1 = k_B$, $\gamma_2 = X_A$, $\alpha_1 = 0$, $\alpha_2 = 1$ and $\beta = k_{AB}/(X_A - k_B)$ so that the decay kinetics of A become monoexponential, those of B remaining biexponential:

$$I_A(\lambda, t) = I_A(\lambda)\exp[-X_A t]$$

$$I_B(\lambda, t) = I_B(\lambda)\beta(\exp[-k_B t] - \exp[-X_A t]) \qquad (4)$$

In this formulation, it appears superficially that the "rise" term for B and the decay of A which feeds it are identical. This will be true experimentally if it so happens that $X_A > k_B$, i.e. if the lifetime of B directly excited is longer than that observed for A. In the converse case, however, the "rise" term must become identified with the (directly excited) decay of B, and the sign change is automatically compensated by simultaneous inversion of the sign of β which contains $X_A - k_B$ in its denominator.

In the reversible case, it may similarly happen that $\tau_1 < \tau_2$ (see Equation 3) and the same compensation will also occur there. In both

cases there also arises the question: what will happen in the event
that the two lifetimes are identical? Although the solution to this,
at least in the irreversible case, is well known in chemical kinetics
(see, e.g., Kynch, 1962), it does not appear to have been dealt with
in the excited-state literature. The result may be expressed (cf.
Dale, 1979):

$$I_B(\lambda, t) = I_B(\lambda) k_{AB} t \exp[-t/\tau] \qquad (5)$$

where τ is the common lifetime which also appears in the monoexponential decay of A, which is identical in both reversible and irreversible cases since, in the former, $\alpha_1 + \alpha_2 = 1$. Equation 5 is most simply
derived by substituting, for example, $\gamma_1 = 1/\tau$, $\gamma_2 = \gamma_1 + \delta$, then
expanding the exponentials and taking the limit for vanishing δ (or,
equivalently, applying l'Hôpital's Rule).

It is easily seen from Equation 4 for an irreversible two-state
excited-state process, and is well known from the theory of radio-
active decay series, that the evolution and decay of the product B
simply represents the convolution of the observed decay of A with
that of B which would obtain were it to be directly excited. As demonstrated by Hauser and Wagenblast (1983), who coined the term
"convolution kinetics" for it, this is also the case in a more complex, irreversible process such as excitation energy transfer in
solution ensembles which involves non-exponential decay forms. Only
recently, however, has it been demonstrated (Lakowicz and Balter,
1982) that it is also true in the two-state reversible case, where it
is not so immediately obvious. The consequence of this from an experimental point of view is that the decay rate of the (directly excited)
product in a two-state excited-state reaction may be obtained via a
monoexponential analysis using the observed kinetics of the progenitor
as the excitation function - "differential wavelength deconvolution"
(Lakowicz and Balter, 1982). Since convolution is commutative, this
applies whether the progenitor function was derived from a true impulse
excitation or is already the result of convolution with an excitation
function of non-negligible width.

Time-scales and heterogeneity

It should be emphasized that the fluorescence decay kinetics observed will depend on the time-scale available for the experiment. The present discussion will be limited to the nanosecond time range. The influence of transitions from higher excited states to the lowest lying excited state will be seen in the fluorescence decay kinetics as data in the picosecond and sub-picosecond range becomes more routinely available (Shank and Ippen, 1975; Beddard and Westby, 1981; Waldeck *et al.*, 1981). Some of the fluorescence changes discussed below have their origin in excited-state interactions that take place on the picosecond time-scale. However, these interactions will manifest themselves in terms of a microheterogeneous mixture of species decaying on the nanosecond time-scale rather than in a more direct influence on the decay kinetics of the emitting state.

Analysis and interpretation of decay data

As with other kinds of kinetics, time-resolved fluorescence decay can be a complex process, requiring suitably sophisticated procedures for data analysis. Additional information is often necessary before proposing suitable mechanistic models for the experimental decay data. Is the decay of the fluorescence intensity influenced directly by excited-state reactions? If so, what is the nature of these reactions, and what information is available regarding the microenvironment? Do the decay kinetics reflect different fluorescent residues in a biological system? Do they reflect different classes or populations of macromolecules? In the case of the decay of the fluorescence emission anisotropy, a complex decay may have its origin in anisotropic rotations of a homogeneous population of probes or in a heterogeneous population of fluorophores, each showing its own rotational behaviour. As answers to the questions raised above become more available, fluorescence spectroscopy will increase in value as a tool for investigations of the static and dynamic structure of biological macromolecules and supra-molecular structures such as proteins and membranes.

The microscopic environment of a fluorophore may be reflected in several measurable fluorescence quantities. The theme of the following discussion is that it is advantageous to attempt to associate different fluorescence decay data. This essentially provides an overdetermination of the system, and may lead to mechanistic proposals that would not be available from an isolated fluorescence decay experiment.

DECAY-ASSOCIATED FLUORESCENCE SPECTRA (DAS)

As was indicated above, a complex decay of the fluorescence intensity may have its origin in heterogeneity or microheterogeneity of the sample. Excited-state interactions may be involved, but, as already indicated, if they are rapid compared to the rates of fluorescence decay, they will not <u>directly</u> influence the decay parameters, and may rather be considered as contributing to the initial heterogeneity observed. Figure 1 illustrates the time-resolved emission of a mixture of two fluorophores with differing decay times and emission spectra. The "blue" emitter has a shorter decay time than the "red" emitter.

Fluorescence emission spectra obtained over a time-slice are referred to as time-resolved emission spectra (TRES) and are easily obtained with most fluorescence decay instrumentation (Ware *et al.*,

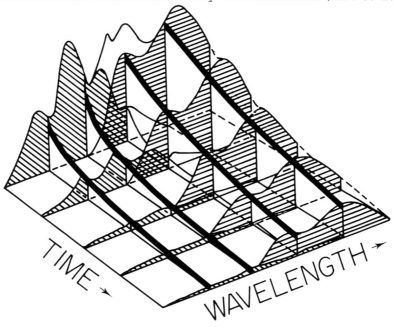

FIG. 1 Decay surface for a typical mixture of fluorescent species. The shorter-wavelength component decays more quickly than the longer-wavelength one (in this case). Thus TRES (time-resolved emission spectra) taken at later times will contain a smaller proportion of the "bluer" spectrum. This is depicted with the shaded time-slices shown. Linear combinations (see text) of these TRES may be used to compute the individual component spectra.

1968; Brand and Gohlke, 1971; Egawa *et al.*, 1971; Ware *et al.*, 1971;
DeToma *et al.*, 1976; Easter *et al.*, 1976; Gafni *et al.*, 1977; Badea
and Brand, 1979; Knorr and Harris, 1981). Likewise decay curves may
be obtained over some wavelength interval ($\Delta\lambda$). Data over the entire
time-energy intensity surface shown in Fig. 1 are thus available in
principle but, for a particular experiment, it is usually possible to
obtain data covering only a limited region on this surface.

Experimental examples

Wahl and his co-workers have obtained decay curves of tryptophan peptides and proteins as a function of emission wavelength. With the aid
of Equation 6 they were able to obtain the decay-associated spectra
$\overline{I}_i(\lambda)$ corresponding to the individual spectra of the folded and
unfolded conformers of *cyclo*-glycyl and *cyclo*-alanyl tryptophan
(Donzel *et al.*, 1974):

$$\overline{I}_i(\lambda) = \overline{I}(\lambda)\alpha_i(\lambda)\tau_i / \sum_i \alpha_i(\lambda)\tau_i \qquad (6)$$

These workers have also resolved the spectra of the tryptophans of the
lac repressor protein of *E. coli.* (Brochon *et al.*, 1977) and the
tryptophans of yeast 3-phosphoglycerate kinase (Privat *et al.*, 1980).
Ross *et al.* (1981b) used the same approach to resolve the tryptophan
emission of horse liver alcohol dehydrogenase into two spectral components. The results of quenching experiments with KI suggested the
assignment of the two spectral components to an outer (exposed)
tryptophan and a buried tryptophan on each subunit of the protein
(Abdallah *et al.*, 1978).

Knutson *et al.* (1982b) used an alternative method to obtain the
decay-associated spectra (DAS) of mixtures of fluorophores and of
alcohol dehydrogenase. A pulse fluorometer was used to collect data
as illustrated in Fig. 2 . Decay curves were collected at 1 nm intervals but with the collection of relatively few counts at each wavelength. The time-dependent spectral data are then divided into quite
broad time windows. The number of windows corresponds to the number
of spectra to be associated. High resolution decay curves were also
obtained at two or more wavelengths. These were analysed and the
individual decay terms were reconvolved with the lamp time profile to
obtain integral mixing coefficients in "convolved" space. These allow

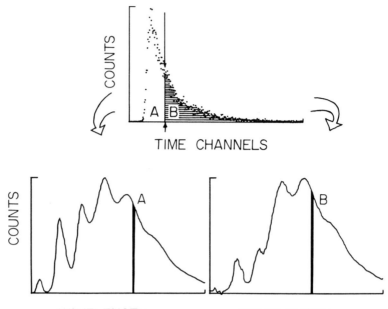

FIG. 2 Method used to collect TRES rapidly, yet retain sufficient accuracy for computing DAS (decay-associated spectra). At each wavelength, a brief collection of the decay histogram (for a flashlamp system, 10 s is a typical dwell time) is made. Then windowed sums are calculated from each region of interest in the decay. These values are added to their respective spectra at the appropriate wavelength. Multiple scans may be added together if the signal is very weak.

computation of the decay-associated spectra from the windowed TRES described above. An example of the use of this method is illustrated in Fig. 3 which shows the individually obtained technical emission spectra of anthracene and 9-cyanoanthracene. The superimposed symbols are the corresponding decay-associated spectra (DAS) obtained from a mixture of the two compounds.

It is of interest to point out that the axis labelled "time" in Fig. 1 may be any other variable that alters the emission distribution between different species. For example, the Stern-Volmer quenching constant is different for anthracene and 9-cyanoanthracene. Thus the same decay-associated spectra shown in Fig. 3 may be obtained with a steady-state spectrofluorimeter by analysing the spectra of a mixture obtained at increasing KI concentrations (Knutson *et al.*, 1983a). Both the decay association and the quenching association approach

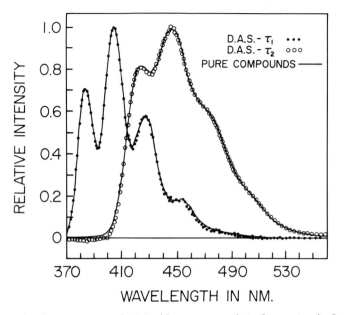

FIG. 3 Typical recovery of DAS (decay-associated spectra) from a heterogeneous fluorescent signal. Linear combinations of the TRES described in previous figures are used to generate a pure component DAS for each decay function found. In this case, they are shown as open or closed symbols. They may be compared to spectra taken for the pure dyes prior to mixing, shown as solid lines. The data displayed are the technical fluorescence emission spectra of anthracene and 9-cyanoanthracene derived separately and from a mixture.

were used to resolve the individual tryptophan emission spectra of horse liver alcohol dehydrogenase. Figure 4 illustrates the technical emission spectrum of the enzyme and the resolution of the two spectral components by decay association (solid curves) and quenching association (symbols) with identical results. The same approach may be used to resolve fluorescence <u>excitation</u> spectra of a mixture or, for example, of different excitation spectral components in the emission of a protein.

Lakowicz and co-workers have modified a phase/modulation fluorometer to achieve phase-sensitive detection (Lakowicz and Cherek, 1981a). They have demonstrated that this instrumental approach may be used to resolve the spectral components in a mixture and have used it to study microheterogeneity in the fluorescence of proteins (Lakowicz and Cherek, 1981b; Lakowicz and Keating, 1983; Lakowicz *et al*., 1983a).

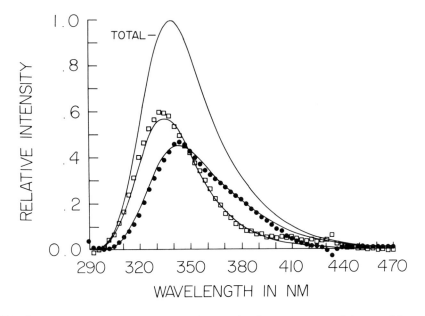

FIG. 4 Decay-associated tryptophan emission spectra of horse liver alcohol dehydrogenase (HLADH) obtained from TRES are shown as solid lines. A similar procedure, using linear combinations of spectra taken under known quenching conditions, may be used to obtain quenching DAS (QDAS). In this case, the iodide-accessible species corresponds to the longer-lived, redder spectrum of the pair. The inaccessible fraction QDAS superimposes a bluer, short-lived DAS.

Horse liver alcohol dehydrogenase/auramine O - a complex example.

Association of spectra with decay constants may aid the investigation of mechanisms underlying complex decay behaviour. In addition, association permits one to focus on the decay of a particular emitting species in a heterogeneous mixture. An example is provided by the studies of Knutson *et al.* (1983b). Previous work had suggested that auramine O binds at the active-site region of horse liver alcohol dehydrogenase, with a large enhancement of dye fluorescence (Conrad *et al.*, 1970; Heitz and Brand, 1971). The free dye shows negligible fluorescence by itself in aqueous solution. Binding of auramine O to the enzyme also quenches the intrinsic tryptophan fluorescence of the enzyme. Since there is good overlap between tryptophan emission and the absorption spectrum of the dye, it has been suggested that resonance energy transfer could account for this decrease in protein fluorescence.

The decay of the tryptophan fluorescence of horse liver alcohol

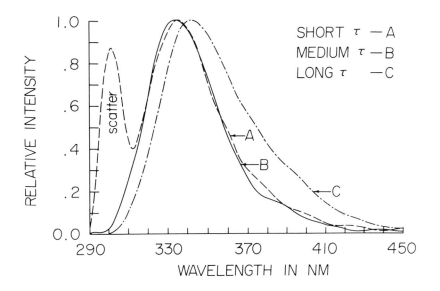

FIG. 5 DAS obtained for the three decay times found in HLADH complexes with the inhibitor auramine O. Addition of this ligand (in
this example, to about 24% occupation) results in the introduction of
a new, shorter-lived decay. The amplitude of this new species increases
with the saturation. The new, short-lived component DAS (curve A) is
found to correspond well with the medium lifetime DAS (previously
found inaccessible to iodide) shown as curve B. Using a minimal 3-
component unmixing matrix, one finds a zero-lifetime scattered light
component added to the subnanosecond species (curve A), while the
long-lived, iodide-accessible tryptophan emission (curve C) is left
unscathed.

dehydrogenase in the presence of non-saturating concentrations of

auramine O can be resolved into three exponential decay terms. A long

and medium decay constant are the same as those observed for alcohol

dehydrogenase in the absence of auramine O. In addition, a short decay

constant of about 0.5 ns is now observed. We questioned whether the

short decay constant has its origin in quenching of the "outer" or the

"inner" or both tryptophan residues.

Figure 5 illustrates the association of the three decay components

for a liver alcohol dehydrogenase-auramine O complex with three spec-

tral components. The spectrum associated with the short decay time is

quantitatively identical with that of the medium component (and also

contains a significant amount of scatter). Thus, the short decay compo-

nent appears to have its origin in quenching the medium decay species

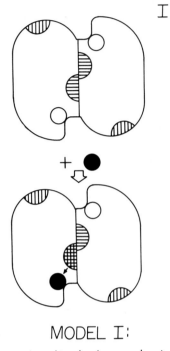

MODEL I:
subunits independent

MODEL II:
cross-subunit quench

FIG. 6 Schematic illustration of the auramine O - HLADH interaction. The first model illustrates a system of independent subunit quenching, i.e., addition of a single ligand acts to dynamically quench only <u>one</u> of the inner tryptophans in the dimer. Model II, in contrast, considers the case where a single ligand quenches both inner tryptophans to a subnanosecond lifetime. The changes (if any) in outer tryptophan intensity are excluded from these models, as they will not alter the distribution of inner tryptophan lifetimes.

which has been assigned to the "inner" tryptophan (Abdallah *et al.*, 1978; Ross *et al.*, 1981b). Horse liver alcohol dehydrogenase is a dimer made up of identical subunits, as is illustrated in Fig. 6. One may now ask: does complex formation only quench the tryptophan on a single subunit (model I), or do the inner tryptophan residues in both subunits respond to binding of the dye to a single subunit (model II)? If quenching occurs only on a single subunit, one expects a linear relation between saturation of sites and the fractional contribution of the medium decay to the sum of the short and the medium decay. In contrast, if binding to one subunit quenches both the inner tryptophan

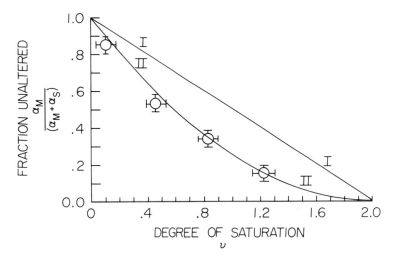

FIG. 7 The effect of auramine O binding to HLADH on the admixture of decays observed for inner (bluer DAS) tryptophan species. The fraction of pre-exponential amplitude that retains a 4 ns lifetime (expressed as the ratio to total of 4 ns-plus-subnanosecond amplitudes) is plotted vs. the degree of saturation (with two sites per dimer maximum). Curves I and II are those expected from the models shown in Fig. 6. Thus, curve I represents the fraction of empty independent subunits, while curve II traces the decline in population of empty dimers. The measured amplitude ratios are plotted as open symbols.

on the same subunit and that on the other subunit, the predicted result is the curvilinear relationship (II) shown in Fig. 7. The experimental results are most consistent with the latter situation. It is of interest to mention that, if this quenching is due to resonance energy transfer, one might expect different transfer efficiencies to the tryptophans on the same and on the other subunit. This would have led to at least four decay times rather than the three obtained. We were, however, unable to resolve the decay into four or more distinct components.

The question now remains, does the binding of auramine O have any effect on the fluorescence of the outer tryptophan residues? The answer is that it does. The fractional value of the <u>pre-exponential</u> term associated with the long ($\sim$7.4 ns) decay component decreases with saturation of sites by the dye. This suggests that static quenching of the outer tryptophan occurs, perhaps due to a conformational change. An alternative explanation is that very efficient energy

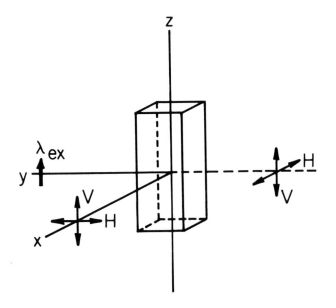

$$R(t) = D(t) / S(t)$$
$$\text{Where}: S(t) = G \cdot V(t) + 2H(t)$$
$$D(t) = G \cdot V(t) - H(t)$$

FIG. 8 Typical optical arrangement used to obtain experimental
vertically [V(t)] and horizontally [H(t)] polarized decay data. The
sample is excited using vertically polarized excitation and observation is at right angles. The time-resolved emission anisotropy R(t),
is related to the time-dependent intensities as shown, the factor G
correcting for any imbalance in the efficiency of detection of V- and
H- polarized emission. Upper case designations are used for intensities
and the anisotropy to indicate that these definitions apply to data,
usually convolved with an excitation function of non-negligible width.

transfer from this tryptophan residue occurs and the decay time of

this quenched species is too short to be resolved by the nanosecond

decay instrument used for these experiments.

The point to be gleaned from these examples is that association of decay terms with individual fluorescence spectra enables one to focus attention on a particular species in a heterogeneous mixture. This makes it possible to successfully address a number of questions that are otherwise unanswerable.

ANISOTROPY DECAY ASSOCIATED FLUORESCENCE SPECTRA (ADAS)

The definition of an experimentally measured fluorescence emission anisotropy decay curve $R(t)$ is given in Fig. 8, together with the optical arrangement typically used to obtain the experimental $V(t)$ and $H(t)$ decay data. In general, the latter represent the convolution products of the corresponding impulse responses $i_V(t)$ and $i_H(t)$ as described in Equation 8 [q.v.].

The impulse response of the emission anisotropy $r(t)$ for a spherical molecule will be described by a single exponential decay:

$$r(t) = \beta \exp[-(t/\phi)] \qquad (7)$$

where $\phi = \eta v/6kT$ is known as the rotational correlation time and $\beta = r_0$, the zero-point anisotropy [q.v.]. For more complex emission anisotropy decays, we will adopt β_j for the pre-exponentials and ϕ_j for the rotational correlation times to distinguish them from the α and τ used in discussing the decay of the fluorescence intensity.

The decay of the emission anisotropy reflects the rotational behaviour of fluorophores during the excited-state lifetime. It may be complex because of anisotropic rotational behaviour. Anisotropic molecules may rotate at different rates about different axes. In addition, the solvent environment may be anisotropic, as is the case with biological bilayer membranes. The anisotropic nature of the "solvent environment" will be reflected in the rotational behaviour of the probe. In the case of proteins, segmental flexibility of a portion of the macromolecule may influence the rotation of both intrinsic and extrinsic fluorophores (Nezlin *et al.*, 1970; Yguerabide *et al.*, 1970; Lakowicz and Weber, 1973a,b; Eftink and Ghiron, 1977; Munro *et al.*, 1979; Hanson *et al.*, 1981; Liv *et al.*, 1981; Reidler *et al.*, 1982; Lakowicz *et al.*, 1983b). Heterogeneity or microheterogeneity may also lead to complex decay of the emission anisotropy.

Anisotropy-decay analysis

Before discussing the advantages of associations as an aid in unravelling anisotropy decay, it may be useful to say a few words about analysis. When data are obtained by pulse fluorometry, the anisotropy decay will be distorted by convolution with the timing profile of the excitation source. Phillipe Wahl first suggested a simple procedure for achieving a deconvolution (Wahl, 1969) and this procedure has been used in our laboratory for a number of years (Badea and Brand, 1979). The $V(t)$ and the $H(t)$ decay curves are obtained as indicated in Fig. 8. The "sum" curve, $S(t)=G\,V(t)+2H(t)$ is computed, a "G"-factor being taken into account to compensate for non-ideality of the instrument (see, e.g., Chen and Bowman, 1965). These data are analysed with the aid of non-linear least squares reconvolution and fitting (Grinvald and Steinberg, 1974), using the appropriate weighting factors (Wahl, 1979, 1983). Other numerical methods such as the method of moments (Isenberg, 1973; Small and Isenberg, 1977a) or Laplace transforms (Gafni *et al.*, 1975; Ameloot and Hendrickx, 1983) may also be used. This provides a model and quantitative values for the parameters describing the decay of the <u>fluorescence</u> <u>intensity</u>. The convolved decay of the fluorescence anisotropy is defined as the ratio of a difference decay to the sum decay, $R(t)=D(t)/S(t)$. For the analysis, one computes the difference curve, $D(t)=GV(t)-H(t)$, again taking the G-factor into account. A second non-linear least squares analysis, again with the appropriate weighting, is now carried out to fit the difference curve. In this, the adjustable parameters are those describing the decay of the emission anisotropy, and the numerical values for the parameters describing the "sum" curve are entered as constants in the analysis. This numerical procedure has given good results in a wide variety of cases. It has the serious, but inherent, disadvantage that the final fit to obtain the anisotropy decay parameters is to the "difference" decay curve which usually has few counts compared with the sum curve. The Laplace transform method has also been adapted to difference curve analysis, at least for simple cases (Dixit *et al.*, 1982).

An alternative procedure is to obtain the anisotropy decay parameters directly from the $V(t)$ and $H(t)$ decay curves. The information content in either of these decay curves alone is usually not sufficient to obtain the required analysis. A procedure for the simultaneous analysis of the V(t) and the H(t) decay curves by means of non-linear least squares has been described by Gilbert (1983). The fitting functions for the V and H decay curves derived by convolution with the excitation pulse E(t) are:

$$V(t) = E(t) \otimes i_V(t)/G$$

$$H(t) = E(t) \otimes i_H(t) \tag{8}$$

where:

$$i_V(t) = (1/3)s(t)[1 + 2r(t)]$$

$$i_H(t) = (1/3)s(t)[1 - r(t)] \tag{9}$$

with:

$$s(t) = \sum_i \alpha_i \exp[-(t/\tau_i)] \tag{10}$$

and:

$$r(t) = \sum_j \beta_j \exp[-(t/\phi_j)] \tag{11}$$

Ameloot and co-workers have described a similar procedure using Laplace transforms (Ameloot and Hendrickx, 1983).

Beechem *et al.* (1984a,b) have developed a global non-linear least squares analysis procedure for use with multiple V(t) and H(t) data sets and applied it to results obtained under a variety of conditions in a rigorous quantitative comparison of various models for the anisotropy decay. For example, a system described by just two intensity decay times and two emission anisotropy decay times may be associative or non-associative (Dale *et al.*, 1977; Beechem *et al.*, 1984b) leading to different fitting functions. The global analysis method developed is quite general and may be used to compare the fits to such different models (Beechem *et al.*, 1984b). This point will be examined in more detail at the end of this contribution.

Rationale of ADAS

Consider a mixture of two fluorophores in solution with total fluorescence [GV(t)+2H(t)] decay courses which are not significantly different. Further, suppose that these molecules rotate isotropically at quite different rates. In such a case, association of spectra with total emission decay times, as described previously, is obviously not appropriate. In contrast, the difference spectra derived from polarized

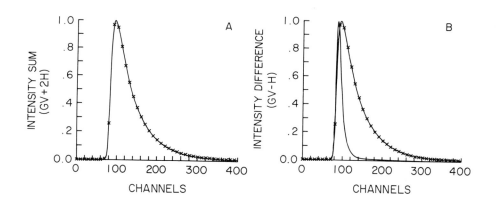

FIG. 9 For a mixture of two fluorophores in solution, with identical decay times but with distinctly different rotational rates, the total fluorescence intensity decay functions (panel A) are identical, while the difference decays (panel B) are significantly different. Using time-windowing, a "late" difference spectrum will reveal an emission associated with a relatively "immobile" probe fraction. The "early" difference spectrum will contain contributions from both components. Channel width = 0.205 ns.

emission intensities <u>will</u> show a time-dependent spectral distribution. The availability of different mixtures of the component spectra in these slices makes it possible to compute ADAS (anisotropy decay associated spectra). The impulse response difference spectrum is defined by both the rotational motions (the rotational correlation times, ϕ) and the total decay parameters of the fluorophore:

$$d(\lambda,t) = (\Sigma_i \alpha_i \exp[-(t/\tau_i)])\{f_1(\lambda)r_{01}\exp[-(t/\phi_1)] +$$
$$f_2(\lambda)r_{02}\exp[-(t/\phi_2)]\}$$

$$= f_1(\lambda)r_{01}\Sigma_i \alpha_i \exp[-t\{(1/\tau_i)+(1/\phi_1)\}] +$$
$$f_2(\lambda)r_{02}\Sigma_i \alpha_i \exp[-t\{(1/\tau_i)+(1/\phi_2)\}] \qquad (12)$$

where $f_1(\lambda)$ is the fraction of the initial intensity associated with species 1 at wavelength λ and $f_1(\lambda)+f_2(\lambda)=1$.

In a manner analogous to the extraction of DAS described above, vertically and horizontally polarized emission decay curves collected as a function of the emission wavelength are selectively integrated

using broad time-windows. Anisotropy and total intensity decay curves
are obtained at a particular emission wavelength. These are analysed
and the individual decay and rotational terms are reconvolved with
the lamp time profile to obtain convolved polarized decay curves V(t)
and H(t). Convolved difference decay curves, D(t)=GV(t)-H(t), are constructed, each of which corresponds to a given rotational correlation
time. Unlike the sum spectra, the difference spectra taken from the
early or late time-windows, will contain different contributions of
the differently rotating species (Fig. 9). The "late" difference spectrum reveals an emission dominated by the relatively "immobile" probe
fraction, since the more "mobile" probes will have rotated out of the
preferred observation direction and become orientationally randomized.
The "early" difference spectrum will contain contributions from both
components. If the rotational heterogeneity is less dramatic than in
the example given here, the "late" difference spectrum will still
contain spectral contributions due to both species. For such cases,
the formalism described above for intensity DAS, can be used to
"unmix" the anisotropy decay associated spectral terms (ADAS).

Experimental examples

An example of such behaviour is supplied by 2,6-TNS (2-toluidinonaph-
thalene-6-sulphonate) which does not show intense fluorescence in
aqueous solution, but fluoresces strongly when adsorbed to horse liver
alcohol dehydrogenase (a protein of MW 84,000) and when adsorbed to
cyclohepta-amylose (β-cyclodextrin), a much smaller molecule (Brand
et al., 1971; Seliskar and Brand, 1971). Figure 10 illustrates the
long anisotropy decay time ($\phi \sim 70$ ns) of the TNS-alcohol dehydrogenase complex as compared to the faster rotational decay observed
for the cyclodextrin complex with 2,6-TNS ($\phi \sim 1$ ns). The fluorescence
emission spectrum of TNS bound to cyclodextrin is red-shifted as compared to the emission spectrum of the dye bound to alcohol dehydrogenase. As expected, the emission spectrum of a mixture of the dye,
the protein, and cyclodextrin is intermediate between the two "pure"
systems. This is illustrated in Fig. 11. In a case where the emission
anisotropies differ markedly (as shown in Fig. 10), the "late" difference spectrum of the mixture reveals the "immobile" species corresponding to the enzyme-dye complex. The early difference TRES is

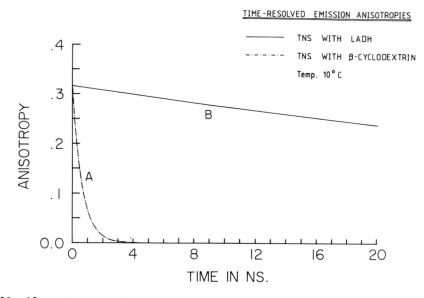

FIG. 10 Anisotropy decays of fluorophores adsorbed to small mobile bodies (curve A) as compared to the same dye bound to large, relatively immobile macromolecules (curve B). The extreme contrast in Brownian rotational rates shown here simplifies ADAS calculation but is not an obligatory requirement. The simplification occurs because the difference value D(t) for the mobile species is miniscule in a late time-window.

multiplied by a factor to normalize the area under the "immobile" difference curve in that window to that found in the late window. The normalized "early" minus "late" difference spectrum then corresponds to the spectrum of the more "mobile" cyclodextrin-dye complex. These results, also shown in Fig. 11, demonstrate that ADAS are able to successfully resolve mixtures of fluorescent dye molecules which are located within two different rotational environments.

   We have also applied the technique of ADAS to the study of microheterogeneity of location of the probe DPH (1,6-diphenyl-1,3,5-hexatriene) in single bilayer vesicles. Partitioning of probes within domains of a bilayer may lead to rotational microheterogeneity and fluorescence decay studies can provide information about this important phenomenon (Hui and Parsons, 1975; Klausner and Wolf, 1980; Klausner *et al.*, 1980). The interpretation of fluorescent-probe depolarization in bilayer membranes has recently been reviewed in considerable

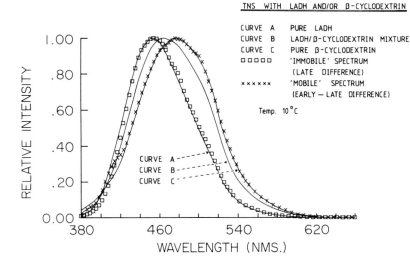

FIG. 11 Anisotropy decay associated spectra (ADAS) and steady state emission spectra for a rotationally heterogeneous system. Curve A is the steady state spectrum of TNS bound to HLADH, curve C is the corresponding spectrum for a complex with cyclodextrin (a much smaller molecule), and curve B is a spectrum observed for a mixture of these two species. This mixture was studied using polarized TRES to compute ADAS. The open squares represent the calculated "immobile" component, the crosses that of the more mobile species.

detail (Badley, 1976; Dale, 1983). The complex decay of the emission anisotropy observed in these systems could originate from either anisotropic rotations, with or without hindrance, or from microheterogeneity, or both. The popular probe DPH shows small, but significant spectral shifts in different solvents (Cehelnik *et al.*, 1975). Thus, it was not surprising to find that the fluorescence emission spectrum of DPH in dimyristoyllethicin (DML) single-bilayer vesicles above the main phase transition shows slight spectral shifts compared to DPH at the same temperature in dipalmitoyllecithin single-bilayer vesicles below their phase transition (Davenport *et al.*, 1982; Knutson *et al.*, 1982a). The rotational behaviour of DPH is quite different in these two systems, and the ADAS procedure described above was used to resolve these spectra in a mixture of the two vesicles containing DPH. It was further observed that DPH in DML single-bilayer vesicles at or near the main melting transition exhibited slight differences in the fluorescence spectra associated with the mobile and immobile probe

fractions. These experiments suggest that there is microheterogeneity
in the distribution of DPH even in "homogeneous" single-bilayer vesicles. Thus the concept of microheterogeneity must be taken into account
(in addition to the possibility of hindered rotations) when modelling
the complex anisotropy decays seen in bilayer vesicles. Similar spectral shifts have been observed by Chong and Weber (1983) when pressure
is used to induce transitions in vesicles.

ANISOTROPIC ROTATIONAL DIFFUSION

Experiments using a variety of physical methods have provided evidence
that some small molecules show anisotropic rotational behaviour even
in homogeneous solvents (Huntress, 1970; Vold *et al.*, 1977; Zinsli,
1977). Such behaviour may be greatly enhanced when the usual "sticking"
boundary condition of hydrodynamics does not apply and a "slipping"
boundary condition may be more appropriate (Hu and Zwanzig, 1974).
Mantulin and Weber (1977) have demonstrated, using differential phase
fluorometry, that unsubstituted aromatic hydrocarbons rotate considerably more rapidly in their own plane than they rotate out of plane.
Although "slipping" boundary conditions may apply to both in- and out-
of-plane rotations, the in-plane rotation will be that much faster
because it does not involve displacement of the solvent. They also
found that aromatics with substituent groups which hydrogen bond with
the solvent undergo essentially isotropic rotations.

For a rigid ellipsoidal body, the decay of the emission anisotropy
is described by a sum of five exponential terms (Belford *et al.*, 1972;
Chuang and Eisenthal, 1972; Ehrenberg and Rigler, 1972). In many cases
no more than three exponentials will be experimentally resolvable
(Small and Isenberg, 1977b). The decay of the emission anisotropy of
ellipsoids of revolution is exactly described by a sum of three exponential terms. The pre-exponentials are geometric factors which depend
on the relative orientation of the absorption and emission dipoles with
respect to the principal diffusion axes of the ellipsoid, and in general depend on the wavelength of excitation (Perrin, 1936; Witholt and
Brand, 1970; Shinitzky *et al.*, 1971; Lakowicz and Knutson, 1980). The
rotational correlation times depend on the principal diffusion coefficients of the ellipsoid (Weber, 1977; Zinsli, 1977; Lakowicz, 1983).

Anisotropic rotations of simple macromolecules

In the case of Brownian rotations of macromolecules, such as many proteins, that may be considered as ellipsoids of revolution in solvents such as water which essentially provide "sticking" hydrodynamic boundary conditions for their rotation, the principal diffusion coefficients and therefrom the rotational correlation times can be calculated from the volume and axial ratio of the macromolecule (including any possible solvation shell) along with the temperature and viscosity of the solvent (Perrin, 1934; see also Koenig, 1975). Explicitly, the time-dependence of the emission anisotropy of a fluorescent probe rigidly attached to such a macromolecule is given by:

$$r(t) = \beta_1 \exp[-t/\phi_1] + \beta_2 \exp[-t/\phi_2] + \beta_3 \exp[-t/\phi_3] \tag{13}$$

where (see Fig. 12):

$$\beta_1 = 0.3\sin2\theta_A \sin2\theta_E \cos\xi$$

$$\beta_2 = 0.3\sin^2\theta_A \sin^2\theta_E \cos2\xi$$

$$\beta_3 = 0.1(3\cos^2\theta_A - 1)(3\cos^2\theta_E - 1) \tag{14}$$

from which the zero-point emission anisotropy r_0 (in this idealized case equivalent to the fundamental anisotropy r_f) is given by:

$$r_0 = \beta_1 + \beta_2 + \beta_3 = 0.2(3\cos^2\omega - 1) \tag{15}$$

and:

$$\phi_1 = (5D_\perp + D_\parallel)^{-1}$$

$$\phi_2 = (2D_\perp + 4D_\parallel)^{-1}$$

$$\phi_3 = (6D_\perp)^{-1} \tag{16}$$

in which:

$$D_\perp = \frac{3\rho[(2\rho-1)S'-\rho]}{2(\rho^4-1)} D$$

$$D_\parallel = \frac{3\rho(\rho-S')}{2(\rho^2-1)} D \tag{17}$$

where D is the rotational diffusion coefficient for a sphere of the same volume and:

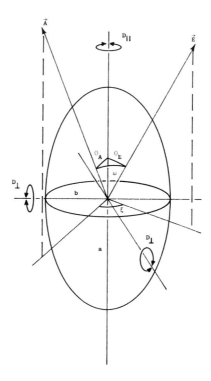

FIG. 12 Relationship of absorption and emission transition moment vectors $\vec{A}$ and $\vec{E}$ to the principal axes of a prolate ellipsoid of revolution (axial ratio $\rho = a/b > 1$) in whose coordinate frame they are rigidly fixed. For convenience of representation, the origin of the transition moments has been translated to coincide with that of the ellipsoid. Θ_A and Θ_E are the polar angles subtended by $\vec{A}$ and $\vec{E}$, respectively, about the unique axis of the ellipsoid, ω is the angle between them and ξ their azimuth in the perpendicular plane. $D_{\parallel}$ and $D_{\perp}$ are the diffusion coefficients for rotations of and about the unique axis and are defined in the text.

$$S' = \begin{cases} (\rho^2-1)^{-1/2} \ln [\rho + (\rho^2-1)^{1/2}] & \rho > 1 \\ \\ (1-\rho^2)^{-1/2} \arctan[(1-\rho^2)^{1/2}/\rho] & \rho < 1 \end{cases} \tag{18}$$

It is also noted that, using the cosine rule:

$$\cos\omega = \cos\theta_A \cos\theta_E + \sin\theta_A \sin\theta_E \cos\xi$$

β_1 and β_2 may be rewritten:

$$\beta_1 = 1.2\cos\theta_A\cos\theta_E(\cos\omega-\cos\theta_A\cos\theta_E)$$

$$\beta_2 = 0.3[2(\cos\omega-\cos\theta_A\cos\theta_E)^2 - \sin^2\theta_A\sin^2\theta_E] \qquad (19)$$

while β_3 depends only on the angles made by $\vec{A}$ and $\vec{E}$ with the unique axis of the ellipsoid of revolution and is independent of the angle between them, i.e. independent of r_0 which cannot in general, therefore, be factorized out of the β_j.

   Inspection of Equations 14 and 19 reveals that any or all of the β_j may take on negative signs (the angular ranges necessary to cover all possibilities are $0 \rightarrow \pi/2$ for any two of the three angles concerned, $0 \rightarrow \pi$ for the third). This predicates the observation under certain conditions of rather unusual and interesting anisotropy decay curves, particularly for prolate ellipsoids with a high axial ratio. Initial increases in the absolute value of the anisotropy before its decay to zero on complete randomization of the initially photoselected partially oriented excited-state population can be expected (Harvey and Cheung, 1972; 1977), initially positive anisotropies may become negative and vice versa, and two cycles of increase or decrease of the anisotropy as a function of time are possible. A number of more or less exotic cases are illustrated in Figs 13 and 14.

   It should be noted that experimental resolution of these effects will depend on the ratio of the fluorescence lifetime and the reduced correlation time $\phi_0 = (6D)^{-1}$ for the sphere of equivalent volume. Reasonable anisotropy data can be obtained only over about ten lifetimes. In general, the lifetime is likely to be on the order of 10 ns which is also the correlation time for fairly small spherical proteins such as apomyoglobin (Kinosita *et al.*, 1975) so that only part of the anisotropy decay curves presented will normally be available on the fluorescence time-scale, e.g. in an experimental example presented by Harvey and Cheung (1977), only the rising part of a positive anisotropy is resolved. The use of triplet probes, such as those discussed in other contributions to this volume, would extend this range. Both phosphorescence and delayed fluorescence anisotropies as well as the absorption anisotropy derived from transient dichroism measurements are described for freely rotating macromolecular ellipsoids of

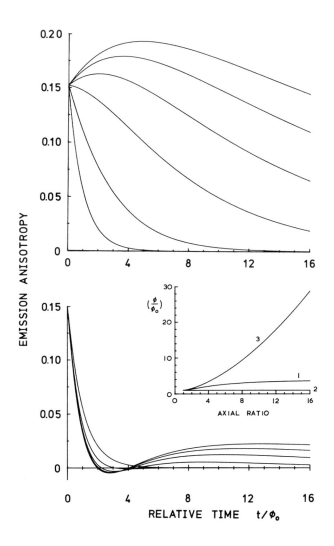

FIG. 13 Anisotropy decays for prolate ellipsoids of revolution of axial ratio $\rho = 1,3,6,9,12$ and 15 as a function of time relative to $\phi_0 = (6D)^{-1}$, the correlation time for a sphere of the same volume. *Upper*: $\theta_A = \theta_E = 20°$, $\xi = 180°$ ($\omega = \theta_A + \theta_E = 40°$ giving $\beta_1 \sim -0.124$, $\beta_2 \sim 0.004$, $\beta_3 \sim 0.272$ and $r_0 \sim 0.152$); ρ increasing from lower left to upper right. *Lower*: $\theta_A = \theta_E = 70°$, $\xi = 180°$ ($\omega = \theta_A + \theta_E = 140°$, the angle between $\vec{A}$ and $\vec{E}$ across the meridional plane, $\pi - \omega = 40°$, giving $\beta_1 \sim -0.124$, $\beta_2 \sim 0.234$, $\beta_3 \sim 0.042$ and $r_0 \sim 0.152$); ρ increasing from upper right to lower left near the origin. *Insert*: Correlation times ϕ_1, ϕ_2 and ϕ_3 relative to ϕ_0, as a function of the axial ratio ρ. $\phi_2/\phi_0 \sim 1$ for all ρ, while the limits of ϕ_1/ϕ_0 and ϕ_2/ϕ_0 for infinite ρ are 4 and 1, respectively, since their reciprocals are dominated by $4D_\parallel$ and $D_\parallel$, and $D_\parallel/D$ has a limit of 3/2 as ρ becomes infinite (see Eqs 16 to 18).

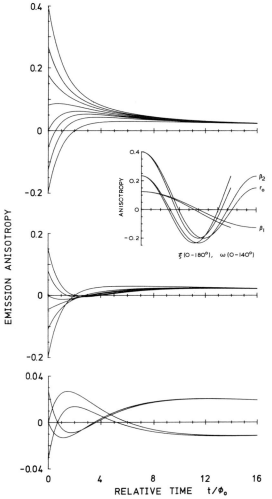

FIG. 14 Anisotropy decays for a prolate ellipsoid of revolution of axial ratio $\rho=15$. *Upper*: $\theta_A=\theta_E=70°$ with, in descending order, $\xi=\omega=0°$ ($r_0=0.4$), $\xi=30°$, $\xi=40°$, $\xi=50°$, $\omega=\arccos(\sqrt{1/3})\backsim54.7°$ (the "magic" angle), $\xi=65°$, $\xi=75°$ and $\omega=90°$ ($r_0=-0.2$). *Middle*: $\theta_A=\theta_E=70°$ with, in ascending order, $\omega=90°$ ($r_0=-2.0$), $\xi=125°$, $\xi=135°$, $\omega=\pi-\arccos(\sqrt{1/3})\backsim$ $125.3°$ (angle between $\vec{A}$ and $\vec{E}$ across the meridional plane "magic"), $\xi=155°$ and $\xi=180°$ ($\omega=\theta_A+\theta_E=140°$; $r_0\backsim0.152$). *Lower*: initially *decreasing* curves - $(\theta_A,\theta_E)=(65°,75°)$ with $\omega=\pi-\arccos(\sqrt{1/3})\backsim125.3°$ ["magic" angle condition, $r_0=0$] and $\omega=128°$ [just off "magic" condition, $r_0\backsim0.027$]; initially *increasing* curves - $(\theta_A,\theta_E)=(45°,65°)$ with $\omega=\arccos(\sqrt{1/3})\backsim54.7°$ ["magic" angle; $r_0=0$] and $\omega=58°$ [just off "magic"; $r_0\backsim-0.032$]. *Insert*: zero point and partial anisotropies for $\theta_A=\theta_E=70°$ as functions of the azimuth ξ and the angle ω ($0\leq\omega\leq\theta_A+\theta_E$); $\beta_3\backsim0.042$ is independent of ξ and ω. Values of $\omega>90°$ correspond to angles of $\pi-\omega$ between $\vec{A}$ and $\vec{E}$ across the meridional plane.

revolution carrying a rigidly oriented probe by the same formalism as in the fluorescence case. Transient dichroism measurements are usually made using the same wavelength for bleaching and for monitoring recovery, in which case one has the equivalent of $\theta_A = \theta_E$, $\omega = 0$ and all terms in the anisotropy decay are positive. If different wavelengths are used, however, such that different transitions, say $\vec{A}$ in bleaching and $\vec{E}$ in monitoring, are employed then, again, the fluorescence anisotropy decay formalism will apply. Finally, if $D_\perp$ in Equations 16 is set equal to zero, the formalism for one-dimensional rotation, such as is effectively the case for proteins embedded in membranes and suited for examination by triplet probe techniques as discussed elsewhere in this volume, is reproduced, and similar possibilities for exotic anisotropy decay behaviour apply.

Free fluorescent probes

The resolution of such effects and their utilization in the accurate quantitation of diffusion coefficients should benefit by <u>overdetermination</u> of the system via the use of more than one excitation wavelength which will elicit changes in the pre-exponential factors β_j without effect on the correlation times ϕ_j. That this is indeed the case has been demonstrated for the free, plate-like aromatic fluorophore perylene (Barkley *et al.*, 1981a,b). It turns out that the "in-plane" rotations of this molecule are very much faster than the "out-of-plane" ones: $D_{ip}(\equiv D_\parallel) \gg D_{op}(\equiv D_\perp)$ (Weber, 1973; Zinsli, 1977; Barkley *et al.*, 1981a,b). The anisotropy decay behaviour then mimics that of a highly elongated rod or prolate ellipsoid of revolution with the transition moments lying in the equatorial plane, and cannot be described by the "sticking" boundary condition formalism of Perrin for an oblate ellipsoid of revolution, since even in the limit of an infinitely thin "plate", $D_\parallel$ and $D_\perp$ do not differ by more than about 20%.

Perylene has two well-defined absorption transitions, both lying in the molecular plane. One, parallel to the emission transition, is in the visible region of the spectrum. The other, perpendicular to the first, is in the ultraviolet region, so that excitation of perylene at 260 nm can give rise to negative values of the fluorescence emission anisotropy: the nanosecond time-resolved emission anisotropy starts at a negative value, rises with time, passes through zero to become

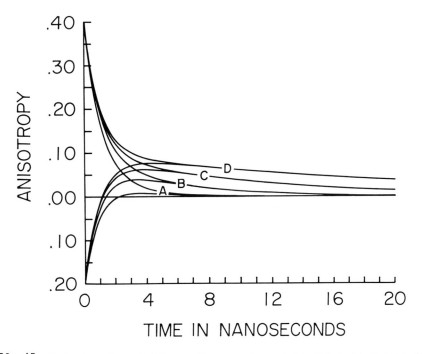

FIG. 15 Anisotropic rotations of a perylene-like molecule result in unusual emission anisotropy decay curves. Pairs of curves from $S_0 \to S_1$ (+) and $S_0 \to S_2$ (-) absorption bands both exhibit fast decay from in-plane motion, while the average of these curves would reveal only the out-of-plane rotations. In this figure, the first rotational correlation time $\phi_1 = (4D_\parallel + 2D_\perp)^{-1}$ is held constant at 1 ns while the second $\phi_2 = (6D_\perp)^{-1}$ increases from 2 ns (curve A) through 5 ns (curve B), 10 ns (curve C) to 20 ns (curve D). Note that the primary change is in the +/- curve average.

positive, reaches a maximum and finally decays towards zero. The simulation of such a decay of the emission anisotropy in an isotropic solvent, according to Equation 20 [q.v.], is shown in Fig. 15. This result implies that the relative magnitudes of the vertically and horizontally polarized components of the fluorescence decay invert with time (Witholt and Brand, 1970; Barkley *et al.*, 1981a). With excitation into a "negative" transition, the initial horizontal intensity is greater than the vertical intensity and so the emission anisotropy is negative. At early times during the decay, the rapid in-plane rotation of the ring moves the emission dipoles on average toward the vertical direction and the anisotropy becomes positive. The slower out-of-plane rotation depolarizes the fluorescence emission at later times. The

steady-state emission anisotropy is the integrated area under the time-
resolved anisotropy weighted by the intensity decay. Thus, the result
shown in Fig. 15 implies that a Perrin plot of $1/r$ vs. T/η (Lakowicz,
1983) is singular, i.e. $1/r$ should decrease towards an infinite negative value at some T/η, reappear from an infinite positive value and
then decrease, as predicted from the theory given by Perrin (1936) -
see also Memming (1961) and Withold and Brand (1970).

Fluorescent probes in anisotropic environments

In contrast with the behaviour in isotropic solvents, consider the
anisotropic rotational motions of disc-shaped molecules in asymmetric
or restricted environments, such as might be expected for the interior
of a bilayer membrane. In these situations, the decay of the emission
anisotropy is expected to be quite complex. Fortunately, the rotating
molecule and its surroundings may each exhibit a degree of symmetry
that can simplify matters. For example, when a probe has D_{4h} or
higher planar symmetries, the oscillators are effectively averaged in-
plane, and only out-of-plane (tumbling) motion will be seen in an
anisotropy experiment. Simplification of the decay functions also
occurs when local order of the environment is symmetric. The general
treatment of these systems and various simplifications have been
discussed by Kinosita *et al.* (1977), Lipari and Szabo (1980) and by
Zannoni and co-workers (Zannoni, 1981; Zannoni *et al.*, 1983).

We have used coronene, a disc-shaped molecule with D_{6h} symmetry,
as a fluorescent probe of lipid bilayer structure (Davenport *et al.*,
1983). The polarized emission intensities from such a highly symmetrical molecule will depend only upon out-of-plane motions and have a
zero-point anisotropy (r_0) of 0.1. However, for coronene, the measured
fluorescence lifetime is quite long ($\sim$250 ns) when compared to the
out-of-plane rotational correlation times (ϕ_{op}) observed for similar
molecules, e.g. perylene. In this particular case, the steady-state
emission anisotropy will arise effectively only from residual out-of-
plane anisotropy $(r_{\infty op})$, as can be confirmed using time-resolved
depolarized measurements. The steady-state anisotropy vs. temperature
plots for coronene incorporated into both DML and DPL vesicles (Fig.
16), show that $r_{\infty op}$ is changing across a 20° temperature span that

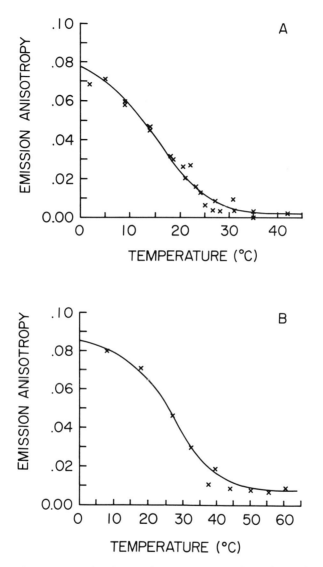

FIG. 16 Steady-state emission anisotropy as a function of temperature for coronene-labelled (A) DML and (B) DPL single-bilayer vesicles. The excitation wavelength was 337 nm, with observation at 448 nm and excitation and emission bandwidths of 4 nm and 6 nm respectively. The gel-liquid crystalline phase transition temperatures measured by differential scanning calorimetry or using the short-lived fluorescence probe DPH fall at about 25 and 40 °C respectively for DML and DPL.

precedes the main transition reported for these phospholipids. It may
be that the motions of coronene reveal slow, subtle structural changes,
such as phospholipid domain melting and reformation taking place on a
similar time-scale to that of emission, occurring well below the complete gel to liquid-crystal transition. Then if, as would be expected,
ϕ_{op} is short (<~10 ns) in fluid domains, and extremely long (>~300 ns)
in gel domains, after about 100 ns, only probes whose <u>entire</u> excited-
state history lies in the gel fraction will effectively contribute anisotropy in the signal (see Equation 12). In the simplest case, the
melting rate constant will be seen as an "apparent" out-of-plane rotational diffusion coefficient $D_{op}=(1/6)\phi_{op}^{-1}$. Long, rather than infinite,
ϕ_{op} values may arise from bulk rotation of the vesicle shell and/or
translational diffusion of the probe over the spherical surface (Dale
et al., 1977). The application of long-lived fluorophores to study
these phenomena in lipid systems is under continuing investigation.

The depolarization arising from rotation in an isotropic environment
of a molecule which behaves hydrodynamically as an ellipsoid of revolution (or other solid of revolution, e.g. rod, disc) and has both its
absorption and emission transition moments perpendicular to the symmetry axis, such as perylene or pyrene, is given by:

$$r(t) = 0.3(2f-1)\exp[-(4D_{\parallel}+2D_{\perp})t] + 0.1\ \exp[-6D_{\perp}t] \qquad (20)$$

where $D_{\parallel}$ and $D_{\perp}$ are the "in-plane" (spinning) and "out-of-plane" (tumbling, wobbling) rotational diffusion coefficients respectively, and
f is the fractional $S_0 \to S_1$ absorption, (1-f) that of $S_0 \to S_2$ when their
transition moments are parellel and perpendicular, respectively, to
that of emission. If tumbling does not occur ($D_{\perp} \to 0$), the long-time
limit of the emission anisotropy, r_∞ is determined by the second term
in Equation 20, i.e. $r_\infty=0.1$. The simulated anisotropy decay curves in
Fig. 15 show the effect of changes in the rate of tumbling for excitation into both the positive (f=1) and the negative (f=0) transitions.
The appearance of the anisotropy decay curves is altered in a very
different way if the rate of tumbling is kept the same, but the rate
of spinning, $D_{\parallel}$, is changed. Figure 17 shows a family of such curves.
The average of the positively and negatively polarized curves gives
the tumbling part of the decay exclusively. In the case of a hindered

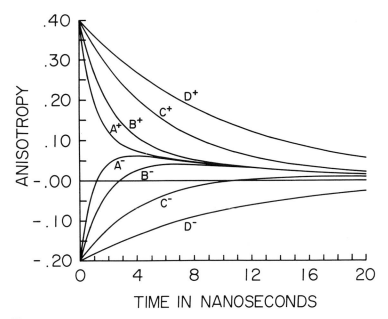

FIG. 17 In this case, unlike that depicted in Fig. 15, the "tumbling" rate ϕ_2 is held constant at 10 ns while ϕ_1 (dominated by the in-plane "spinning" rate) is raised from a perylene-like value of 1 ns towards more symmetric values ($\phi_1 = \phi_2$ for a sphere). The ϕ_1 values are 1 ns (curve A), 2 ns (curve B), 5 ns (curve C) and 10 ns (curve D). Note the absence of any change in the +/- average.

environment, the same kind of averaging can be used to visualize tumbling as opposed to spinning restrictions.

The model of Lipari and Szabo (1980) applied to rotation of such molecules in an asymmetric environment allows for restriction on the tumbling motion. Examples of this restriction are shown in Fig. 18. Note that $r_\infty = 0.1 S^2$ is independent of the absorption transition excited and is always positive (S is the order parameter about a local director coincident with the axis of symmetry that is tumbling, see e.g. Heyn, 1979; Kooyman et $al.$, 1983; Zannoni et $al.$, 1983). As mentioned above, total restriction of tumbling motion leads to $r_\infty = 0.1$.

Early work by Wahl has provided a model for rotational motion, where wobble is totally restricted ($D_\perp = 0$) and in-plane rotation is partially restricted (Wahl, 1975, 1983). Figure 19 shows simulated emission anisotropy decays according to this model. Here $r_\infty = 0.1$ when in-plane motion is complete since wobble is completely restricted. If the in-plane motion is also restricted, $r_\infty = 0.1 \pm \Delta$, with $0 \leq \Delta \leq 0.3$,

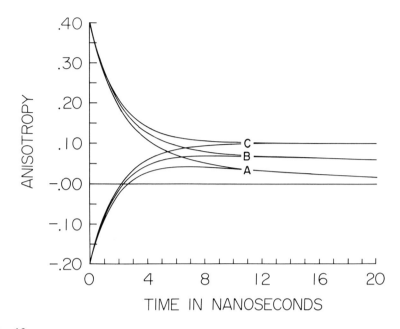

FIG. 18 In this figure, a slow initially isotropic out-of-plane "tumbling" (ϕ_2=10 ns) is restricted in such a way that r_∞=0.1S^2, while fast in-plane motions are left constant (ϕ_1=2 ns). The S^2 values are 0 (curve A), 0.5 (curve B) and 1.0 (curve C). Note that restriction of the tumbling motion alone leads to a +/- average value equal to r_∞ at long times.

providing a model in which a negative r_∞ is expected for excitation into a negative transition.

A more recent model developed by Attila Szabo (personal communication) permits restriction both of, and around, the axis through the plane of a disc (the C_∞ axis), with r_∞=0.1S^2±0.3$S^2\sigma^2$, where σ^2 is a "transverse" order parameter for in-plane restrictions. As mixing of the previous models intuitively suggests, this also leads to the possibility of a negative value for r_∞.

The origins of r_∞ may be found in inhomogeneous ordering potentials (restrictions on rotational motion). At the same time, it must be kept in mind that a non-zero r_∞ may also have its origin in hetero- geneity between the environments of the probe in the bilayer membrane. The simplest example of this situation is the case of an <u>immobile</u> <u>subfraction</u> of probes for which r_∞=r_0. As discussed above, evidence

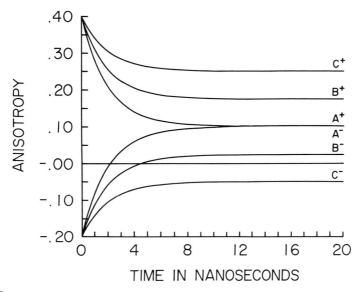

FIG. 19 In this simulation, no tumbling at all is permitted, but the remaining in-plane mobility may be restricted. The greater the restriction, the more r_∞ will diverge for the + and - curves. In the examples shown, ϕ_1 is held approximately constant (ϕ_2 is infinite). r_∞ takes the form $(0.1\pm\Delta)$ with Δ set at 0 (curve A), 0.25 (curve B) and 0.5 (curve C). Note how the in-plane restriction separates the + and - curves without altering the average of 0.1.

for probe rotational heterogeneity can be obtained with the aid of anisotropy-decay-associated excitation or emission spectroscopy.

In Fig. 20, the steady-state emission anisotropy of perylene in DML (dimyristoyllecithin) single bilayer vesicles as a function of temperature (Kowalczyk *et al.*, 1982) is displayed. Data were obtained with excitation into both a positively and a negatively polarized transition. Data for excitation in a spectral region where $r_0=0.1$ (314 nm) is also shown. As indicated above, excitation at 314 nm will yield anisotropy values which depend only on out-of-plane mobility. These results suggest that, below the phase transition, there is little change in out-of-plane motion or restriction. Through and above the phase transition, there <u>does</u> appear to be a significant change in out-of-plane motion and restriction. In contrast, the changes in emission anisotropy with temperature observed at excitation wavelengths of 400 nm and 258 nm suggest that there is a significant moderation of in-plane rotation below the phase transition.

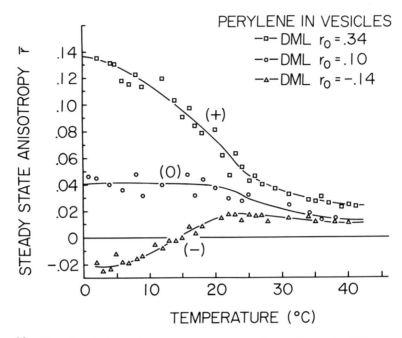

FIG. 20 Steady-state emission anisotropy of perylene in DML single
bilayer vesicles. The probe was excited in the positive band
($r_0 = 0.34$), in the negative band ($r_0 = -0.14$), and at a wavelength where
equal $S_0 \rightarrow S_1$ and $S_0 \rightarrow S_2$ contributions occur ($r_0 = 0.1$). The latter anisotropies are less precise, due to weak absorption at ~314 nm, but are
(importantly) sensitive to out-of-plane motions only. Comparison of
these curves gives a qualitative idea of the in-plane vs. out-of-
plane average mobility, especially as regards the different responses
to the lipid phase transition. The out-of-plane motion apparently
becomes less active on cooling through 24 °C (the transition temperature), but remains almost constant on further cooling. In contrast,
the in-plane rotations of perylene among the hydrophobic tails become
progressively more restricted with further cooling. Thus, the in-plane
mobility responds to structural changes at lower temperatures than
does the out-of-plane mobility.

Figure 21 shows the time-dependence of the emission anisotropy of

perylene in DML single bilayer vesicles at 3, 10 and 24°C for excita-

tion into both the positive and negative transitions. There is a large

difference in r(t) at late times between the positive and negative

curves. At 3°C, the data indicate the presence of a small but signi-

ficant negative r_∞. It therefore appears that perylene may experience

restrictions to its in-plane as well as its tumbling motion. However,

the possibility of alternative or parallel influences of microhetero-

geneity must also be considered.

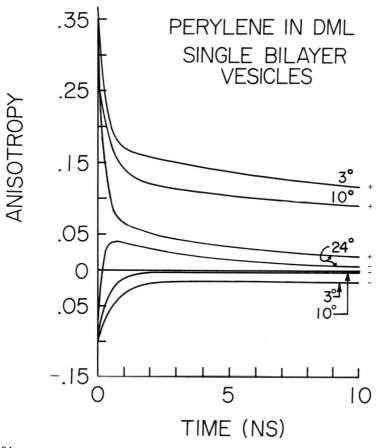

FIG. 21 Decays of emission anisotropy for perylene embedded in the hydrophobic region of small unilamellar DML vesicles. At 24°C, the motion appears to be only slightly hindered, with an in-plane rotational rate much faster than that out-of-plane [$(\phi_2/\phi_1)>20$]. As the temperature is decreased, both out-of-plane restriction (average r_∞) and in-plane hindrance (differences in r_∞ for + and - bands) are seen to increase. At 3°C, a small but significantly negative r_∞ is found for excitation into the negative band. Thus, both in-plane and out-of-plane restrictions must be present. The data is not yet sufficiently precise to rule out a heterogeneous origin for these restrictions.

While the details of the rotational motion of perylene in bilayers remain to be fully elucidated, it appears that anisotropic rotors may be of particular value in helping to define the anisotropic nature of bilayer membranes. Moreover, it should not come as a surprise if the course of steady-state emission anisotropy "melting-curves" taken at a particular excitation wavelength does not appear to agree with those obtained at other wavelengths or, for example, with gel to liquid-crystalline transition curves obtained by other methods.

ASSOCIATIONS BETWEEN INTENSITY DECAY AND ANISOTROPY DECAY

So far, only associations between spectral distributions and either intensity or anisotropy decay parameters have been discussed. Associations between intensity and anisotropy decay parameters, independent of spectral distribution, also make up an important class of correlations. Consider first the situation of a homogeneous distribution of probe molecules. It might exhibit biexponential intensity decay kinetics whose origin is in an excited-state reaction, as discussed earlier. It might also exhibit a biexponential decay of the emission anisotropy due to anisotropic rotations of the type described above for perylene. The decay of the impulse responses of the polarized components of emission in this system would then be described by:

$$i_V(t) = (1/3) \sum_{i=1}^{2} \alpha_i \exp[-(t/\tau_i)]$$

$$+ (2/3) \sum_{i=1}^{2} \alpha_i \exp[-(t/\tau_i)] \sum_{j=1}^{2} \beta_j \exp[-(t/\phi_j)]$$

$$i_H(t) = (1/3) \sum_{i=1}^{2} \alpha_i \exp[-(t/\tau_i)]$$

$$- (1/3) \sum_{i=1}^{2} \alpha_i \exp[-(t/\tau_i)] \sum_{j=1}^{2} \beta_j \exp[-(t/\phi_j)] \tag{21}$$

and the impulse response emission anisotropy decay reduces simply to:

$$r(t) = [i_V(t) - i_H(t)]/[i_V(t) + 2i_H(t)] = \sum_{j=1}^{2} \beta_j \exp[-(t/\phi_j)] \tag{22}$$

i.e. the "original" biexponential time-course.

In contrast, consider a situation where a probe exhibits biexponential decay of both fluorescence intensity and emission anisotropy because the probe is located in two microenvironments, e.g. in a membrane. In this case:

$$i_V(t) = (1/3) \sum_{i=1}^{2} \alpha_i \exp[-(t/\tau_i)] + (2/3) \sum_{i=1}^{2} \alpha_i \exp[-(t/\tau_i)] \beta_i \exp[-(t/\phi_i)]$$

$$i_H(t) = (1/3) \sum_{i=1}^{2} \alpha_i \exp[-(t/\tau_i)] - (1/3) \sum_{i=1}^{2} \alpha_i \exp[-(t/\tau_i)] \beta_i \exp[-(t/\phi_i)]$$

$$(23)$$

and the impulse response anisotropy decay is non-exponential:

$$r(t) = \sum_{i=1}^{2} \alpha_i \beta_i \exp[-t\{(1/\tau_i)+(1/\phi_i)\}] / \sum_{i=1}^{2} \alpha_i \exp[-(t/\tau_i)] \qquad (24)$$

The previous case may be referred to as non-associative (or homogeneous), while the latter is associative (or heterogeneous), since each intensity decay time is selectively associated with a specific rotational behaviour, here with one particular correlation time. The decay curves described by Equations 21 and 23 may differ sufficiently to be distinguished by a good analysis procedure.

    In general, for n intensity and m anisotropy decay components:

$$i_V(t) = (1/3) \sum_{i=1}^{n} \alpha_i \exp[-(t/\tau_i)]$$

$$+ (2/3) \sum_{i=1}^{n} \alpha_i \exp[-(t/\tau_i)] \sum_{j=1}^{m} \beta_j \exp[-(t/\phi_j)] L_{ij}$$

$$i_H(t) = (1/3) \sum_{i=1}^{n} \alpha_i \exp[-(t/\tau_i)]$$

$$- (1/3) \sum_{i=1}^{n} \alpha_i \exp[-(t/\tau_i)] \sum_{j=1}^{m} \beta_j \exp[-(t/\phi_j)] L_{ij} \qquad (25)$$

leading to:

$$r(t) = \sum_{i=1}^{n} \sum_{j=1}^{m} \alpha_i \beta_j \exp[-t\{(1/\tau_i)+(1/\phi_j)\}] L_{ij} / \sum_{i=1}^{n} \alpha_i \exp[-(t/\tau_i)] \qquad (26)$$

where L_{ij} are the terms of an association matrix. These are equal to unity if the intensity decay component i contributes light to the anisotropy decay component j, and are zero otherwise. In the examples given above, then, n = m = 2 and L_{ij} = 1 for all [i,j] in the case described by Equation 22., whereas $L_{ij} = 1$ if $i = j$ and 0 if $i \neq j$ in Equation 24.

As a simple comparative example, consider the model heterogeneous system:

$$i_1(t) = 1.5\exp[-(t/4)] , \qquad r_1(t) = -0.2\exp[-t/5)]$$

$$i_2(t) = 0.7\exp[-(t/15)] , \qquad r_2(t) = 0.35\exp[-(t/11)] \qquad (27)$$

A convolved emission anisotropy decay curve for this model is shown in Fig. 22 together with the convolved decay for the non-associative (but in this case also non-homogeneous with respect to rotation) analogue involving the same basic decay terms:

$$i(t) = 1.5\exp[-(t/4)] + 0.7\exp[-(t/15)]$$

$$r(t) = -0.2\exp[-(t/5)] + 0.35\exp[-(t/11)] \qquad (28)$$

There is a significant difference in the shape of the curves at "early" times. For additional figures showing associative-type effects in real data, see Wolber and Hudson (1982) and Ludescher (1984).

The results of analyses of the simulated <u>associative</u> data using an algorithm described below are given in Table 1. Both associative and non-associative analysis yielded accurate recovery of the total intensity decay parameters. Only the associative analysis returns the correct anisotropy decay parameters. Although the fits for the associative and non-associative cases could not be differentiated on a strictly statistical basis, the non-associative model returned a value for one of the parameters of the anisotropy decay that was

TABLE 1 Analyses of convolved data simulated using the model heterogeneous system described by Equation 27.

ANALYSIS	a_1	τ_1	a_2	τ_2	β_1	ϕ_1	β_2	ϕ_2	χ^2
Associative	1.49	3.98	.70	14.95	-.195	4.5	.33	11.4	1.2
Non-associative	1.49	3.96	.70	14.92	-.36	4.9	.33	12.0	1.3

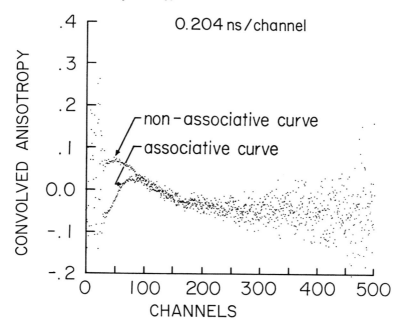

FIG. 22 Associative and non-associative convolved anisotropies for
emission components 1.5exp[-(t/4ns)] and 0.7exp[-(t/15ns)] with
anisotropy decay components -0.2exp[-(t/5ns)] and 0.35exp[-(t/11ns)].
The intensity of the short lifetime component (4 ns) is associated
with the rotational correlation time of 5 ns in the associative model.
Therefore "early" emission, which predominantly comes from the 4 ns
component is negatively polarized. As its intensity component decays,
the anisotropy increases with time reflecting the change in the
distribution of light to the 15 ns component associated with a
rotational correlation time of 11 ns.

"non-physical": the pre-exponential anisotropy decay term corresponding
to the short correlation time was, at -0.36, well outside the range of
-0.2 to +0.4 consistent with physical reality. Similar effects have
been noted with real data, but in the opposite direction, i.e. physically unrealistic values of both pre-exponential factors and correlation times were obtained in associative analyses (Dale *et al.*, 1977).

GLOBAL ANALYSIS

The theme of this contribution has been that the characteristics of
the intensity decay and the emission anisotropy decay can be quite
complex. Thus, associations between measurable quantities and fluorescence decay parameters obtained under differing conditions (i.e.

by varying excitation or emission wavelengths) can be helpful in
interpreting the complexity. Although it has been shown in other
areas of biophysics that it is advantageous to analyse multiple data
sets together when they have one or more parameters in common
(Edelstein and Schachman, 1967; Ackers *et al.*, 1975; Johnson *et al.*,
1976; Johnson *et al.*, 1981; Sheiner and Beal, 1981; Turner *et al.*,
1981; Nagle *et al.*, 1982; Laue *et al.*, 1984), it has usually been the
custom in the fluorescence field to analyse decay curves one at a
time. Subsequent analysis, usually graphical, of decay constants has
been used to provide mechanistic information about the system. In
many cases this does not allow sufficient resolution to unravel systems of the type discussed here. Eisenfeld has pointed out the theoretical advantages of a vectorial analysis approach in moment transform space (Eisenfeld and Ford, 1979; Eisenfeld, 1983). The advantages of simultaneous non-linear least squares analysis of V(t) and
H(t) decay curves to obtain the parameters of the time-dependent
emission anisotropy have also been indicated (Gilbert, 1983).

Beechem *et al.* (1983a,b; 1984a,b) and Knutson *et al.* (1983c) have
described the use of a non-linear least squares technique for "global"
analysis of fluorescence decay and of the decay of the emission
anisotropy. In normal non-linear least squares analysis, one decay
curve is analysed at a time and all the parameters are optimized. If
additional information is available, the values of some parameters
may be fixed, aiding in the recovery of the others. In "global"
analysis, on the other hand, several decay curves are fitted simultaneously and, while values of parameters are not necessarily fixed,
the values of any of the parameters may be "linked" between different
decay curves. Up to 150 decay curves with 250 independent parameters
have been analysed for an internally consistent set of decay parameters. For this purpose, a matrix-mapping subroutine (Knutson *et
al.*, 1983c) added to a standard non-linear least squares programme,
allows the analysis of <u>decay</u> <u>surfaces</u> instead of sequential two-
dimensional slices of experimental data. We have found that this procedure greatly decreases the amount of covariance between parameters
and increases the ability to recover closely-spaced decays.

With the aid of global analyses, the ability to discriminate between different models and to recover quantitative values for decay parameters is enhanced. This approach to data analysis leads to a reasoned design of experiments, wherein multiple decay curves are obtained under selected conditions. With simultaneous analysis it should prove feasible in many cases to discriminate between the various models for excited-state reactions, heterogeneity and anisotropic rotations discussed in this contribution.

The nature of luminescence decay is multidimensional. This allows possibilities for overdetermination and system-wide modelling that should not be ignored. The exploitation of these features will be facilitated as numerical procedures for analysis of multidimensional data surfaces become more widely available.

ACKNOWLEDGEMENTS

We thank Mary D. Barkley and J.B.A. Ross for helpful discussions. They also participated in some of the studies cited in this review. We thank Nancy Beechem for outstanding help with the art work and Ms Shermina and Ms Ward for assistance with typing. The work discussed here was supported by N.I.H. grant No. GM 11632.

REFERENCES

Abdallah, M.A., Biellman, J.F., Wiget, P., Joppich-Khun, R. and Luisi, P.L. (1978). *Eur. J. Biochem.* 89, 397.

Ackers, G.K., Johnson, M.L., Mills, F.C., Halverson, H.R. and Shapiro, S. (1975). *Biochemistry* 14, 5128.

Ameloot, M. and Hendrickx, H. (1983). *Biophys. J.* 44, 27.

Badea, M.G. and Brand, L. (1979). *Methods in Enzymology* 61H, 378.

Badley, R.A. (1976). *In* "Modern Fluorescence Spectroscopy" (Ed. E.L. Wehry) Vol. 2, p. 91.

Barkley, M.D., Kowalczyk, A.A. and Brand, L. (1981a). *J. Chem. Phys.* 75, 3581.

Barkley, M.D., Kowalczyk, A.A. and Brand, L. (1981b). *In* "Biomolecular Stereodynamics" (Ed. R.H. Sarma) Vol 1, p. 391, Adenine Press, New York.

Beddard, G.S. and Westby, M.J.(1981). *J. Chem. Phys.* 57, 121.

Beechem, J.M., Knutson, J.R., Ross, J.B.A., Turner, B.W. and Brand, L. (1983a). *Biochemistry* 22, 6054.

Beechem, J.M., Knutson, J.R. and Brand, L. (1983b). *Photochem. Photobiol.* Abstr. S20.

Beechem, J.M., Knutson, J.R. and Brand, L. (1984a). *Biophys. J.* 45, 127A.

Beechem, J.M., Knutson, J.R. and Brand, L. (1984b). *Photochem. Photobiol.* In press.

Belford, G.C., Belford, R.L. and Weber, G. (1972). *Proc. Natl Acad. Sci. USA* 69, 1392.

Birks, J.B. (1970)."Photophysics of Aromatic Molecules". Wiley, New York.

Brand, L. and Gohlke, J.R. (1971). *J. Biol. Chem.* 17, 2317.

Brand, L., Seliskar, C.J. and Turner, D.C. (1971). *In* "Probes of Structure and Function of Macromolecules and Membranes, 1, Probes and Membrane Function" (Ed. B. Chance) p. 17. Academic Press, New York.

Brochon, J.C., Wahl, Ph., Charlier, M., Maurizot, J.C. and Hélène, C. (1977). *Biochem. Biophys. Res. Commun.* 79, 1261.

Cehelnik, E.D., Cundall, R.B., Lockwood, J.R. and Palmer, T.F. (1975). *J. Phys. Chem.* 79, 1369.

Chang, M.C., Petrich, J.W., McDonald, D.B. and Fleming, G.R. (1983). *J. Amer. Chem. Soc.* 105, 3819.

Chen, R.F. and Bowman, R.L. (1965). *Science* 147, 729.

Chong, P.L.-G. and Weber, G. (1983). *Biochemistry* 22, 5544.

Chuang, T.-J. and Eisenthal, K.B.(1972). *J. Chem. Phys.* 57, 5094

Conrad, R.H., Heitz, J.R. and Brand, L. (1970). *Biochemistry* 9, 1540.

Dale, R.E., Chen, L.A. and Brand, L. (1977). *J. Biol. Chem.* 252, 7500.

Dale, R.E. (1979). *In* "Applications of Synchrotron Radiation to the Study of Large Molecules of Biological Interest" (Eds I.H. Munro and R.B. Cundall) Daresbury Laboratory Report DL/SC1/R13, p. 94.

Dale, R.E. (1983). *In* "Time-Resolved Fluorescence Spectroscopy in Biochemistry and Biology" (Eds R.B. Cundall and R.E. Dale) p. 555. Plenum Press, New York.

Davenport, L., Knutson, J.R. and Brand, L. (1982). *Amer. Soc. Photobiol.* 10. 69a.

Davenport, L., Markby, D.W., Knutson, J.R. and Brand, L. (1983). *Amer. Soc. Photobiol.* 37, S20.

DeToma, R.P., Easter, J.H. and Brand, L. (1976). *J. Amer. Chem. Soc.* 98, 5001.

DeToma, R.P. and Brand, L. (1977). *Chem. Phys. Lett.* 47, 231.

Dixit, B.P.S.N., Waring, A.J., Wells, K.O.,III, Wong, P.S., Woodrow, G.V.,III and Vanderkooi, J.M. (1982). *Eur. J. Biochem.* 126, 1.

Donzel, B., Gauduchon, P. and Wahl, Ph. (1974). *J. Amer. Chem. Soc.* 96, 801.

Easter, J.H., DeToma, R.P. and Brand, L. (1976). *Biophys. J.* 16, 571.

Edelstein, S.J. and Schachman, H.K. (1967). *J. Biol. Chem.* 242, 306.

Eftink, M.P. and Ghiron, C.A. (1977). *Biochemistry* 16, 5546.

Egawa, K., Nakashima, N., Mataga, N. and Yamanaka, C. (1971). *Bull. Chem. Soc. Japan* 44, 3287.

Ehrenberg, M. and Rigler, R. (1972). *Chem. Phys. Lett.* 14, 539.

Eisenfeld, J. and Ford, C.C. (1979). *Biophys. J.* 26, 73.

Eisenfeld, J. (1983). *In* "Time-Resolved Fluorescence Spectroscopy in Biochemistry and Biology" (Eds R.B. Cundall and R.E. Dale) p. 233. Plenum Press, New York.

Gafni, A., DeToma, R.P., Manrow, R.E. and Brand, L. (1977). *Biophys. J.* 17, 155.

Gafni, A.R., Modlin, R.L. and Brand, L. (1975). *Biophys. J.* 15, 263.

Gafni, A. and Brand, L. (1978). *Chem. Phys. Lett.* 58, 346.

Gilbert, C.W. (1983). *In* "Time-Resolved Fluorescence Spectroscopy in Biochemistry and Biology" (Eds R.B. Cundall and R.E. Dale) p. 605. Plenum Press, New York.

Grinvald, A. and Steinberg, I.Z. (1974). *Analyt. Biochem.* 59, 583.

Hanson, D.C., Yguerabide, J. and Schumaker, V.N. (1981). *Biochemistry* 20, 6842.

Harvey, S.C. and Cheung, H.C. (1972). *Proc. Natl Acad. Sci. USA* <u>69</u>,
    3670.
Harvey, S.C. and Cheung, H.C. (1977). *Biochemistry* <u>16</u>, 5181.
Hauser, M. and Wagenblast, G. (1983). *In* "Time-Resolved Fluorescence
    Spectroscopy in Biochemistry and Biology" (Eds R.B. Cundall and
    R.E. Dale) p. 463. Plenum Press, New York.
Heitz, J.R. and Brand, L. (1971). *Biochemistry* <u>10</u>, 2695.
Heyn, M.P. (1979). *FEBS Lett.* <u>108</u>, 359.
Hu, C.M. and Zwanzig, R. (1974). *J. Chem. Phys.* <u>60</u>, 4354.
Hui, S.W. and Parsons, D.F. (1975). *Science* <u>190</u>, 383.
Huntress, J.,Jr (1970). *Adv. Magn. Reson.* <u>4</u>, 1.
Isenberg, I. (1973). *J. Chem. Phys.* <u>59</u>, 6596.
Jameson, D.M. (1983). *In* "Time-Resolved Fluorescence Spectroscopy in
    Biochemistry and Biology" (Eds R.B. Cundall and R.E. Dale) p. 623.
    Plenum Press, New York.
Johnson, M.L., Halverson, H.R. and Ackers, G.K. (1976). *Biochemistry*
    <u>15</u>, 5363.
Johnson, M.L., Correia, J.C., Yphantis, D.A. and Halverson, H.R.
    (1981). *Biophys. J.* <u>36</u>, 575.
Kinosita, K., Jr, Mitaku, S. and Ikegami, A. (1975). *Biochim. Biophys.
    Acta* <u>393</u>, 10.
Kinosita, K., Jr, Kawato, S. and Ikegami, A. (1977). *Biophys. J.* <u>20</u>,
    289.
Klausner, R.D., Kleinfeld, A.M., Hoover, R.L. and Karnovsky, M.J.
    (1980). *J. Biol. Chem.* <u>255</u>, 1286.
Klausner, R.D. and Wolf, D.E. (1980). *Biochemistry* <u>19</u>, 6199.
Knorr, F.J. and Harris, J.M. (1981). *Analyt. Chem.* <u>53</u>, 272.
Knutson, J.R., Davenport, L. and Brand, L. (1982a). *Biophys. J.* <u>37</u>,
    203a.
Knutson, J.R., Walbridge, D.G. and Brand, L. (1982b). *Biochemistry*
    <u>21</u>, 4671.
Knutson, J.R., Baker, S.H., Cappuccino, A.G., Walbridge, D.G. and
    Brand, L. (1983a). *Photochem. Photobiol.* <u>37</u>, Abstr. S21.
Knutson, J.R., Walbridge, D.G. and Brand, L. (1983b). *Biophys. J.*
    <u>41</u>, 268A.
Knutson, J.R., Beechem, J.M. and Brand, L. (1983c). *Chem. Phys. Lett.*
    <u>102</u>, 501.
Koenig, S.H. (1975). *Biopolymers* <u>14</u>, 2421.
Kooyman, R.P.H., Vos, M.H. and Levine, Y.K. (1983). *Chem. Phys.* <u>81</u>,
    461.
Kowalczyk, A.A., Knutson, J.R., Barkley, M.D., Christy, R. and Brand,
    L. (1982). In "Abstr. Fourth Conference in Luminescence", p. 187.
    Szeged, Hungary.
Kynch, G.J. (1962). "Mathematics for the Chemist" (revised edition)
    p. 230. Butterworths, London.
Lakowicz, J.R. and Weber, G. (1973a). *Biochemistry* <u>12</u>, 4161.
Lakowicz, J.R. and Weber, G. (1973b). *Biochemistry* <u>12</u>, 4171.
Lakowicz, J.R. and Knutson, J.R. (1980). *Biochemistry* <u>19</u>, 905.
Lakowicz, J.R. and Cherek, H. (1981a). *J. Biochem. Biophys. Methods*
    <u>5</u>, 19.
Lakowicz, J.R. and Cherek, H. (1981b). *J. Biol. Chem.* <u>256</u>, 6348.
Lakowicz, J.R. and Balter, A. (1982). *Biophys. Chem.* <u>16</u>, 223.
Lakowicz, J.R. (1983). "Principles of Fluorescence Spectroscopy".
    Plenum Press, New York.

Lakowicz, J.R. and Keating, S. (1983). *J. Biol. Chem.* 258, 5519.
Lakowicz, J.R., Townson, R.B. and Cherek, H. (1983a). *Biochim. Biophys. Acta* 734, 295.
Lakowicz, J.R., Maliwal, B.P., Cherek, H. and Balter, A. (1983b). *Biochemistry* 22, 1741.
Laue, T.M., Johnson, A.E., Esmon, C.T. and Yphantis, D.A. (1984). *Biochemistry* 23, 1339.
Laws, W.R. and Brand, L. (1979). *J. Phys. Chem.* 83, 795.
Lipari, G. and Szabo, A. (1980). *Biophys. J.* 30, 489.
Liv, B.W., Cheung, H.C. and Mestecky, J. (1981). *Biochemistry* 20, 1997.
Ludescher, R.D. (1984). *Biophys. J.* 45, 379a.
Mantulin, W.W. and Weber, G. (1977). *J. Chem. Phys.* 66, 4092.
Memming, R. (1961). *Z. Phys. Chem. (Frankfurt)* 28, 168.
Munro, I., Pecht, I. and Stryer, L. (1979). *Proc. Natl Acad. Sci. USA* 76, 55.
Nagle, J.F., Parodi, L.A. and Lozier, R.H. (1982). *Biophys. J.* 38, 161.
Nezlin, R.S., Zagyansky, Y.A. and Tumerman, L.A. (1970). *J. Mol. Biol.* 50, 569.
Perrin, F. (1934). *J. Phys. Radium* 5, 497.
Perrin, F. (1936). *J. Phys. Radium* 7, 1.
Petrich, J.W., Chang, M.C., McDonald, D.B. and Fleming, G.R. (1983). *J. Amer. Chem. Soc.* 105, 3824.
Porter, G.B. (1972). *Theoret. chim. Acta* 24, 265.
Privat, J.P., Wahl, Ph. and Auchet, J.C. (1980). *Biophys. Chem.* 11, 239.
Rayner, D.M. and Szabo, A.G. (1978). *Can. J. Biochem.* 56, 743.
Reidler, J., Oi, V.T., Carlsen, W., Minh Vuong, T., Pecht, I., Herzenberg, L.A. and Stryer, L. (1982). *J. Mol. Biol.* 158, 739.
Robbins, R.J., Fleming, G.R., Beddard, G.S., Robinson, G.W. and Thistlethwaite, P.J. (1980). *J. Amer. Chem. Soc.* 102, 6271.
Ross, J.B.A., Rousslang, K.W. and Brand, L. (1981a). *Biochemistry* 20, 4369.
Ross, J.B.A., Schmidt, C.J. and Brand, L. (1981b). *Biochemistry* 20, 4361.
Seliskar, C.J. and Brand, L. (1971). *Science* 171, 799.
Shank, C.V. and Ippen, E.P. (1975). *Appl. Phys. Lett.* 26, 62.
Sheiner, L.B. and Beal, S.L. (1981). *In* "Kinetic Data Analysis" (Ed. L. Endreynyi) p. 271. Plenum Press, New York.
Shinitzky, M., Dianoux, A.-C., Gitler, C. and Weber, G. (1971). *Biochemistry* 10, 2106.
Small, E.W. and Isenberg, I. (1977a). *J. Chem. Phys.* 66, 3347.
Small, E.W. and Isenberg, I. (1977b). *Biopolymers* 16, 1097.
Szabo, A.G. and Rayner, D.M. (1980). *J. Amer. Chem. Soc.* 102, 554.
Turner, B.W., Pettigrew, D.W. and Ackers, G.K. (1981). *Methods in Enzymology* 76, 596.
Vold, R.L., Vold, R.R. and Canet, D. (1977). *J. Chem. Phys.* 66, 1202.
Wahl, Ph. (1969). *Biochim. Biophys. Acta* 175, 55.
Wahl, Ph. (1975). *Chem. Phys.* 7, 210.
Wahl, Ph. (1979). *Biophys. Chem.* 10, 91.
Wahl, Ph. (1983). *In* "Time-Resolved Fluorescence Spectroscopy in Biochemistry and Biology" (Eds R.B. Cundall and R.E. Dale) p. 497. Plenum Press, New York.

Waldeck, D., Cross, A.J., McDonald, D.B. and Fleming, G. (1981). *J. Chem. Phys.* <u>74</u>, 3381.

Ware, W.R., Chow, P. and Lee, S.K. (1968). *Chem. Phys. Lett.* <u>2</u>, 356.

Ware, W.R., Lee, S.K., Brant, G.K. and Chow, P.P. (1971). *J. Chem. Phys.* <u>54</u>, 4729.

Weber, G. (1973). *In* "Fluorescence Techniques in Cell Biology" (Eds A.A. Thaer and M. Sernetz) p. 5. Springer-Verlag, New York.

Weber, G. (1977). *J. Chem. Phys.* <u>66</u>, 4081.

Witholt, B. and Brand, L. (1970). *Biochemistry* <u>9</u>, 1948.

Wolber, P.K. and Hudson, B.S. (1982). *Biophys. J.* <u>37</u>, 253.

Yguerabide, J., Epstein, H.F. and Stryer, L. (1970). *J. Mol. Biol.* <u>51</u>, 573.

Zannoni, C. (1981). *Mol. Phys.* <u>42</u>, 1303.

Zannoni, C., Arcioni, A. and Cavatorta, P. (1983). *Chem. Phys. Lipids* <u>32</u>, 179.

Zinsli, P.E. (1977). *Chem. Phys.* <u>20</u>, 299.

Fluorescence Spectroscopy in the Time Domain Using Synchrotron Radiation

I.H. MUNRO

INTRODUCTION

The past decade has seen the introduction and utilization of a wide
variety of new experimental techniques in the time domain applied in
the areas of physics, chemistry and the biological sciences which
make use of synchrotron radiation from storage rings and synchrotrons,
and these have been summarized in various review articles (Lopez-
Delgado *et al.*, 1974; Monahan and Rehn, 1978; Schwentner *et al.*, 1979;
Munro and Sabersky, 1980; Munro and Schwentner, 1983; Mills, 1984.
The methods fully exploit the characteristics of synchrotron radiation,
whose principal advantages are the emission of a broad continuum of
electro-magnetic radiation of high brightness, which is highly linearly polarized and modulated at high (Megahertz) frequencies. The
most significant scientific progress to date using synchrotron radiation sources has probably been in the areas of high-resolution,
element-specific X-ray spectroscopy, crystal structure studies and
small angle scattering; all methods which yield valuable structural
information. No doubt further significant progress will continue in
these fields. However, the development and construction of "purpose-
built" storage rings from the beginning of the 1980s appears to be
directed, at least in part, to two new major areas where synchrotron
radiation sources promise exciting scientific developments particularly in the biosciences. The new areas are concerned with high
spatial resolution imaging (for example tuneable sub-micron X-ray
microscopy, trace element analysis and X-ray imaging), and in the

SPECTROSCOPY AND THE DYNAMICS
OF MOLECULAR BIOLOGICAL SYSTEMS

broad field of time-resolved studies including subnanosecond time-
resolved fluorescence spectroscopy.

Measurements in the time domain are becoming established at some
storage ring centres and reviews of the significance of past work with
likely prospects for the future have been given (Munro and Schwentner,
1983; Mills, 1984). Most time-resolved excitation and emission fluorescence studies of atoms and molecules to date have been undertaken
with a resolution of at least 1 ns on a range of materials, notably
rare gas solids and low pressure, low quantum yield vapours in the
near vacuum ultraviolet region. Individual timing measurements have
covered an extremely broad subject area at an exploratory level. For
example, quantum coherence effects have been observed in rare gases,
kinetic processes associated with exciton formation and energy migration have been observed in crystals and studies have been made of X-
ray scattering and diffraction from the nanosecond to the millisecond
time domain.

The exploitation in particular, of the highly polarized nature of
synchrotron radiation has allowed measurements to be made of time-
resolved fluorescence anisotropy on a considerable variety of materials, including amino acids, with a best time resolution of about
±50 ps using standard photon-counting techniques. A considerably
higher time resolution ($\sim$ picoseconds) is possible using phase or
amplitude comparison methods at single frequencies. Measurements of
this kind may become of value in determining changes in chemical
environment, local geometry in proteins, membranes and other important
biochemical systems via an analysis of the time-dependence of fluorescence decay of intrinsic or extrinsic fluorescent labels.

THE CHARACTERISTICS OF STORAGE RINGS AT TIME-MODULATED LIGHT SOURCES

The principal merit of synchrotron radiation sources is that they
provide high brightness (and small source size) over an extraordi-
narily wide wavelength range, which can extend from the hard X-ray
($\geq 0.1\mathring{A}$) to the centimetre wave region as shown in Fig. 1 (Cundall
and Munro, 1979). At any particular source point in a typical storage
ring, the beam cross section will be of the order of, or rather larger
than, $1\,mm^2$. The time modulation (defined by the longitudinal size of
the electron bunches) of the emitted radiation is imposed primarily

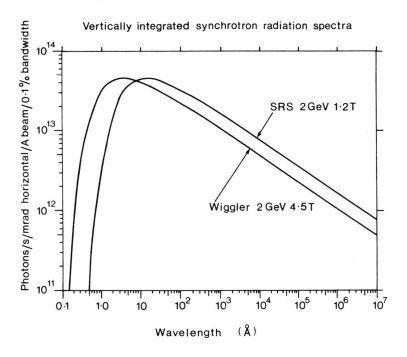

FIG. 1 The source spectrum from the SRS at Daresbury Laboratory operating at 2 GeV.

by the radio-frequency field which is used to accelerate the electron beam and to maintain a stable equilibrium orbit. Different storage rings often use rather different radio frequencies for acceleration, leading to quite different electron bunch separations within different accelerators. A list of parameters for several storage rings which have been used for timing experiments is presented in Table 1.

The minimum inter-bunch separation is defined by the time period of the radio-frequency system. The maximum inter-bunch spacing is defined by the dimensions of the storage ring and is, of course, equal to the time required for a single bunch of electrons to rotate once around the storage ring, a time which is usually less than one

TABLE 1 Storage ring parameters

	Ring period (ns)	No. of bunches	RF period (ns)	Pulse width (ns)
ACO (Orsay) [0.5 GeV]	73.5	2	36.7	0.8 -1.4
BESSY (Berlin) [0.7 GeV]	208.2	26 } 104	8.0 } 2.0	0.01-0.13
SRS (Daresbury) [2.0 GeV]	302.5	160	2.0 [500 MHz]	0.2
DORIS (Hamburg) [3.5 GeV]	961.5	481	2.0	0.13
SPEAR (Stanford) [4.0 GeV]	781.2	279	2.8	0.08-0.4
CHESS (Cornell) [8.0 GeV]	2222	1111	2.0	0.1

microsecond (see Table 1). Suggestions have been made, but never
implemented, to introduce electron deflecting devices to increase
this maximum single bunch period. The bunch duration is related to
machine parameters in a calculable way and will tend to a minimum
value as the momentum conpaction factor and the ring energy are
reduced and as the applied radio-frequency voltage is increased. In
fact the electron bunch length (and hence the photon pulse duration)
σ_z is given by:

$$\sigma_z \simeq [\alpha T_0 E^3 / (\frac{dV}{dt})]^{\frac{1}{2}} \tag{1}$$

where α is the ring momentum compaction factor, T_0 and E are the ring
period and energy, respectively, and V is the applied radio-frequency
voltage.

A major obstacle in the exploitation of storage rings for timing
studies is the difficulty in achieving high circulating current in
the storage ring under single bunch conditions, particularly for short
(≤ 100 ps) bunch durations. Current-dependent bunch lengthening has
been seen to occur in several storage rings. Typical operating parameters, in the Daresbury SRS, for example, would be currents of
≤ 300 mA at 2 GeV in the filled, multibunch mode and of ≤ 30 mA in
the single bunch mode. In fact, in the limit, when σ_z tends to a few

millimetres, the circulating current will tend to zero. Nevertheless, such conditions at the SRS give from 10^4 to 10^7 photons per pulse at the sample (depending on the bandwidth selected for excitation), and these values could be increased by at least a factor of ten when the undulator magnet is installed (NIM, 1983a).

The pulse shape provided by a storage ring for sample excitation is exceedingly stable in amplitude, shape (a good approximation to a Gaussian) and in repetition frequency. Of the greatest importance is the fact that all wavelengths radiated must possess identical time registration. This property is of special benefit when minimizing errors associated with the wavelength responsivity of detectors, particularly when fluorescence decay profiles lie close to the time resolution of the apparatus.

EXPERIMENTAL TECHNIQUES FOR BIOCHEMICAL FLUORESCENCE SPECTROSCOPY

The successful exploitation of the intrinsic time modulation of syn-chrotron radiation is still somewhat limited both by constraints imposed by the design and operation of synchrotron radiation sources, and also by a limited appreciation by the scientific community of users, of what measurements are now feasible. The variety of experimental results which have been published demonstrate the potential merits of time-domain spectroscopy, particularly in the near VUV, for photophysical and photochemical investigations of gases and solids, in terms of time-of-flight studies for electron and ion energy and mass analysis and even for time-resolved X-ray diffraction in the nanosecond range (Mills, 1984). Synchrotron radiation is an almost ideal excitation source for photochemical and biochemical fluorescence excitation and emission spectroscopy and for time-resolved measure-ments of polarized fluorescence (Rapp. D'Activ., 1980-1984). The measurement of the time decay of fluorescence anisotropy of fluores-cent probes in proteins, membranes and other biochemical systems is well matched to the properties of the source.

Pulsed methods

Storage-ring radiation has been studied using a fast detector with a sampling oscilloscope (response function FWHM $\sim$ 100 ps) and, at Hamburg and Daresbury, using a streak camera both in the single shot

and in the repetitive mode. It is to be noted that the limited dynamic
range and the noise level of the streak camera combined with the
relatively low "pulse power" available from a storage ring (only a
few watts, which is many orders of magnitude below the pulse power
available from lasers), may limit the role of the streak camera to
the study of storage ring bunch conditions with broad band radiation.
Conventional delayed-coincidence methods using single-photon counting
are ideal for time-dependent excitation and for fluorescence and
fluorescence anisotropy evolution and decay studies with synchrotron
radiation because of the appropriate match of the detector (microchannel plate or fast photomultiplier tube) to the pulse width, which
is usually ∿100-200 ps but may be as narrow as 50 ps (SPEAR, BESSY).
Measurements using apparatus at the SRS will accumulate data at rates
$\lesssim$50 kHz and will yield a best time resolution from a microchannel
plate of $\gtrsim$20 ps. Count rates of this magnitude allow experiments to
be completed routinely using 1024 channels and the accumulation of
∿50000 counts in the peak channel within a small fraction of an hour.
With the development of improved detectors and the exploitation of
digital counter/timers, in place of a time-to-amplitude converter
and a multi-channel analyser, it is envisaged that data collection
rates approaching 1 MHz will soon be feasible.

Frequency domain measurements

The "Gaussian" electron bunch distribution within the ring is equivalent in the frequency domain to a "Gaussian" frequency envelope of
harmonics of the ring fundamental frequency (3.2 MHz for the SRS)
extending to very high frequencies (several GHz defined by the pulse
width (∿200 ps for the SRS). Such an extended range of "pure" frequencies enables the prospective user to make measurements either of
phase delay or amplitude reduction (demodulation) at very many individual frequencies or over any selected group of frequencies. Measurements of phase shift at single frequencies have been recorded at a
number of storage rings. An output from a photodetector giving
$\geq$20 µv is sufficient to enable phase comparisons of differences as
small as 0.1° to be derived between reference and fluorescence signals at frequencies of up to 500 MHz. The equivalent time delay from

measurements of this kind will be around a few picoseconds and the
measurement would be simple and convenient to carry out in time-
dependent excitation spectroscopy. Analogous measurements can be made
using analogue comparisons between signal and reference at selected
frequencies. The use of a power spectrum analyser can provide valuable
information in the understanding of instabilities arising from single
bunch operation. This is effectively a frequency-domain analogue of
the streak camera, although phase information cannot be obtained.

The advantages of the phase and amplitude methods have recently
been combined to achieve the measurement and analysis of heterogeneous
emissions using a multi-frequency cross-correlation phase and modulation fluorometer technique which can yield picosecond resolution,
in conjunction with the storage ring at Frascati, Italy (Jameson *et
al.*, 1984). This elegant technique will be of value in understanding
complex decay systems in the subnanosecond region and also in studying
the most rapid molecular motions via the observation of the decay of
fluorescence anisotropy.

The technical prospects for time-resolved studies with synchrotron
radiation are good. Some of the new generation of storage rings, for
example the VUV ring at BESSY (Berlin), possess relatively low energy,
small electron beam emittance and a high fundamental frequency leading
to "minimum" predicted useful bunch lengths of 50 ps or less. The
installation of undulators in future storage rings will, to a good
approximation, not provide any significant time deterioration (i.e.
no measurable increase in pulse width) although the simultaneous
accumulation of radiation from up to one hundred or more magnet poles
in series will increase the intensity per pulse by a corresponding
amount. Eventually, pulsed coherent radiation sources will become
available through the development of free electron laser devices in
conjunction with recirculating electron storage rings (NIM, 1983b).

TIME-RESOLVED FLUORESCENCE DEPOLARIZATION STUDIES IN SYSTEMS OF BIOCHEMICAL SIGNIFICANCE

The extensive research programme at the ACO storage ring (Rapp.
D'Activ., 1980-1984) illustrates clearly the benefits of the use of
synchrotron radiation to study both steady-state and time-resolved
fluorescence spectroscopic properties of a very wide range of organic

compounds quickly and effectively. Synchrotron radiation sources should be able to make a particularly valuable contribution in the field of time-resolved fluorescence depolarization, where the time-registration properties, pulse width, intrinsically high polarization and good level of tuneable intensity are an excellent match to the requirements of those experiments.

Protein fluorescence

From the behaviour and movements of intrinsic or extrinsic fluorescent probes in proteins and peptides one can obtain unique information about changes in local conformation around the probe site, about the flexibility of large molecules and also a numerical indication of the overall size of a protein (including its hydration envelope) according to the particular circumstances.

The fluorescent probe (extrinsic or intrinsic) can be conjugated covalently or bound by non-covalent ionic or hydrophobic bonds to sites of interest. All intrinsic probes are readily spectroscopically accessible using synchrotron radiation, although, of course, the quality of interpretation of the final data will depend critically on an awareness of the specific site or sites responsible for fluorescence and on the absence of any processes which could simulate geometrical depolarization of fluorescence (for example, inter- or intra-protein energy transfer processes). Although little has been published relating to intrinsic probe fluorescence from proteins using synchrotron radiation (Munro et al., 1979), it is known to the author that a number of preliminary results have been obtained from single-tyrosine-containing compounds (simple peptides), which could yield information on hydrogen bonding and proton transfer, and on single-tryptophan-containing compounds which exhibit considerable environmental sensitivity both in fluorescence (red shift) and in the time-dependence of their emission anisotropy. Phenylalanine (excitation at ∿260 nm) is a less promising probe, but is of interest in hormones and ionophores. When the fluorescence lifetime is short and the quantum yield small, the capacity to provide variable excitation wavelength is particularly important.

A broad range of single-tryptophan-containing compounds exists, and valuable data will undoubtedly be derived also from multi-tryptophan-

containing proteins (e.g. lactate dehydrogenase) when they are sufficiently well understood for the principal emitting residue(s) to be
defined. Preliminary measurements have also been made on hydrated
protein crystals (e.g., of lysozyme).

Nucleic acids

In general, the nucleic acid bases do not emit under physiological
conditions (i.e. the quantum yield is $\lesssim 10^{-4}$) or they possess complex
excited-state properties. Nevertheless, spatial relationships between
segments of these molecules and also estimates of their flexibility
have been inferred from observations on the "odd" bases in t-RNA. To
date, it has been seen to be necessary to introduce specific modifi-
cations which yield fluorescent derivatives, for example, by intercalation of acridine range or ethidium bromide. The intensity, the
tuneability and the sufficiently short pulse duration have proved
adequate for wavelength-resolved lifetime studies of emissions from
DNA to be identified using the ACO storage ring (Ballini et $al.$,
1982). Decay studies have been carried out on polyadenylic acid as a
function of wavelength and also on cytidine, thymidine, guanosine and
adenosine where lifetimes are short - typically $\lesssim 100$ ps. Lifetimes
measured at low temperatures (77°K) were seen to be anomalously long.
Other studies have been undertaken using tryptophan as the fluorescent probe in peptide-nucleic acid complexes. Two decay components
were readily detected, and the mean fluorescence lifetimes are
unchanged on binding to nucleic acids despite a reduction in quantum
efficiency (Montenay-Garestier, Brochon and Hélène, 1981). Further
research will undoubtedly follow using polarized synchrotron radiation
to observe the dynamics of peptide-nucleic acid complexes (Montenay-
Garestier, Fidy et $al.$, 1981).

Membrane systems

Membrane systems have been very extensively researched using "quasi-
intrinsic" probes which are usually linear polymers such as retinal,
carotenoid derivatives or diphenylhexatriene and its many derivatives.
Measurements have also been made on linear polyene fatty acids such
as cis- and trans-parinaric acids. Despite the abundance of data on
fluorescence spectra, fluorescence lifetimes and steady-state

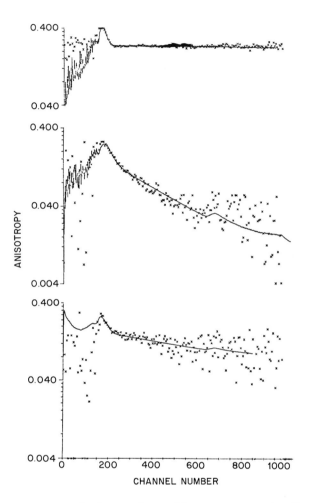

FIG. 2 Fluorescence anisotropy decay time measurements for DPH derivatives in DPPC lipid vesicles. The time scale factor is 0.027 ns per channel; weighted non-linear least squares fitting started at channel 150.

Upper: DPHTG   30°C; $r_0 = 0.244$, $\phi = 228$ns   (highly scattering sample)
                                                                                    $\chi_\nu^2 = 2.21$

Middle: DPHTG   50°C; $\beta_1 = 0.203$, $\phi_1 = 0.53$ns
                                 $\beta_2 = 0.161$, $\phi_2 = 6.8$ns
                                 —————————
                                 $r_0 = 0.364$ $\chi_\nu^2 = 1.82$

Lower: TMAEPC   20°C; $\beta_1 = 0.193$, $\phi_1 = 0.21$ns
           1570 Bar $\beta_2 = 0.165$, $\phi_2 = 23$ns
                                 —————————
                                 $r_0 = 0.358$ $\chi_\nu^2 = 0.94$

depolarization, measurements of time-resolved fluorescence depolarization are relatively scarce. The specific advantage of a synchrotron radiation source, such as the SRS at Daresbury Laboratory, is that individual anisotropy measurements can be made extremely rapidly with no requirement for time-shifting the data in the software data-fitting routines.

Of course, proper understanding of the behaviour of "model" membrane systems must be preceded by a comprehensive understanding of the excited-state behaviour of the probe molecule itself. This understanding is of the greatest importance in probe molecules related to diphenylhexatriene, which can introduce apparent anisotropy (geometrical) changes resulting solely from photophysical changes in the probe.

Some preliminary results are given in Fig. 2 from a series of experiments to investigate the behaviour of bilayer lipid membranes under a variety of conditions of temperature and pressure. The probes were synthesized from 2-carboxyethyl-4'-diphenyl-1,3,5-hexatriene to give triacylglycerol (DPHTG) and lecithin (DPHPC) derivatives. DPHPC and an analogous probe, the trimethylammonium DPH derivative (TMAEPC), partition into the phospholipid bilayer with the charged groups in the polar head region and the probe aligned with the acyl chains. When such different molecules are dispersed at low concentrations within a phospholipid bilayer, then differences in fluorescence anisotropy behaviour will be seen if constraints on probe motion or orientation do not apply equally. Interpretation of data can be comprehensively valid only when the location and orientation of the probe are known, for example, in ordered lipid bilayers (van der Meer *et al.*, 1982). The possibility of phase separation with the formation of domains rich in probe, for example, will be difficult to detect and would obscure the interpretation of the general fluorescence information.

The anisotropy decay of DPHTG in bilayers of dipalmitoyllecithin (DPPC) at 30° shows a rapid depolarization followed by a residual static polarization indicative of a restricted motion of the probe within a "cone angle" defined by the orientating effect of the bilayer membrane in the gel phase. Above the phase transition temperature (40°C) of the membrane, probe motion is much less hindered as shown by the data at 50 °C in Fig. 2b. Preliminary measurements, without temperature control or birefringence corrections, indicate that rotation becomes more restricted as the pressure is elevated (Fig. 2c). Other measurements of this type at Daresbury Laboratory have revealed significant differences between different probes (such as DPH and the phosphatidylcholine derivative of DPH), and have identified detailed changes in two-component lifetimes and in rotational correlation times for DPHPC in DPPC over a range of temperatures. Further work has revealed the effects of pressure on unilamellar vesicles, with and without a membrane stiffener such as cholesterol, illustrated the "fluidizing" effects of ethanol (an anaesthetic) and will give some insight into the antagonistic relationship between pressure and anaesthetics in model membrane systems (Cundall *et al.*, 1983).

Steady-state measurements in this field are, of course, severely limited by the dramatic changes in the multi-exponential decays of the probe which arise with changing temperature and pressure, and also by local environmental changes with added solvent, cholesterol, etc. Finally, time-dependent depolarization measurements permit an independent (but simultaneous) observation of fluorescence decay and anisotropy decay.

FUTURE PROSPECTS

Studies of fluorescence lifetimes and of the decay of fluorescence anisotropy using synchrotron radiation are still relatively few in number, but already extensive in scope. The speed with which experiments can be undertaken, the wavelength excitation capability for all probes and the high quality of the data, particularly in the subnanosecond domain, will clearly produce very active programmes of research associated with photophysics, photochemistry and the dynamics of molecules of biochemical significance.

REFERENCES

Ballini, J.P., Daniels, M. and Vigny, P. (1982). *J. Luminescence* 27, 389.
Cundall, R.B., Jones, G., Kelly, M.M., Mant, G.R., Morgan, C., Munro, I.H. and Shaw, D. (1983). Unpublished work.
Cundall, R.B. and Munro, I.H. (Eds)(1979). "Application of synchrotron radiation to the study of large molecules of chemical and biological interests". Daresbury Report DL/SCI/R13.

Jameson, D.M., Gratton, E. and Hall, R.D. (1984). *Appl Spectrosc. Rev.* 20, 55.

Lopez-Delgado, R., Tramer, A. and Munro, I.H. (1974). *Chem. Phys.* 5, 72.

Mills, D.M. (1984). *Physics Today* 37, 22.

Monahan, K.M. and Rehn, V. (1978). *Nucl. Instrum. Methods* 152, 225.

Montenay-Garestier, T., Brochon, J.C. and Hélène, C. (1981). *Int. J. Quantum Chem.* 20, 41.

Montenay-Garestier, T., Fidy, J., Brochon, J.C. and Hélène, C. (1981). *Biochimie* 63, 937.

Munro, I.H., Pecht, I. and Stryer, L. (1979). *Proc. Natl Acad. Sci. USA* 76, 56.

Munro, I.H. and Sabersky, A.B. (1980). *In* "Sunchrotron Radiation Research" (Eds H. Winick and S. Doniach), Chapter 9. Plenum Press, New York.

Munro, I.H. and Schwentner, N. (1983). *Nucl. Instrum. Methods* 208, 819.

Nucl. Instrum. Methods (1983). 208, Part II and references therein.

Nucl. Instrum. Methods (1983). 208, Part III and references therein.

Pecht, I. (1982). *In* "Uses of Synchrotron Radiation in Biology" (Ed. H.B. Stuhrmann), p. 71. Academic Press, London.

Rapports D'Activé (1980-1984). L.U.R.E., C.N.R.S. Université Paris-Sud, 91405 Orsay, France.

Schwentner, N., Hahn, U., Einfeld, D. and Mulhaupt, G. (1979). *Nucl. Instrum. Methods* 167, 499.

van der Meer, B.W., Kooyman, R.P.H. and Levine, Y.K. (1982). *Chem. Phys.* 66, 39.

Time-resolved X-ray Scattering Studies of Microtubule Assembly Using Synchrotron Radiation

E. MANDELKOW, E.-M. MANDELKOW, W. RENNER and J. BORDAS

GENERAL ASPECTS OF SYNCHROTRON RADIATION STUDIES

Experimental

Synchrotron radiation is emitted when electrons or positrons are accelerated and forced into circular orbits. This happens in particle accelerators used by high energy physicists for whom synchrotron radiation represents merely a waste of energy. However, there are now an increasing number of applications of synchrotron radiation to other branches of science, ranging from solid-state physics to biology, which has led to the current construction of new synchrotron laboratories in several countries. The potential uses have been described in a number of reviews (e.g. ESF proposal, 1979; Kunz, 1979; Winick and Doniach, 1980; Stuhrmann, 1983), including several reports on biological applications, e.g. to muscle structure and dynamics (Rosenbaum and Holmes, 1980; Huxley et al., 1980) and structural transitions of proteins in solution (Mandelkow et al., 1980a; Bordas and Mandelkow, 1983; Bordas, 1983; Koch et al., 1983).

Synchrotron radiation has several unique properties which include the following:

(i) the wavelength spectrum is continuous, ranging from the X-ray region to the infrared (Fig. 1);

(ii) the radiation is highly focussed and polarized in the plane of the electron orbit;

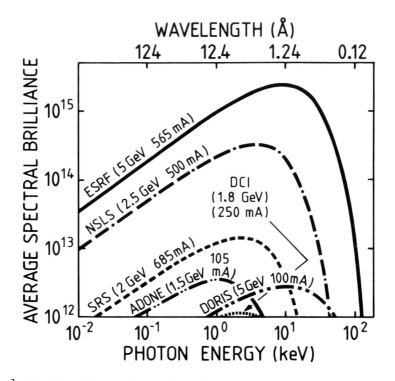

FIG. 1 Wavelength spectrum of synchrotron radiation and comparison
of several synchrotrons in operation or projected. Note the differences in high energy cut-off and maximum brilliance. From ESF proposal, 1979.

'iii) the radiation has a time structure since it is emitted by
discrete bunches of particles;

(iv) there is a large gain in spectral brilliance (defined as
photons/unit time/unit area/unit solid angle/unit wavelength bandpass) compared with conventional X-ray sources.

For X-ray scattering experiments, the high intensity is the most
useful property. For example, in the case of DORIS operated at 5 GeV
and 100 mA there are 2.9 x 10^{12} photons of 10 KeV energy (wavelength
0.124 nm). This is considerably better than conventional X-ray
generators and opens up the possibility of time-resolved X-ray studies.

When dealing with biological structures we wish to record data in
resolution ranges from atomic to supramolecular. The specimens usually
scatter weakly and require a high incident flux. Moreover, the samples

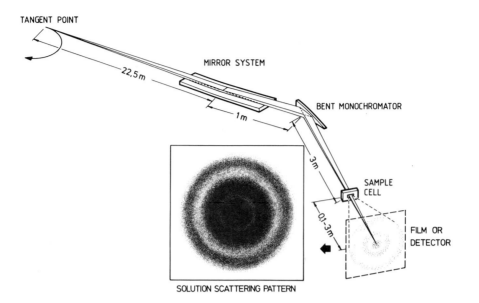

TANGENT POINT

MIRROR SYSTEM

22.5 m

1 m

BENT MONOCHROMATOR

3 m

SAMPLE
CELL

0.1-3 m

FILM OR
DETECTOR

SOLUTION SCATTERING PATTERN

FIG. 2 Diagram of EMBL instrument X13 (Hendrix *et al.*, 1979; Bordas *et al.*, 1980). The radiation emitted by the positrons is focussed and monochromatized by a set of mirrors and a Germanium crystal inside the shielding wall. Maximum specimen to detector distance is about 3 m. The solution scattering pattern consists of a set of circularly symmetric fringes. A new instrument (X33, Koch and Bordas, 1983) has now been installed in the HASYLAB hall with a basically similar design, except that the positions of mirror and monochromator have been interchanged, and the specimen-to-detector distance may be varied up to 8 m.

are often only available in small amounts so that the irradiated volume must be kept small (for details on the general requirements see Bordas and Koch, 1979). One way to meet the experimental needs is illustrated in Fig. 2 (EMBL instrument X13, see Hendrix *et al.*, 1979; Bordas *et al.*, 1980). The X-rays are monochromatized and focussed by a mirror-plus-crystal arrangement which evolved from X-ray studies of muscle (Huxley and Brown, 1967; Rosenbaum *et al.*, 1971). The resolution range may be varied by choosing camera lengths up to 8 m in the case of EMBL instrument X33 (Koch and Bordas, 1983).

Static X-ray patterns may be recorded on film, but for time-resolved studies one requires a position-sensitive detector. A number of different designs are being developed (see Boulin *et al.*, 1982). In our

experiments we used a linear detector in which positional information
was obtained by the delay line method (Gabriel, 1977). This is adequate for solution scattering patterns which are circularly symmetric
so that a radial scan contains all available information (apart from
the loss of counting statistics, compared with the full two-dimensional
patterns). The detector and its associated data acquisition system
is limited to the recording of approximately 3×10^5 photons/s, with
a 20% dead-time loss and a spatial resolution of about 0.4 mm (Bordas
et al., 1980; Boulin *et al.*, 1982).

The intensities contained in the 256 detector channels, as well as
other information pertinent to the experiment (e.g. temperature of the
sample, incident X-ray flux, etc.) is accepted by the data acquisition
system up to 10^6 events per second). Under optimal storage ring
conditions, the incident beam delivers about 10^{11} to 10^{12} photons per
second, of which about 10^7 are scattered. Since the detection system
collects only on the order of 10^5 events about 99% of the data must
be discarded. This illustrates that experiments are often limited by
the detectors rather than the available X-ray intensity.

Reactions in the solution may be induced by temperature jump or
rapid mixing. At the same time the cells must be equipped with X-ray
transparent windows which precludes the use of commercially available
chambers designed for UV or visible light detection. The apparatus
used for studying microtubule assembly is described elsewhere (Renner
et al., 1983). In general, different types of experiment require
different chambers, depending on the rates of the reactions, sample
concentrations, etc.

The data are stored on magnetic discs, ready for further analysis.
Due to the large amount of data, special programmes are required for
handling and displaying the results as an aid to their interpretation
(Koch and Bendall, 1981; Golding, 1982).

Interpretation of time-resolved solution scattering patterns

Since the particles are randomly oriented in solution, their scattering can be analysed in terms of small angle scattering theory (see
Guinier and Fournet, 1955). The intensity is proportional to the
incident beam, number of particles, square of excess electron density,
and a shape function containing the structural information. We recall

a few aspects important for time-resolved experiments of self-assemb-
ling proteins (see Bordas and Mandelkow, 1983).

(i) The incident intensity decreases with time, due to the loss
of particles in the storage ring. In order to relate patterns
taken at different times, the primary beam must be monitored
continuously so that the scattering traces can be normalized.
This is achieved by placing an ion chamber immediately before
the specimen.

(ii) In classical solution scattering work one tries to work with
monodisperse solutions. This is not the case when studying self-
assembly, since it is the transition between different states which
is of interest.

(iii) If the structures observed during the reaction are built from
(nearly) identical subunits, the scattering traces can be expressed
in terms of the total subunit concentration which remains constant
during the experiment. This leads to the following formula
(Mandelkow *et al.*, 1980a):

$$I(S,t) = const * I_0 c_0 \sum_k x_k(t) p_k i_k(S) \tag{1}$$

where $S = 2\sin\theta/\lambda$ is the Bragg spacing, I_0 the incident beam, c_0
the concentration of subunits, the index k refers to the different
types of structures present in the solution, x_k is the fraction of
subunits incorporated into aggregates of type k, p_k is their
degree of polymerization, and i_k is the shape function.

In general, there is no direct way to obtain the shape from the
scattering intensity. The standard practice is to compute model
scattering curves and compare them with the observed ones. Even with
a homogeneous solution the results of this approach are often ambiguous, and this holds even more for mixtures. However, there are ways
to circumvent the problem. Firstly, one must make use of structural
information obtained by other methods, mainly electron microscopy and
analytical ultracentrifugation. Second, the interpretation is greatly
facilitated if there are known X-ray patterns for a reference, e.g.
that of oriented microtubules (Mandelkow *et al.*, 1977a). Third, since

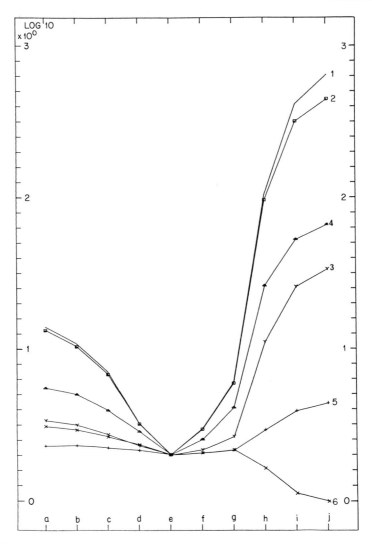

FIG. 3 Comparison of several scattering signals for a model reaction:
(a) 40% "rings", 60% "subunits"; (e) 100% "subunits"; (j) 100% microtubules"; the others are intermediate mixtures. Abscissa, reaction
parameter; ordinate, log (intensity). All traces have been normalized
to state (e). Curve (1), forward UV light scattering ($0°$); (2) UV
light scattering at $90°$; (3) UV "turbidity", i.e. light scattered
outside the entrance slits of the photomultiplier; (4) "central X-ray
scattering" at 0.015 nm^{-1} (just outside beam stop); (5) X-ray scattering at 0.5 nm^{-1} (first subsidiary maximum of microtubule pattern);
(6) X-ray scattering at 0.03 nm^{-1} (around first subsidiary maximum of
ring pattern, roughly coincident with first minimum of microtubule
pattern). Note that the relative signals are greatest in curves 1,
2 and 4. They become considerably smaller with UV turbidity (3) and
with increasing angle in the X-ray pattern (5, 6), especially during
the initial part of the model reaction where the aggregates are small.

one is mainly interested in structural transitions, it is often convenient to analyse intensity differences between certain time points;
this eliminates the background arising from those particles which do
not participate in the reaction.

   In planning a time-resolved experiment, one must decide on the
desired time resolution at which structural changes are to be observed,
given a certain spatial resolution. The two kinds of resolution
counteract each other, i.e. in order to see fine detail one has to
watch longer, and vice versa. The detectability of a structural
change is given by the relative contrast $C_r = (I_1-I_2)/I_1$, where I_1
and I_2 are the intensities measured at different time points at a
given spacing. The minimum exposure time is given roughly by:

$$\Delta t = const/(IC_r^2) \qquad\qquad\qquad (2)$$

where the constant is about 25 (determined by the requirement that
the signal should be visible above noise). I is the mean intensity
which decreases rapidly with scattering angle, so that the time
resolution deteriorates concomitantly.

   The contrast expected from several types of scattering experiments
is illustrated in Fig. 3. We have assumed a simplified model reaction
inspired by some features of microtubule assembly. The initial solution
contains 40% ring aggregates and 60% soluble subunits (state a). The
rings fall apart gradually until all protein is in the form of subunits
(state e, minimum degree of polymerization). After that, tubules grow
by end-wise addition of subunits until the protein is fully polymerized
(state 1; note that the x-axis does not represent time but some arbitrary reaction parameter). Curve 1 represents the theoretical forward
scattering of UV light at 350 nm, while curve 2 is the UV light scattering at right angles. Curve 3 is the UV "turbidity", calculated in
terms of the apparent absorption caused by light scattered away from
the forward direction; this is often used to monitor microtubule
assembly (Berne, 1974). Curves 4, 5 and 6 represent the X-ray

scattering at 0.015, 0.03 and 0.05 nm^{-1}, corresponding to the "central"
scatter (just outside the beam stop) and the first subsidiary maxima
of rings and microtubules, respectively. We note that the contrast
(i.e. sensitivity towards changes in structure) decreases as the
resolution increases. Curves 1-4 all are measures of the overall
degree of polymerization, the least sensitive being turbidity, especially when the structures are small (a-e). This explains in part why
the pre-nucleation events which are readily detected with X-ray or
light scattering are less apparent when measuring turbidity (cf.
Mandelkow *et al.*, 1980b; Palmer *et al.*, 1982).

When studying solutions by temperature jump and UV light detection
a common artifact is the initial apparent decrease of scattering.
This is caused by the "lens" effect due to inhomogeneous heating, and
it may mask the signal from a change in state of polymerization. This
problem does not arise with X-rays since they are insensitive to
temperature-dependent changes in refractive index.

Radiation damage

For X-rays of 0.15 nm wavelength the radiation dose is given by:

$$D = 2.3 \times 10^{-9} \, N_{inc} \, [rad],$$

so that, for an incident beam of 4×10^{11} photons per second, we have
a dose rate of 1000 rad per second. For comparison, doses around
10^9 rad destroy the structures of organic molecules, while about 10^6
rad kill living cells (see Sayre *et al.*, 1977). It has been reported
that the polymerizability of microtubules is strongly inhibited after
irradiation with about 50,000 rad by conventional X-ray sources (Coss
et al., 1981; Zaremba and Irwin, 1981). This would be equivalent to
an exposure of 50 seconds in the synchrotron beam. However, in some
experiments we have exposed solutions of microtubule protein for up
to two hours, with total doses around 500,000 rad, without major
radiation damage. This suggests that biological specimens tolerate
considerably higher doses (about an order of magnitude in this case)
in synchrotron radiation experiments. This experience seems to be
fairly general (see Bordas and Randall, 1978) and may be related to
the higher dose rate.

X-RAY STUDIES OF MICROTUBULE STRUCTURE AND ASSEMBLY

Microtubule structure

We shall illustrate the application of synchrotron X-radiation by examples taken from the studies of microtubules. These are fibrous protein structures, found in most eukaryotic cells, which are involved in a variety of functions such as mitosis, axonal transport, flagellar beating, etc. They consist of globular subunits, tubulin, of molecular weight 50000 (Ponstingl *et al.*, 1981). The chemical building block is a heterodimer of two somewhat different species, alpha and beta tubulin. They self-assemble into hollow cylinders of diameter about 25-30 nm, composed of 13 longitudinal protofilaments (Fig. 4).

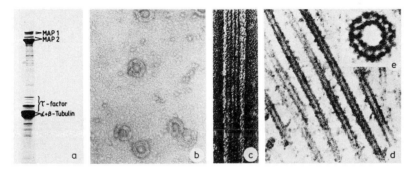

FIG. 4   (a) SDS-polyacrylamide gel pattern of microtubule protein showing the doublet of alpha and beta tubulin and a variety of microtubule-associated proteins (e.g. Tau, MAP1, MAP2). (b) Electron micrograph of negatively-stained microtubule protein at 4°C, showing rings of mean diameter around 35-40 nm and other aggregates. The rings account for about 40% of the protein and contain the MAP fraction; the remainder is mainly in the form of alpha-beta-heterodimers, as judged by analytical ultracentrifugation. (c) Negatively-stained microtubule of diameter about 25 nm, showing longitudinal striations of protofilaments separated by 5.0-5.5 nm. (d) Thin section showing associated proteins attached to the surface of microtubules. Inset: Cross-section through microtubule showing 13 protofilaments in projection.

In the cell, many microtubules are in a dynamic equilibrium with their subunits. This enables them to assemble and disassemble when necessary. The surface is coated with several kinds of associated proteins which may be responsible for the physiological function, commonly termed microtubule-associated proteins, MAPs (Sloboda *et al.*, 1975).

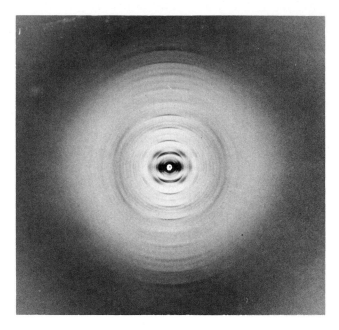

FIG. 5 X-ray pattern of a pellet of oriented microtubules, obtained
on an Elliott GX6 rotating anode generator with double mirror camera
and specimen to film distance of about 10 cm (Harmsen, 1980). Exposure times for such samples are several days. Note the series of
layer lines at orders of 4 nm, extending to a resolution of about
0.4 nm.

Assembly and disassembly may be studied *in vitro* by changing the
temperature between 4 and 37°C (Weisenberg, 1972). This forms the
basis of the temperature-cycle method of preparation used in most
studies (Borisy *et al.*, 1975; Shelanski *et al.*, 1973). Cold microtubule protein ($\equiv$ tubulin and MAPs) consists of tubulin heterodimers
(sedimentation constant 6S) and ring structures containing tubulin
and MAPs (Borisy and Olmsted, 1972). Microtubule protein is capable
of forming a number of polymorphic forms, some of which have been
implicated in the pathway of assembly (reviewed by Kirschner, 1978
and by Amos, 1979). The surface lattice of negatively-stained microtubules has been determined by image processing of electron micrographs (Amos and Klug, 1974), and X-ray patterns of oriented microtubules have been obtained showing reflections up to 0.4 nm resolution
(Cohen *et al.*, 1971; Mandelkow *et al.*, 1977a; see Fig. 5). The general

aim of the experiments to be described is to determine the structures of the initial, final, and intermediate states during assembly, and to characterize the transitions between them in order to distinguish between various models of microtubule assembly.

Figure 6 illustrates the relationship between the oriented fibre pattern of microtubules, the static low angle patterns of solutions containing microtubules or rings, and the temperature-induced transitions between them. As a rough guide one can subdivide the patterns into several angular regions of different structural information. The central scattering peak (inside the beam stop and just outside it) is sensitive to the average degree of polymerization and higher modes of aggregation (e.g. bundling, cross-linking). The subsidiary peaks of Bessel order 0 (see Fig. 6) indicate the concentration and diameter of microtubules or rings. Higher angle peaks (e.g. Bessel terms 13, 3 and 10) arise from the subunit structure and arrangement within the aggregates. In solution the peaks are disoriented into circles and tend to overlap. In principle, an unambiguous interpretation is possible only for the innermost region (see Finch and Holmes, 1967; Federov et al., 1977). In the present case the higher resolution pattern can also partly be disentangled by reference to the known oriented fibre pattern (see below).

Static X-ray experiments

A particularly powerful application of synchrotron radiation is to cases of weakly scattering and dilute samples, especially when the protein is as labile as tubulin which loses its activity within hours at room temperature.

A basic parameter obtainable from small angle scattering without the aid of model calculations is the radius of gyration, which is a measure of the rotationally-averaged particle size. Figure 7 shows a Guinier plot (see Guinier and Fournet, 1955) of tubulin subunits purified by phosphocellulose chromatography. The slope of the curve indicates a radius of gyration of 3.1 nm (Bordas et al., 1983). This value is somewhat larger than one would have expected, considering the molecular weight of tubulin and the monomer separation of 4 nm in the microtubule wall. It could be explained if unassembled tubulin had a shape or conformation different from the assembled state. The

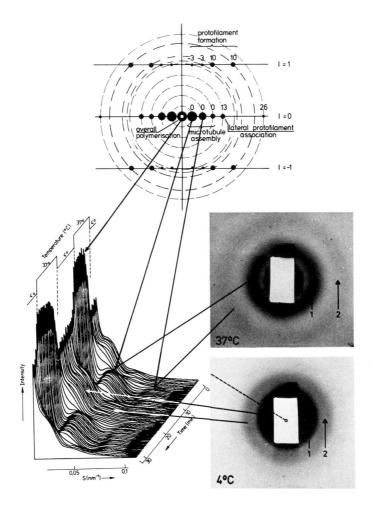

FIG. 6 Top: Diagram of inner region of the oriented fibre pattern out to 2.5 nm resolution, with Bessel term assignment (cf. Mandelkow *et al.*, 1977a) and indication of structural interpretation. This pattern serves as a reference for the solution patterns in which the reflections are smeared out into circles (dashed). Right: Solution patterns of microtubule protein taken on EMBL camera X13, showing subsidiary maxima typical of rings (4°C) and microtubules (37°C). The position of the linear detector is indicated by the dashed line (from Mandelkow *et al.*, 1980b). Left: Scattering traces during two cycles of assembly and disassembly (from Mandelkow *et al.*, 1980a). Note the change in height of the central scatter and in the subsidiary maxima. Long arrows connect corresponding peaks in the oriented fibre and solution scattering patterns.

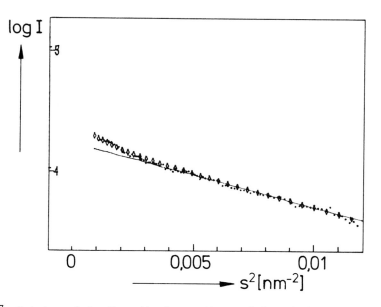

FIG. 7 Guinier plot of scattering pattern of 6S tubulin heterodimers.
The linear slope indicates a radius of gyration of 3.1 nm. The rise
towards the centre shows the presence of protein oligomers. From
Bordas *et al.* (1983).

rise towards smaller scattering angles comes from the tendency of the

protein to oligomerize. The effect was always observed, even after

removing aggregates by high speed centrifugation. This behaviour of

the protein is probably important for the assembly of microtubules

(see below). The size of the tubulin subunit obtained in this experi-

ment is used when comparing the observed scattering traces at differ-

ent states of aggregation with model scattering curves.

A second example of static experiments is the comparison of micro-

tubules and rings in the low and intermediate angle regions. The

results may be compared with various models of ring structure derived

from electron microscopy. Some of the structural models have been

used to imply different pathways of microtubule assembly (reviewed by

Kirschner, 1978). The basic differences may be summarized as follows:

in a microtubule the long axis of the tubulin dimers points in the

direction of the longitudinal protofilaments, while in rings the dimer

axis could in principle point along the ring periphery (Voter and

Erickson, 1979), in a radial direction (Frigon and Timasheff, 1975),

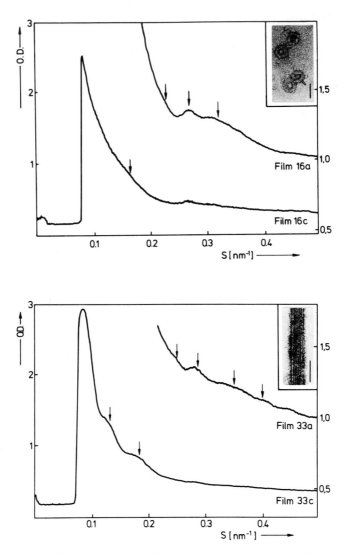

FIG. 8 Intermediate angle scattering patterns of solutions of (a)
microtubules at 37°C and (b) rings at 4°C, taken on X-ray film and
processed by densitometry and circular integration. The bottom trace
shows the high intensity region revealing the equatorial J_0 and J_{13}
maxima in (a) which are absent in (b). The top traces show the weaker
intermediate angle scattering patterns. The maxima arise mainly from
the subunit packing. The similarity indicates that rings may be considered as coiled protofilaments. From Mandelkow *et al.* (1983a).

or out of the plane of the ring (Scheele and Borisy, 1978). In the
first case, the ring would be a coiled protofilament, in the third it
would be related to one of the helices running around a microtubule.
Since the lateral and longitudinal intersubunit vectors are different
the cases can be distinguished by solution scattering of rings (Fig.
8). The X-ray patterns support the first possibility (coiled protofilaments), and any assembly model has to be compatible with this
structure (Mandelkow et al., 1983a).

Temperature-jump experiments

Having established the static patterns of the initial and final states,
one would like to observe the transitions between them. This may be
done by raising the temperature from cold (0-4°C) to warm (37°C) in a
time short compared to the assembly process, and by monitoring the
scattering pattern in short time-frames. This yields the rate constants for those structural transitions which contribute to the scattering region under observation. For example, all transitions by
which the degree of polymerization is altered will contribute to the
central scatter, whereas the region around 0.5 nm^{-1} is particularly
sensitive to microtubule formation. Figure 6 (left) shows an example
of two cycles of assembly and disassembly. The sample cell is contained in a thermostatted copper block connected to four water baths.
Two of them are set at the desired initial and final temperatures,
the other two at higher and lower settings (70 and -20°C respectively).
This allows the final equilibrium to be obtained quickly by using
a brief pulse of overshoot temperature. The current temperature is
recorded by a thermoresistor inside the solution and stored along
with the scattering data.

The temperature jump induces dramatic changes in the scattering
pattern (Mandelkow et al., 1980a). After a brief drop the central
scatter rises steeply, and new maxima and minima are formed at higher
scattering angles. Most of the rising phase is due to the formation
of microtubules. However, the initial undershoot indicates net disassembly. This reaction may be analysed by subtracting the scattering
at 30 seconds after the T-jump from the initial one, which procedure
should yield the contribution of species which disappear during this
phase. The difference plot (Fig. 9) is in good agreement with the

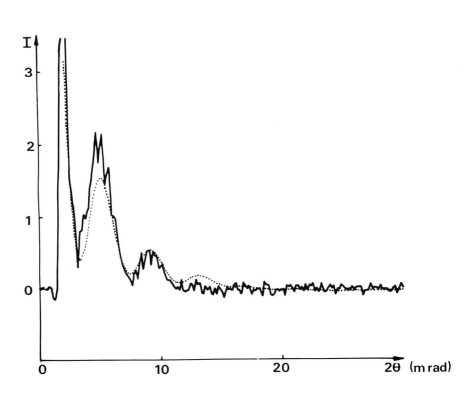

FIG. 9 Difference pattern obtained by subtracting the pattern at 30 seconds after the temperature jump from the initial cold pattern (see experiment of Fig. 5, left). The model curve was calculated for tubulin rings. This shows that rings tend to break apart prior to microtubule assembly. From Mandelkow *et al.* (1980a).

theoretical scattering expected of tubulin rings. In other words, rings tend to break apart before microtubule assembly. This means that the rings are not likely to act as nucleation centres, as assumed in some assembly models. That role, if any, must be transferred to the ring fragments, i.e. tubulin oligomers and/or subunits, whose interaction is somehow responsible for the following phases of nucleation and microtubule elongation. Since rings consist of coiled protofilaments (see above), the ring fragments are short pieces of protofilaments.

Near-equilibrium temperature scans

If the temperature is raised gradually rather than by a jump, the reactions take place at or close to equilibrium. This enables one to

study the sequence of structural changes with better statistical accuracy (Bordas *et al.*, 1983). From a practical point of view, the
lowest heating rates were around $1^{\circ}C$/minute during main-user shifts
at DORIS, equivalent to a total exposure time of around 100 minutes
for a complete cycle ($0^{\circ}C$ - $37^{\circ}C$ - $0^{\circ}C$). The limitation is mainly
caused by instabilities in the beam rather than by radiation damage,
at least at particle energies below 5 GeV. *A priori*, the assembly
pathways observed during temperature scans are not necessarily the
same as those of temperature jumps, especially when nucleation is the
limiting step. For example, the slowness of a scan affects the relative rates of nucleation and elongation (Hinz *et al.*, 1979). The
influence of heating rate was noticed in the X-ray experiments. However, as shown below, the types of structural transitions were similar
for fast and slow temperature changes.

Figure 10a shows the central scatter as a function of temperature
for one complete cycle. During the rising phase (lower branch) there
is an undershoot indicating net depolymerization. Assembly starts at
point C, passes through a transition at D and reaches its final equilibrium at F. During the temperature drop (upper branch), the intensity stays roughly constant until point H and then decreases. In this
representation, the most noticeable feature is the hysteretic behaviour
of assembly: the midpoints of the assembly and disassembly branches
are displaced by about $12^{\circ}C$. This shows that the reaction is still
kinetically limited, even at slow heating rates. The limitation lies
mainly in the reaction taking place between C and D. The final level
of the central scatter is the same as the initial level, indicating
that the average degrees of polymerization are the same before and
after the cycle. This appearance is deceptive, however. If one plots
the intensities at other scattering angles, the hysteresis loops fail
to close. Thus, the structures in the two states are not identical.

The structural interpretation of the initial phase A-C is similar
to that of the temperature jump, i.e. there is a biphasic breakdown
of rings and other aggregates. The period following D may be interpreted in terms of microtubule elongation by endwise addition of subunits. The events between C and D are more complex; in fact, one
first has to establish point D as a separate transition. This may be

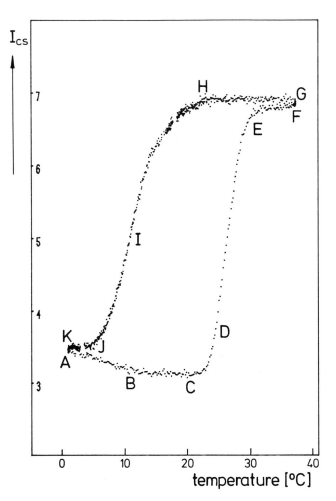

FIG. 10 Data derived from slow temperature scans (heating rate 1-2°C/ minute). (a) Central scattering intensity as a function of temperature; lower branch is assembly, upper branch is disassembly. The hysteresis loop arises because assembly is nucleation limited, even at slow heating rates. Stages A-C = prenucleation events (e.g. ring dissolution), C-D = nucleation, D-F = elongation and post-assembly events, G-I = microtubule disassembly, I-K = further disassembly and reformation of rings.

achieved by cross-correlating different parts of the scattering pattern (Fig. 10b). If there is a transition between the two states, the correlation diagram is expected to show a straight line. Conversely, intermediate reactions are revealed by kinks in the correlation plot. The diagram shows that D represents a distinct transition, in addition

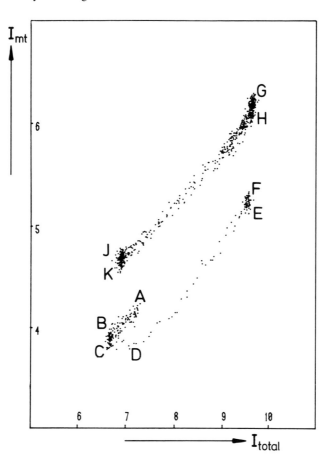

FIG. 10 (b) Cross-correlation of the total intensity received by the detector and the intensity at 0.5 nm^{-1} (first subsidiary maximum of microtubule pattern). The breakpoints correspond to the transition from one type of reaction to another and serve to define the stages of assembly. Assembly and disassembly branches are displaced vertically.

to B and C. The scattering patterns at this point are difficult to measure accurately because the transition is short-lived, even during temperature scans. However, the average of several experiments shows that it is best explained by a mixture of short elongated species equivalent to laterally associated protofilament fragments. The most plausible interpretation is that the period C-D represents microtubule nucleation by lateral association of protofilament fragments.

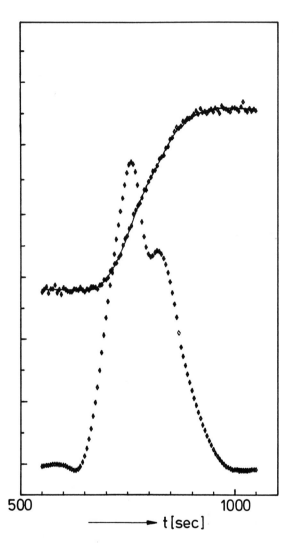

FIG. 10 (c) Central scatter (top) and its derivative (bottom) with respect to time or temperature. The two maxima coincide roughly with the midpoints of nucleation and elongation. From Bordas *et al.* (1983).

An alternative method of separating the phases of nucleation and elongation is illustrated in Fig. 10c which shows the derivative of the central scatter with respect to temperature. If assembly followed a simple isodesmic mechanism where the association reactions of subunits are equal and independent of the size of the aggregate, one would expect one maximum in the derivative curve (see, for example,

Sturtevant *et al.*, 1981). In the case of the central scatter there
are two clear peaks, indicating first the assembly of nuclei and then
that of microtubules. If a derivative plot is constructed using the
intensity at 0.5 nm^{-1} we observe only one peak, corresponding to the
elongation phase, because this scattering angle is dominated by the
presence of complete microtubules.

Stopped-flow experiments

Assembly by temperature-jump requires the presence of GTP which is
hydrolysed during the reaction. An alternative is to mix the protein
rapidly with GTP at a constant temperature of $37^{\circ}C$. This type of
experiment is presumably more closely related to the events taking
place in the cell. In Fig. 11 we compare the time course of the central scatter following a temperature jump from 4 to $37^{\circ}C$ or after
mixing with GTP at $37^{\circ}C$. The T-jump experiment shows the usual brief
undershoot (ring dissolution), followed by a steep rise (nucleation,
growth of microtubules). The mixing experiment first shows an apparent sudden intensity drop caused by the filling of the chamber with
solution after passing the pre-mixing chamber (dashed portion of the
curve). After that we also observe a brief drop and a steep rise.
The structural interpretation is the same as in (a), as deduced from
the scattering traces. This means that the structural transitions
accompanying microtubule assembly depend on assembly conditions, but
- to a first approximation - they do not depend on how one reaches
assembly conditions, i.e. by adding heat at constant GTP, or by adding
GTP at constant temperature, provided that the change in conditions
is rapid compared with the reaction rates. We conclude that the
experimentally simpler temperature jump method yields valid information on microtubule assembly. Conversely, if microtubule assembly in
the cell differs from that *in vitro,* it is probably not due to the
constant temperature but to some other factors.

Comparison with electron microscopy

Since the X-ray data often cannot be interpreted unambiguously, it is
desirable to obtain as much independent evidence as possible. For
example, many states of assembly of tubulin have been detected by
electron microscopy. Compared with X-rays, this method has the

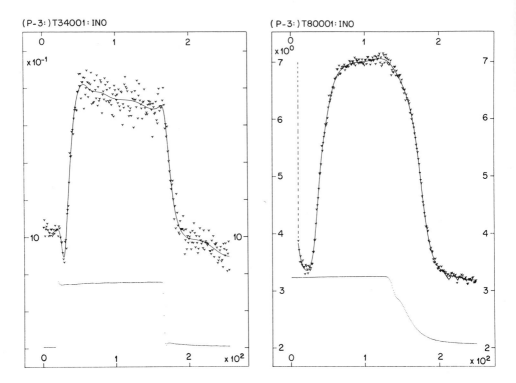

FIG. 11 Central scatter of solutions of microtubule protein after
(a) T-jump and (b) rapid mixing time. Data were collected in 256
time frames of 2 seconds each. Dots, temperature; crosses, intensity;
solid line, intensities smoothed by cubic spline function. (a) Assembly and disassembly were induced by T-jumps from 4 to 37°C and back
to 4°C. Note the initial undershoot, followed by an intensity overshoot. (b) Assembly was induced by mixing the protein with GTP at
37°C, disassembly by a slow temperature drop to 4°C. Note the undershoot after filling of the cell (dashed).

obvious advantage of providing images rather than scattering patterns.
The disadvantages include not only the well-known artifacts of preparation, but, more to the point in the present context, the difficulty
of obtaining time resolution, the low protein concentration required
for negative stain work (which causes disassembly of the structures
one tries to observe), the insensitivity towards small and non-perio-
dic aggregates (which tend to be disregarded as "background"), and
finally the problem of obtaining quantitative information about the
average state of assembly. We therefore found it necessary to use
X-ray scattering and electron microscopy in parallel on the same or

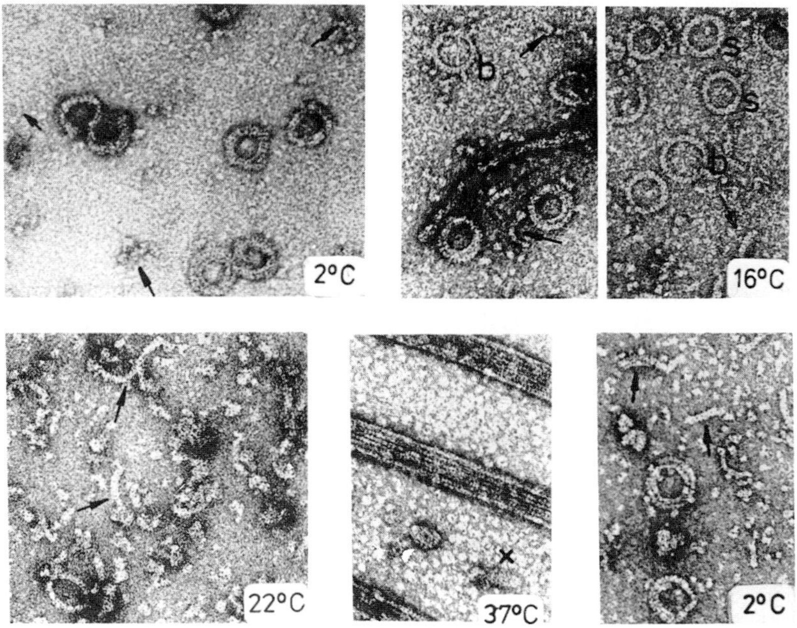

FIG. 12 Stages of microtubule assembly monitored by electron microscopy. (a) and (b), rings (mostly double concentric) and other aggregates observed at 2°C; (c) at 16°C there are fewer rings, the remaining ones are often single, with diameters corresponding to the turns
of the double rings in (a); (d) ring fragments = tubulin oligomers at
22°C; (e) microtubules at 37°C; (f) rings, incomplete rings, and
oligomers after return to 2°C. From Bordas et al. (1983).

similar protein preparations. Information on kinetics and weight
averages of structural transitions were obtained from X-ray experiments, whereas individual particles were observed by electron microscopy.

Figure 12 shows a temperature scan similar to that of Fig. 10
monitored by electron microscopy (Bordas et al., 1983). It shows the
rings present in the initial cold solution, their gradual breakdown
into fragments and subunits, the microtubules at 37°C, and the final
cold state after reversing the scan, showing many half-finished rings.
The images illustrate that there is no direct transition between the
major states of aggregation seen in the electron microscope (i.e.
rings and microtubules) so that there must be smaller building blocks
(oligomers and subunits) from which the higher forms of aggregation
are constructed.

Overshoot assembly

From theoretical considerations of nucleated assembly, one expects an
initial phase of formation of nuclei, followed by a phase of elongation during which almost no further nucleation occurs. This phase
should therefore show a nearly exponential approach to saturation,
since the number of polymer ends stays constant and all subunits are
assumed to have identical probabilities of binding to the ends. This
behaviour has been demonstrated for microtubules (Gaskin *et al.*,
1974; Bryan, 1976; Engelborghs *et al.*, 1976). The second phase is
particularly clear with seeded assembly where the nucleation phase is
bypassed (Johnson and Borisy, 1977). However, careful analysis of
the progress curves shows that they often deviate from the ideal
behaviour. In part this may often be explained by the mode of detection. For example, turbidity is not a strictly linear function of
the degree of polymerization as long as the particles are smaller than
the wavelength of light (Fig. 3), and even when this condition is
satisfied, one has to apply corrections (Carlier and Pantaloni, 1978).
This problem is inherently less severe with X-rays since their wavelength is always much smaller than the particle size.

If assembly is induced by temperature jump, the approach to equilibrium is often faster than expected for an exponential time course,
indicating that there may be more reactions than microtubule growth
(see Fig. 11a). Under certain assembly conditions (e.g. by varying
cation concentration or ionic strength) this behaviour can be amplified. Figure 13 depicts an example where the intensity shows a large
overshoot following the temperature jump, then relaxes to an intermediate level followed by another rise. All of these events take
place at the constant temperature of 37°C. Analysis of the scattering
curves shows that, during the initial overshoot period, there is a
mixture of microtubules and incomplete microtubule walls which fail
to close into cylinders. The intensity decrease is mainly caused by
the disappearance of the incomplete aggregates, and the second rise
shows the formation of more microtubules. Thus, the overall assembly
process may be subdivided into two parts: one is the usual assembly
of intact microtubules, the second is the transient overshoot assembly
of incomplete microtubules followed by a redistribution phase in which

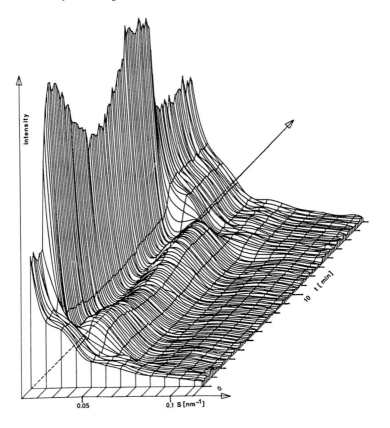

FIG. 13 Temperature jump experiment showing pronounced kinetic overshoot. Note the central scatter (left) which rises rapidly after a
small undershoot following the T-jump to 37°C. The intensity then
drops and rises again at constant temperature. Finally, disassembly
is induced by a reverse T-jump to 4°C. From Mandelkow *et al.* (1983b).

they are dismantled and their subunits re-used to form intact microtubules.

ASSEMBLY MODEL FOR MICROTUBULES AND POLYMORPHIC FORMS

Experiments like that of Fig. 13 can be explained within the framework of nucleation-condensation assembly (Oosawa and Asakura, 1975)
if the usual assumption of linear or helical assembly is dropped in
favour of a two-dimensional assembly scheme. With one-dimensional
assembly, overshoot can only take the form of a non-equilibrium length
distribution. This has been observed, for example, for actin (Oosawa,
1970), tobacco mosaic virus protein (Schuster *et al.*, 1979), and also

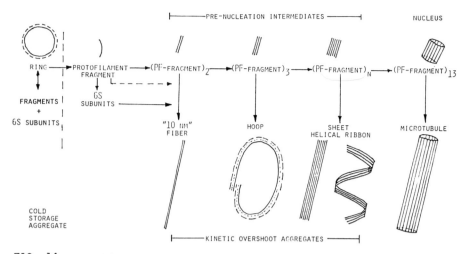

FIG. 14 Assembly scheme for microtubules and polymorphic forms. The
initial cold state contains rings in equilibrium with oligomers and
6S subunits. After raising the temperature the rings break apart into
ring fragments (= short protofilaments) and subunits. Nucleation
takes place by lateral association of protofilament fragments (top
row). Protofilaments can be elongated by subunits at any stage. If
nucleation is slowed down relative to elongation one observes a variety
of polymorphic overshoot aggregates whose intersubunit bonds are similar to microtubules, but their gross shape depends on slight conformational differences (e.g. fibres, hoops, sheets, twisted ribbons).
They tend to disintegrate during a redistribution phase so that their
subunits can be incorporated into microtubules. From Mandelkow *et al.*
(1983b).

with microtubules (Johnson and Borisy, 1977). With two-dimensional

assembly, overshoot can generate new polymorphic forms, depending on

the relative rates of lateral and longitudinal growth and the nuclea-

tion conditions (Mandelkow *et al.*, 1983b). Microtubules are a case in

point because their subunits form a two-dimensional lattice constrained

to the surface of a cylinder (cf. Erickson and Pantaloni, 1980).

As described elsewhere, the effective rate constants can be

changed in several ways, e.g., by solution conditions, protein

concentration, heating rate, etc. In the electron microscope, one

observed a number of polymorphic aggregates represented diagram-

matically in Fig. 14. Although they differ in overall shape, they

share two important properties: (a) the surface lattices of

tubulin subunits are similar to that of microtubules, and

(b) they occur preferentially at early times in assembly, under conditions where nucleation is slow (e.g. at low protein concentrations), and they tend to disappear again with time. This behaviour is observed most prominently with tubulin hoops which are nucleated by a triplet of protofilaments. Such triplets therefore seem to play a special role during assembly, possibly because they are associated with particular microtubule-associated proteins (Mandelkow et al., 1977b). In these two respects, the polymorphic forms considered here differ from others which are induced by microtubule poisons such as vinblastine or zinc and do not have a microtubule-like surface lattice (reviewed by Amos, 1979).

The structural properties of the tubulin aggregates and the kinetic data are consistent with the scheme of Fig. 14. In it, the rings are considered as cold-storage aggregates in equilibrium with fragments and subunits, in agreement with analytical ultracentrifugation results (Marcum and Borisy, 1978; Bayley et al., 1982). As the temperature is raised, the rings progressively disappear. Their breakdown products (i.e. short protofilament fragments) may further disassemble into subunits, but they may also bind laterally. This is the start of the nucleation sequence which eventually leads to a small microtubule cylinder (the shortest lengths observed are on the order of 40-100 nm). From then on, further assembly takes place by elongation of protofilaments via subunits (or possibly oligomers). Note that the processes of both nucleation and elongation are considered non-helical, although the final products (microtubules) have a near-helical structure.

Under "normal" assembly conditions, nucleation would be fast compared to elongation. This would ensure the formation of properly closed microtubules. However, if nucleation is slowed down in relative terms, e.g. by reducing the concentration of nucleating species (tubulin oligomers) relative to the elongating species (subunits), then elongation may become appreciable before cylindrical closure of the tubule. This results in polymorphic forms whose shapes depend on slight differences in the conformation of tubulin (e.g., they often have axial as well as radial curvature). The appearance of these structures depends on the kinetics of assembly rather than on differences in the binding constants. Thus microtubules may be considered

as the most stable form, having the largest number of intersubunit
bonds. As a result there will be a redistribution phase during which
the polymorphic forms disappear in favour of microtubules. This is the
two-dimensional analogue of the length redistribution phase observed
with linear or helical polymers (Oosawa and Asakura, 1975). Note that
in this scheme there is in general no direct conversion of the kinetic
overshoot aggregates into microtubules, other than by disassembly
(such transitions have been postulated in some assembly models).

Many details of the assembly scheme have yet to be elucidated.
However, the results described here should serve as an illustration
of how one can combine the structural and kinetic results from time-
resolved X-ray scattering with data from other methods in order to
arrive at a better understanding of protein self-assembly.

ACKNOWLEDGEMENTS

We thank our colleagues C. Boulin, A. Gabriel, J. Hendrix, M. Koch and
the staff at EMBL Hamburg for their substantial contributions to the
development of the facilities which were used in these studies, as
well as A. Harmsen at the MPI Heidelberg for his involvement during
the early stages of this work, and K. Maier for building some of the
equipment for the kinetic studies.

REFERENCES

Amos, L.A. and Klug, A. (1974). *J. Cell Sci.* 14, 523-549.
Amos, L.A. (1979). *In* "Microtubules". (Eds K. Roberts and J.S. Hyams),
    pp. 1-64, Academic Press, New York.
Bayley, P.M., Charlwood, P.A., Clark, D.C. and Martin, S.R. (1982).
    Eur. J. Biochem. 121, 579-585.
Berne, B.J. (1974). *J. Mol. Biol.* 89, 755-758.
Bordas, J. and Randall, J.T. (1978). *J. Appl. Cryst.* 11, 434-441.
Bordas, J. and Koch, M.H.J. (1979). *In* "Applications of Synchrotron
    Radiation to the Study of Large Molecules of Chemical and Biologi-
    cal Interest". (Eds R.B. Cundall and I.H. Munro), pp.
    Science Research Council, Daresbury Laboratory DL/SCI/R13, U.K.
Bordas, J., Koch, M.H.J., Clout, P.N., Dorrington, E., Boulin, C.
    and Gabriel, A. (1980). *J. Phys. E: Sci. Instrum.* 13, 938-944.
Bordas, J. and Mandelkow, E. (1983). *In* "Fast Methods in Physical
    Biochemistry and Cell Biology". (Eds R.I. Sha'afi and S.M. Fernan-
    dez), pp. 137-172. Elsevier/North-Holland Biomedical Press.
Bordas, J. (1983). *In* "Uses of Synchrotron Radiation in Biology".
    (Ed. H. Stuhrmann). Academic Press, New York.
Bordas, J., Mandelkow, E.-M. and Mandelkow, E. (1983). *J. Mol. Biol.*
    164, 89-135.
Borisy, G.G. and Olmsted, J.B. (1972). *Science* 177, 1196-1197.

Borisy, G.G., Marcum, J.M., Olmsted, J.B., Murphy, D.B. and Johnson, K.H. (1975). *Ann. N.Y. Acad. Sci.* 253, 107-132.

Boulin, C., Dainton, D., Dorrington, E., Elsner, G., Gabriel, A., Bordas, J. and Koch, M.H.J. (1982). *In* "X-Ray Detectors for Synchrotron Radiation". (Eds J. Bordas, R. Fourme and M.H.J. Koch), Nucl. Instrum. Meth. 201, 209-220.

Bryan, J. (1976). *J. Cell Biol.* 71, 749-767.

Carlier, M.F. and Pantaloni, D. (1978). *Biochemistry* 17, 1908-1915.

Cohen, C., Harrison, S.C. and Stephens, R.E. (1971). *J. Mol. Biol.* 59, 375-380.

Coss, R.A., Bamburg, J.R. and Dewey, W.C. (1981). *Radiation Res.* 85, 99-115.

Engelborghs, Y., Heremans, K., de Maeyer, L. and Hoebeke, J. (1976). *Nature* 259, 686-689.

Erickson, H.P. and Pantaloni, D. (1981). *Biophys. J.* 34, 293-309.

European Science Foundation (ESF) Proposal for the European Synchrotron Radiation Facility (ESRF) (1979). Suppl. I: The Scientific Case (Eds Y. Farge and P.J. Duke), Suppl. II: The Machine (Eds G. Thompson and M. Poole), Suppl. III: Instrumentation (Eds B. Buras and G.V. Marr), ESF, 1 Quai Lezay-Marnesia, Strasbourg, France.

Federov, B.A., Shpungin, I.L., Gelfand, V.I., Rosenblat, V.A., Dama-Schun, G., Damaschun, H. and Papst, M. (1977). *FEBS Lett.* 84, 153-155.

Finch, J.T. and Holmes, K.C. (1967). *In* "Methods in Virology", Vol. 3, (Eds K. Maramorosch and H. Koprowski), pp. 351-474, Academic Press, New York.

Frigon, R&P% and Timasheff, S.N. (1975). *Biochemistry* 14, 4567-4573.

Gabriel, A. (1977). *Rev. Sci. Instrum.* 48, 1303-1305.

Gaskin, F., Cantor, C.R. and Shelanski, M.L. (1974). *J. Mol. Biol.* 89, 737-758.

Golding, F. (1982). *In* "X-Ray Detectors for Synchrotron Radiation". (Eds J. Fourme, J. Bordas and M.H.J. Koch), Nucl. Instrum. Meth. 201, 231-235.

Guinier, A. and Fournet, G. (1955). "Small Angle Scattering of X-rays". Wiley, New York.

Harmsen, A. (1980). Ph.D. Thesis, University of Heidelberg.

Hendrix, J., Koch, M.H.J. and Bordas, J. (1979). *J. Appl. Cryst.* 12, 467-472.

Hinz, H.-J., Gorbunoff, M.J., Price, B. and Timasheff, S.N. (1979). *Biochemistry* 18, 3084-3089.

Huxley, H.E. and Brown, W. (1967). *J. Mol. Biol.* 30, 383-434.

Huxley, H.E., Faruqi, A.R., Bordas, J., Koch, M.H.J. and Milch, J. (1980). *Nature* 284, 140-143.

Johnson, K.A. and Borisy, G.G. (1977). *J. Mol. Biol.* 117, 1-31.

Kirschner, M.W. (1978). *Int. Rev. Cytol.* 54, 1-71.

Koch, M.H.J. and Bendall, P. (1981). "Proc. Digital Equipment Computer User Society", pp. 13-16. Warwick, UK.

Koch, M.H.J., Stuhrmann, H.B., Vachette, P. and Tardieu, A. (1983). *In* "Uses of Synchrotron Radiation in Biology". (Ed. H.B. Stuhrmann), pp. 223-253. Academic Press, New York.

Koch, M.H.J. and Bordas, J. (1983). *Nucl. Instrum. Meth.*, in press.

Kunz, Chr. Ed. (1979). *In* "Synchrotron Radiation Techniques and Applications". Springer-Verlag, New York.

Mandelkow, E., Thomas, J. and Cohen, C. (1977a). *Proc. Natl. Acad. Sci. USA* 74, 3370-3374.
Mandelkow, E.M., Mandelkow, E., Unwin, P.N.T. and Cohen, C. (1977b). *Nature* 265, 655-657.
Mandelkow, E.M., Harmsen, A., Mandelkow, E. and Bordas, J. (1980a). *Nature* 287, 595-599.
Mandelkow, E., Harmsen, A., Mandelkow, E.M. and Bordas, J. (1980b). *In* "Microtubules and Microtubule Inhibitors 1980" (Eds M. DeBrabander and J. DeMey), pp. 105-117, Elsevier, Amsterdam.
Mandelkow, E., Mandelkow, E.M. and Bordas, J. (1983a). *J. Mol. Biol.* 167, 179-196.
Mandelkow, E., Mandelkow, E.M. and Bordas, J. (1983b). *Trends in Bioch. Sci.* 8, 374-377.
Marcum, J.M. and Borisy, G.G. (1978). *J. Biol. Chem.* 253, 2825-2833.
Oosawa, F. (1970). *J. Theoret. Biol.* 27, 69-86.
Oosawa, F. and Asakura, S. (1975). *In* "Thermodynamics of the Polymerisation of Protein". Academic Press, London.
Palmer, G.R., Clark, D.C., Bayley, P.M. and Satelle, D.B. (1982). *J. Mol. Biol.* 160, 641-658.
Postingl, H., Krauhs, E., Little, M. and Kempf, T. (1981). *Proc. Natl. Acad. Sci. USA* 78, 2757-2761.
Renner, W., Mandelkow, E.-M., Mandelkow, E. and Bordas, J. (1983). *Nucl. Instrum. Meth.* in press.
Rosembaum, G., Holmes, K.C. and Witz, J. (1971). *Nature* 230, 434-437.
Rosenbaum, G. and Holmes, K.C. (1980). *In* "Synchrotron Radiation Research". (Eds. H. Winick and S. Doniach), pp. 533-564. Plenum Press, New York.
Sayre, D., Kirz, J., Feder, R., Kim, D.M. and Spiller, E. (1977). *Ultramicroscopy* 2, 337-349.
Scheele, R.B. and Borisy, G.G. (1978). *J. Biol. Chem.* 253, 2846-2851.
Schuster, T.M., Scheele, R.B. and Khairallah, L.H. (1979). *J. Mol. Biol.* 127, 461-485.
Shelanski, M.L., Gaskin, F. and Cantor, C.R. (1973). *Proc. Natl. Acad. Sci. USA* 70, 765-768.
Sloboda, R.D., Rudolph, S.A., Rosenbaum, J.L. and Greengard, P. (1975). *Proc. Natl. Acad. Sci. USA* 72, 177-181.
Stuhrmann, H.B. Ed. (1983). *In* "Uses of Synchrotron Radiation in Biology". Academic Press, New York.
Sturtevant, J.M., Velicelebi, G., Jaenicke, R. and Lauffer, M.A. (1981). *Biochemistry* 20, 3792-3800.
Voter, W.A. and Erickson, H.P. (1979). *J. Supramol. Struct.* 10, 419-431.
Weisenberg, R.C. (1972). *Science* 177, 1104-1105.
Weisenberg, R.C. and Rosenfeld, A. (1975). *Ann. N.Y. Acad. Sci.* 253, 78-89.
Winick, H. and Doniach, S. Eds (1980). *In* "Synchrotron Radiation Research". Plenum Press, New York.
Zaremba, T.G. and Irwin, R.D. (1981). *Biochemistry* 20, 1323-1332.

Iodine Laser Temperature-jump: Relaxation Processes in Phospholipid Bilayers on the Picosecond to Millisecond Time-scale

J.F. HOLZWARTH, V. ECK and A. GENZ

INTRODUCTION

The investigation of fast dynamic processes in biological systems
depends strongly on suitable experimental techniques giving good time-
resolved signals, which can be interpreted without the requirement of
dominating model assumptions. Most chemical equilibria are connected
with an increase in enthalpy caused by a rise in temperature. This
fact can be utilized in the powerful temperature jump (T jump) kinetic
method, provided that the temperature jump is achieved in a very short
time without producing other unwanted effects like transient electric
fields, shockwaves or photochemical reactions. We have therefore
developed a laser T-jump arrangement for aqueous solutions using the
absorption of photons in the near infrared region of the most important biological solvent, water. The apparatus has been used to investigate the main phase transition, T_m, of vesicles composed of phospholipids, the simplest cell-like aggregates, to demonstrate the potential of our method for the characterization of fast dynamic phenomena
in biological systems.

The relevance of the temperature-dependent order to disorder transition of phospholipid bilayers for their function in membranes of
living cells as transport barriers, protein matrices and enzyme activity switches, has already been demonstrated. In the following, an
attempt is made to contribute to the knowledge about the dynamics

SPECTROSCOPY AND THE DYNAMICS
OF MOLECULAR BIOLOGICAL SYSTEMS

JOULE HEATING

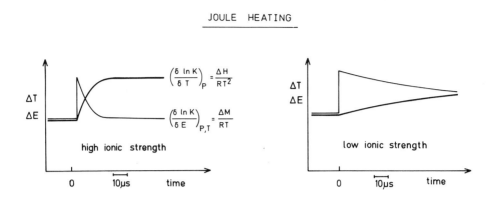

FIG. 1 Time dependence of the energy in the electric field E and the temperature T in T-jump arrangements which use the discharge of a capacitor through the sample as energy source.

which are responsible for these control functions, by offering an extremely flexible biophysical technique and its application to membrane-like aggregates. The difficulties arising from commercially available T-jump arrangements are indicated in Fig. 1. The strong transient electric field across the sample which produces the T-jump by driving ions between two electrodes (Joule-heating) induces unwanted physical and chemical changes, superimposed upon the temperature effect, which are especially crucial in biological systems. These side-effects can be tolerated only in simple systems where they decay fast enough to their former state of equilibrium, leaving the subsequent temperature effect undisturbed. Besides these difficulties, the time resolution of Joule-heating depends on the ionic strength of the sample given by the product resistance times capacity (RC/2). All these restrictions made it essential to develop a better T-jump technique (Holzwarth, 1979).

IODINE-LASER TEMPERATURE-JUMP TECHNIQUE

Many attempts have been made to use a laser as the energy source to produce fast T-jumps in aqueous solutions. These experiments were

performed with commercial systems like neodymium or rubidium lasers.
None of the arrangements presented in the literature produced more
than partially satisfactory results (Holzwarth, 1979). They either
used dyes which were photodecomposed as energy absorbers, or the
laser wavelength was shifted by stimulated Raman emission into an
infrared region where the absorption of water is too strong for homogeneous temperature rises in a layer of one millimetre or more. An
absorption of 0.7 to 1 cm^{-1} would be optimal to heat a layer of 2 mm
with the laser beam passing through the sample twice. In Fig. 2 we
have summarized the problems resulting from the absorption of water
in the near infrared region. Only the iodine laser with a wavelength
of 1.315 μm is capable of producing less than 10 per cent difference
in temperature rise between the entrance and the exit windows of the
measuring cell. By passing the laser beam twice through a water layer

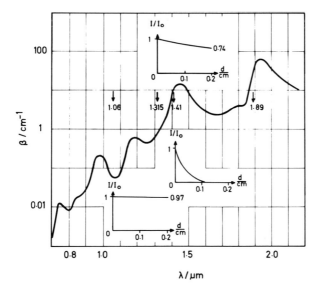

FIG. 2 Absorption spectrum of water and the distance dependence of
the intensity for three wavelengths to demonstrate the different
temperature gradients caused in water samples exposed to photons of
the energies indicated.

TABLE 1   The modular iodine laser (top) and its threshold phenomena (bottom)

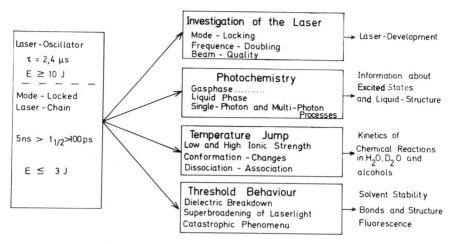

Laser - Oscillator		
$\tau = 2{,}4 \ \mu s$		
$E \geq 10 \ J$		
Mode - Locked Laser - Chain		
$5 ns > t_{1/2} > 100 \ ps$		
$E \leq 3 \ J$		

Investigation of the Laser
Mode - Locking
Frequence - Doubling
Beam - Quality
→ Laser - Development

Photochemistry
Gasphase ....
Liquid Phase
Single - Photon and Multi - Photon
                                    Processes
Information about
Excited States
and Liquid - Structure

Temperature   Jump
Low and High Ionic Strength
Conformation - Changes
Dissociation - Association
Kinetics of
Chemical Reactions
in H_2O, D_2O and
alcohols

Threshold   Behaviour
Dielectric Breakdown
Superbroadening of Laserlight
Catastrophic Phenomena
Solvent Stability
Bonds and Structure
Fluorescence

Normal   Behaviour :   Excitation of Rotations, Vibrations, (Electrons)
                         and their Relaxation

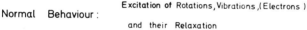

Dielectric   Breakdown
Avalanche   Ionization
(multiphoton - ionization) :
with tunnelling mechanism

Hot   Electrons
Thermalised   Electrons
Solvated   Electrons
Different Kinds of Ions

Recombination
Processes

Superbroadening :   ⟶   White   Lightpulse   "Very Fast
                                                    Flashlamp"

Self - Focusing :   ⟶   Catastrophic   Phenomena   Yablonovitsch
                            (Liquids are self healing!)   Pulse

of 2 mm by the use of a reflecting mirror, a temperature gradient of
less than 6 per cent inside the sample can be achieved. In this way
it is possible to avoid shockwaves which would disturb the observation
channel in the microsecond to millisecond time range.

A summary of the performance and possible applications of the
iodine laser is given in Table 1. In the following part we describe
only the iodine laser T-jump and its principal limitations. Details
about the laser itself are to be found in the literature (Holzwarth,
1979; Brederlow et al., 1983). The long time limit of a T-jump is
given by the deviation of temperature from its value immediately after
the T-jump. This time is around 1 sec for a temperature decrease of
10 per cent with water surrounding the heated volume. If special care
is taken and the heated sample is well isolated, much longer times
can be achieved.

The highest possible time resolution is given by the emission time
of the laser and the thermalization of the absorbed photon energy.
The shortest emission time of an iodine laser is around 10^{-10} s. The
lifetime of the excited rotational-vibrational states of water is
around 10^{-13} s (Goodall and Greenhow, 1971), but the temperature
equilibration between all water molecules takes longer. Figure 3 gives
a schematic model for this process, and Table 2 summarizes the results.
If we assume a hopping model for equilibration, 1.5×10^{-10} s are
needed for a homogenous energy distribution throughout a sample of
140 µl, compared with 9×10^{-11} s calculated from the thermal conductivity of water, and we can conclude that the shortest available
emission time of the iodine laser is about the same as the physical
barrier set by thermodiffusion in water.

This barrier for energy equilibration in the sample coincides with
the power threshold of water, 1.5×10^{10} W/cm^2 (Natzel et al., 1982),
which is equivalent to 1 J emitted in 10^{-10} s into an area of 0.7 cm^2.
Following discussion of the physical limits of an iodine laser T-jump,
we should mention the limitations set by electronic registration.
The best oscilloscopes have a bandwidth of 10^9 Hz and a sensitivity
of 10 mV/cm. If suitable photomultipliers or photodiodes are used,
rise times of 5×10^{-10} s are measurable (Frisch et al., 1979). For
faster signals a streak camera and special oscilloscopes like the

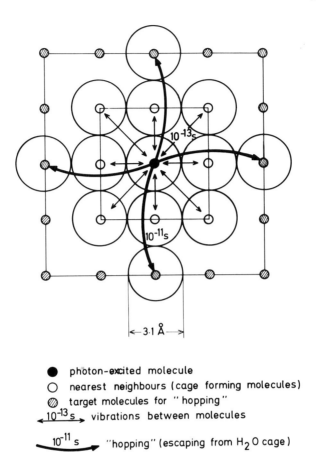

● photon-excited molecule
○ nearest neighbours (cage forming molecules)
◉ target molecules for "hopping"
◄─10⁻¹³s─► vibrations between molecules

─────10⁻¹¹ s────► "hopping" (escaping from H_2O cage)

FIG. 3 Two dimensional model of photon energy transfer in water by
vibrations and diffusion.

Thompson CSF with a bandwidth of 6×10^9 Hz are needed, but such
instruments are very expensive.

In Fig. 4 a schematic diagram of the iodine laser T-jump with optical detection is shown. The use of a differential method has some
advantages in compensating for slow fluctuations of the detection
light, but should not be used for relaxation processes faster than
10^4 to 10^5 Hz. The transient digitizer which allows digital registration of the signals, offers the possibility of sampling several

TABLE 2 Summary of the physical characteristics of the iodine laser and two calculations to estimate the time necessary to reach thermal equilibrium after an iodine laser pulse.

TIME RESOLUTION: IODINE-LASER-T-JUMP

1. Laser

    Wavelength   : 1.315 μm $\triangleq$ 0.943 eV $\triangleq$ 2.3 x 10^{14} Hz

    Emission time   : $\geq 10^{-10}$ s

    Photon - energy: = 1.5 x 10^{-19} J

    Photon number
     per pulse   : ~ 7 x 10^{18} $\triangleq$ 1 J

2. Probe

    Heated Volume H_2O: 140 x 10^{-3} cm^3 $\triangleq$ 140 μl

    Temperature-jump : $\Delta T = 1^{\circ}C$

    1 photon per 1300 molecules H_2O $\triangleq$ $\Delta T = 1^{\circ}C$ in 140 μl

3. Temperature - Equilibration in H_2O

    Thermodiffusion: Hopping across 1 H_2O in 10^{-11} s $\triangleq$ 6.2 $\overset{\circ}{A}$
     (escaping of H_2O from the cage of its neighbours)

    Relaxation time: 5 jumps in the 3 directions of space
                          $\triangleq$ volume with 11^3 molecules of H_2O

$$\boxed{\Delta t = 1.5 \text{ x } 10^{-10} \text{ s}}$$

4. Heat - Transfer in H_2O

    Thermal conductivity   : k = 6.1 x 10^{-3} J/s cm $^{\circ}C$ $(25^{\circ}C)$

    Heat quantity transported   : $Q = \dfrac{k \cdot area \cdot time \cdot \Delta \, ^{\circ}C}{distance}$

    Time needed to thermalize a
    photon (Q=1.5 x 10^{-19} J) among: $t = \dfrac{Q \cdot distance}{k \cdot \Delta \, ^{\circ}C \cdot area}$
    1300 molecules

$$\boxed{t = 8 \text{ x } 10^{-11}s} = \frac{1.5 \text{ x } 10^{-19} \text{ x } 31 \text{ x } 10^{-8}}{6.1 \text{ x } 10^{-3} \text{ x } 1 \text{ x } 31^2 \text{ x } 10^{-16}}$$

5. Threshold for Dieletric Breakdown on H_2O

    Power - threshold : P = 1.5 x 10^{10} W/cm^2

    Laser - power   : L = 1 J per 10^{-10} s $\triangleq$ 10^{10} W

6. Result

    With a laser emission time of 10^{-10} s and a laser energy of 1 J

    the physical barrier for T-jump experiments is reached.

$$10^0 \, \text{s} \geqq \text{Time resolution} \geqq 10^{-10} \, \text{s}$$

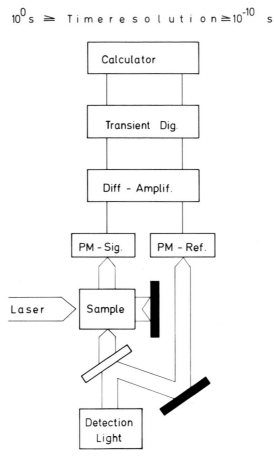

FIG. 4 Schematic arrangement of the iodine laser T-jump.
Laser T-jump technique: (1) independent of additives, (2) no E-field
effects, (3) time-resolution <1 ns, (4) sample-volumes 10 μl–100 μl,
(5) flexible detection system (optical or conductance), (6) time range
from seconds to picoseconds.

signals and in this way increasing the signal-to-noise ratio. The
subsequent on-line calculator is useful for fast signal processing
and flexible calculation of the results. In Fig. 5 we present three
relaxation traces: a very fast one with a relaxation time of 4 ns, a
second one with a decay time of 320 ns, and a third one of 14 μs.
The last signal clearly demonstrates the lack of shockwaves in the
microsecond region which would drive the signal in the opposite direction. The sensitivity used in this experiment was 1 mV, and the

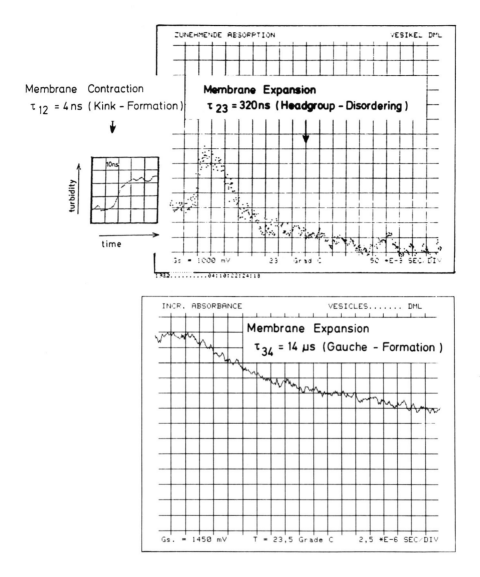

FIG. 5 Three relaxation signals from phospholipid vesicles measured with the iodine laser T-jump.

signal to noise ratio (S/N) of 5×10^3 at a bandwidth of 10^6 Hz was achieved by the superposition of four signals. Higher S/N could easily be achieved by sampling more single traces. A review of the

many applications of iodine-laser T jump has recently become available
(Bannister *et al.*, 1984).

CHARACTERIZATION OF THE CHEMICAL SYSTEM

Chemicals

The phospholipids dimyristoylphosphatidylcholine (DMPC or DML), dimy-
ristoylphosphatidylethanolamine (DMPE) and dipalmitoylphosphatidyl-
choline (DPPC or DPL) as well as the buffer tris-(hydroxymethyl)amino-
methane(Tris) were of purissimum grade from Fluka, Switzerland and used
without further purification since no impurities could be detected by
thin-layer chromatography. Water was triply distilled from a quartz
apparatus and all other salts or solvents used were of suprapure grade
from Merck, Federal Republic of Germany. The probe molecules diphenylhexatriene (DPH) and trimethylaminodiphenylhexatriene (TMADPH$^+$) were
obtained from Molecular Probes, Texas or Fluka, Switzerland.

Preparation of unilamellar vesicles

The vesicles were prepared according to the injection method developed
by Batzri and Korn (1973) and further modified by Kremer *et al.* (1977).
Details of the preparation which resulted in an approximately 10^{-3} M
solution of lipids in a Tris-buffer of pH 7.5 are given by Eck and
Holzwarth (1983). An electron-micrograph of one of the preparations
is given in Fig. 6.

Phase diagrams for the phospholipid-water mixtures of DMPC and
DPPC have been given by Holzwarth *et al.* (1982) and by Eck and Holzwarth (1983). Most of the measurements which we report here were
made on unilamellar vesicles composed of one kind of phospholipid with
a narrow size distribution between 25 and 60 nm diameter. However, in
one of the equilibrium experiments a solution of multilamellar aggregates was investigated to demonstrate its different behaviour.

Turbidity measurements

The term turbidity as it is used here refers to the light scattering
phenomena caused by vesicles in aqueous solutions. Turbidity was
always measured as a loss of incident light intensity after traversing
the sample in the direction of the detection beam within a cone angle
of approximately 60 degrees determined by a lens system. It strongly
depends on the size and density of the aggregates measured as a

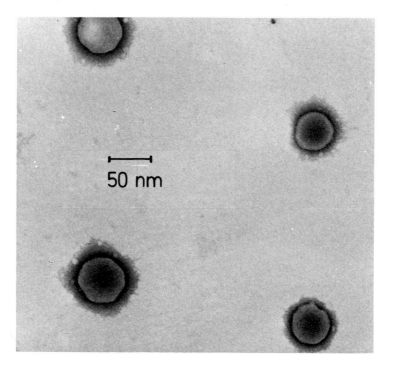

FIG. 6 Electron micrograph of vesicles of DMPE, stained with phosphotungstic acid on Polioform film.

refractive index, as well as on the wavelength of observation (see below). A theory of turbidity has been given by Chong and Colbow (1976). The sample was thermostatted to $\pm 0.1^{\circ}C$ and the temperature was continuously shifted during the static measurements by a Haake thermostat with a temperature time controller.

In Fig. 7 the equilibrium dependence of turbidity on temperature for single-bilayer vesicles of DMPC measured at two wavelengths is shown. The measurements clearly demonstrate the pronounced decrease of the signal at the so-called main phase transition temperature T_m at $23.5^{\circ}C$. The strong wavelength dependence of turbidity allows measurements of light absorption or fluorescence above 400 nm if the aggregates are unilamellar. Multilamellar vesicles or liposomes (elongated structures) show stronger light scattering (turbidity) than small single bilayer vesicles, and give hysteresis effects as well as

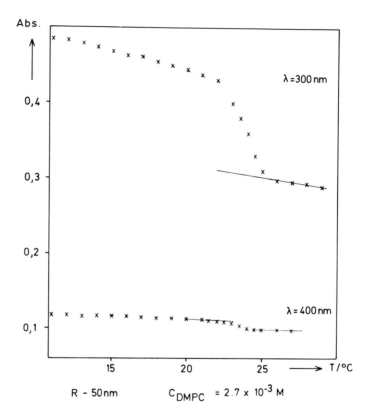

FIG. 7 Equilibrium turbidity-temperature dependence of DMPC vesicles measured at 300 nm and 400 nm to demonstrate the wavelength dependence of turbidity and its strong decrease around the main phase transition temperature, T_m.

exhibiting a pretransition (Fig. 8). Our kinetic measurements were therefore only performed with unilamellar vesicles of narrow size distribution which showed reversible temperature behaviour.

Fluorescence measurements

Turbidity measurement gives a signal integrated over the whole vesicle and, alone, does not allow the resolution at molecular dimensions of effects in different positions within the phospholipid bilayer. Fluorescence measurements, on the other hand, provide information from single molecules. To use this effect for the investigation of the phase transition in vesicles it is necessary to incorporate probe molecules with high fluorescence yields into the bilayer. One of the

Abs./arb.units

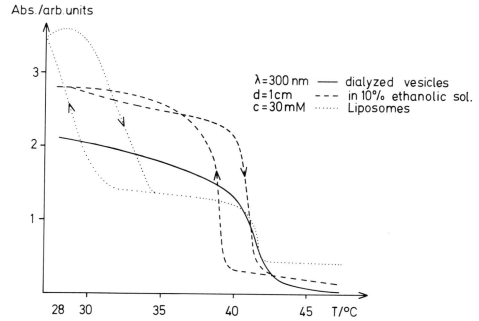

FIG. 8 Equilibrium turbidity-temperature dependence of DPPC vesicles
and liposomes treated differently to show the influence of concentration, size of aggregates, and addition of ethanol.

most widely used probes is DPH which is completely localized in the
hydrophobic part (hydrocarbon chain region) of the phospholipid
bilayer. This is indicated by the fluorescence quantum yield of
almost unity in solutions of vesicles (Shinitzky and Barenholz, 1974);
in an aqueous environment this quantum yield is essentially zero. If
a charged headgroup like trimethylamine is incorporated into one of
the phenyl rings in the para-position, the probe TMADPH[+] is obtained.
This probe associates with one of the negatively charged phosphate
groups in the headgroup area of the phospholipids.

A model for the behaviour of DPH and TMADPH[+] in membranes is shown
in Fig. 9. The position of DPH in the figure does not imply that it
can only reside in the upper or lower hydrocarbon chain area:
Prendergast (1981) suggests that positions between the two layers
are also possible. Our model-figure should not exclude such positions, but detailed consideration of this point is beyond the scope
of the present article. The figure is only to indicate that the

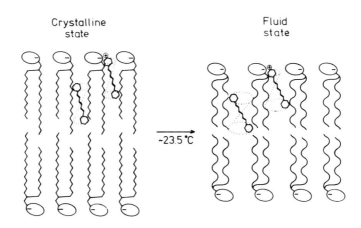

Cone Angle in DMPC

$\theta_c \approx 20°$ (DPH + TMA DPH) $\theta_c \approx 50°$ (DPH)

$\theta_c \approx 30°$ (TMA DPH)

FIG. 9 Model of the space available for the motion of two labels (DPH and TMADPH$^+$) below and above the main phase transition temperature of DMPC bilayers.

space which is available for the movement of the probe molecule depends on its position inside the bilayer, and on the density of the phospholipid in the aggregates. If polarized light is used for the excitation of fluorescence, two types of experiments are possible. One measures either the loss of polarization of fluorescence during the lifetime of the excited state (real-time fluorescence depolariza- tion, e.g. Dale *et al.*, 1977) or the steady-state fluorescence aniso- tropy, both of which provide information about the overall space available for the movement of the probe molecule. The latter experi- ments are carried out with continuous excitation and a time of obser- vation of the fluorescence which is much longer than the lifetime of the excited state. The more space there is available for movement, the lower is the resulting fluorescence anisotropy, reflecting a more randomized distribution of the electric dipole directions associated with fluorescence. Figure 10 shows the experimental arrangement used to measure the fluorescence anisotropy. If the X-Y-recorder is

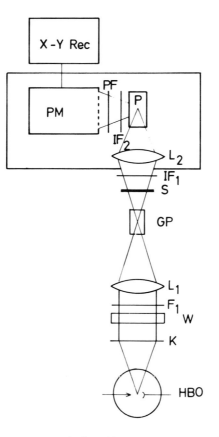

FIG. 10 Schematic arrangement for the measurement of the parallel
and perpendicular polarized components of fluorescence. HBO: mercury
lamp, K and W: heat-filters, F_1: cut-off filter, L_1 and L_2: focussing
lenses, GP: Glan prism, S: shutter, IF_1 and IF_2: interference filters,
P: sample, PF: polarizing filter, PM: photomultiplier, and X-Y recorder.

replaced by a fast digital detection system like that in Fig. 4, time-
resolved measurements are possible, as will be described later in the
kinetic section.

Equilibrium results

In Fig. 11 our equilibrium fluorescence anisotropy-temperature
measurements are shown for DPH and TMADPH$^+$ incorporated into single-
bilayer vesicles of DPPC. DMPC would give similar curves.   The
temperature dependence of the fluorescence anisotropy clearly
shows a strong decrease of the fluorescence anisotropy around the
main phase transition temperature of DPPC vesicles at 40 °C, and

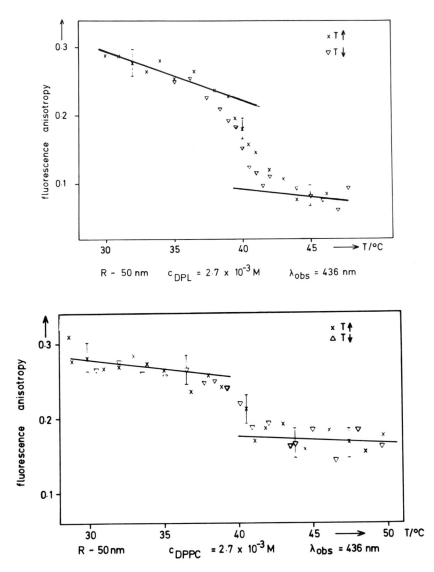

FIG. 11 Equilibrium fluorescence anisotropy-temperature dependence of DPPC vesicles containing the fluorescence probe DPH or TMADPH$^+$ to show the strong change around the main phase transition temperature, T_m.

the system is reversible with respect to temperature. That the value of 0.32 for the fluorescence anisotropy below T_m at 30°C observed pre-viously (Prendergast *et al.*, 1981) is not reached indicates that the

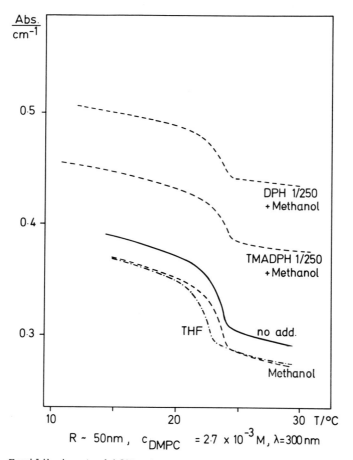

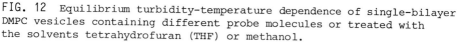

FIG. 12 Equilibrium turbidity-temperature dependence of single-bilayer DMPC vesicles containing different probe molecules or treated with the solvents tetrahydrofuran (THF) or methanol.

excitation light may not be fully polarized. This is of no great moment for our kinetic measurements, giving simply a small reduction of the large change around T_m, and this disadvantage is compensated for in our dynamic experiments by the better signal-to-noise ratio achieved through a higher overall intensity of fluorescence.

In Fig. 12 we show turbidity-temperature curves for pure DMPC vesicles, vesicles containing probes, and vesicles to which the 2 per cent solvent used to incorporate the probe molecules was added. All preparations were exposed to air for 1 h at a temperature 10 °C above the T_m to remove the added solvent. As Fig. 12 demonstrates, there are some persistent differences in turbidity which could not be removed by longer air exposure. Tetrahydrofuran was not further used in our measurements because it shifted T_m by almost 1 °C to lower temperatures. Methanol, TMADPH$^+$ and DPH did not shift T_m but the probe molecules did decrease the change of turbidity around T_m by 15 per cent. When higher concentrations, up to 1/135 (probe/lipid), were used this effect increased to 30 per cent. The absolute values of the absorbances in Fig. 12 changed between different samples by 10 per cent, but the decrease in turbidity caused by methanol and THF and the decrease in the turbidity-jump around T_m caused by the addition of probe molecules was always reproducible. We therefore believe that any addition of probes like DPH or TMADPH$^+$, even at very low concentrations, causes some change inside the bilayer. Its nature is not fully understood, but a clustering of phospholipids around the probes may be responsible.

Kinetic results

For our kinetic experiments we always used single bilayer vesicles of narrow size distribution (±15 per cent) which showed a reversible behaviour around T_m. Two types of detection techniques were used: either turbidity, or fluorescence from probe molecules like DPH or TMADPH$^+$. A fast signal from vesicles of DMPC and DPPC using turbidity as the detection parameter has already been reported by Gruenewald *et al.* (1981). This publication attributed the relaxation at 4 ns to the formation of rotational isomers like kinks, resulting in an increase in the optical density of the bilayer. Slower signals from vesicles of DMPC observed by turbidity changes using the iodine laser T-jump have also been observed (Eck and Holzwarth, 1983). Holzwarth *et al.* (1982) used a pressure jump arrangement with optical detection for signals around 1 ms.

Measurements with other techniques like absorption of ultrasound, Joule-heating and time-resolved NMR are discussed in papers already cited and will only be mentioned again where relevant to the rest of our measurements discussed here.

Turbidity detection Three examples of relaxation traces are included in Fig. 5 which were obtained with the arrangement of Fig. 4. In Fig. 13 we have summarized the relaxation amplitudes A and relaxation

R - 60 nm $C_{DPPC} = 2.5 \times 10^{-3}$ M

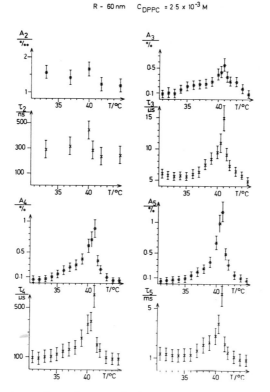

FIG. 13 Temperature dependence of the amplitudes A and relaxation times τ around the main phase-transition temperature of four of the signals from DPPC vesicles with a radius of 60 nm. (Copyright 1984, *Colloid & Polymer Sci.*, Holzwarth and Rys.)

times τ of four signals from 300 ns, to some milliseconds for the pro-cesses designated 2, 3, 4 and 5. These signals were all well separated from each other in time by at least one order of magnitude. Processes 3, 4 and 5 exhibit pronounced maxima of both τ and A around the T_m of DPPC. In the case of signal 2, this maximum is not marked. Similar behaviour in single-bilayer vesicles of DMPC has been noted by Eck and Holzwarth (1983). On addition of probe molecules like DPH the signals 3, 4 and 5 show a time-temperature or amplitude-temperature behaviour in turbidity measurements similar to that in the same samples containing no probe molecules, as documented in Fig. 14. In summary, the time-resolved turbidity measurements with single-bilayer vesicles of DMPC or DPPC containing probe molecules up to a probe/lipid ratio of 1/135, or without any added probe, yield five well-

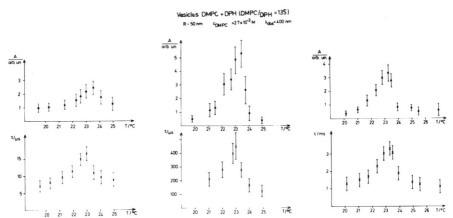

FIG. 14 Temperature dependence of amplitudes A and relaxation times τ of signals (3 (∿10 μs), 4 (∿300 μs) and 5 from DMPC vesicles containing the fluorescence probe DPH measured as a change of turbidity. (Copyright 1984, *Colloid & Polymer Sci.*, Genz and Holzwarth.)

separated relaxation signals. Signals 3, 4 and 5 show strong maxima for the relaxation times, τ, as well as for their amplitudes, A, around T_m while signals 1 and 2 show only weak maxima at T_m.

Fluorescence detection Time-resolved fluorescence measurements were performed using an arrangement combining those depicted in Figs. 4 and 10, except that the detection system utilized different photomultipliers with a risetime of 1 μs, which did not allow measurement of the two fastest signals by fluoresence. The time-resolved fluorescence measurements are summarized in Fig. 15. The most striking difference compared with the turbidity experiments is the lack of signal 4. The fluorescence signals from TMADPH$^+$ all decrease with increasing temperature. In the case of DPH, fluorescence polarized perpendicularly to the polarization of the incident light increases whereas the parallel component decreases. These changes in the intensity of the parallel (∥) and perpendicular (⊥), polarized components of fluorescence can be explained by a decreasing quantum yield of the overall fluorescence (∥+2⊥) with increasing temperature, which is superimposed on increasing signals with ⊥ polarization and decreasing signals with ∥ polarization. All of the signals exhibit maxima in time and amplitude at a temperature slightly above T_m. The slope of the time-resolved fluorescence-temperature dependence curve is smaller below T_m than above T_m where all signals change strongly. This is

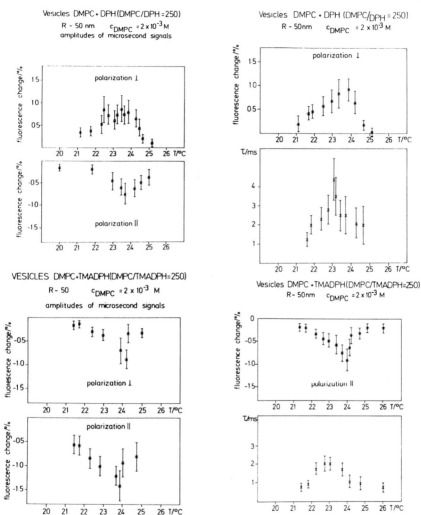

FIG. 15 Temperature-fluorescence dependence of the parallel (‖) and the perpendicular (⊥) polarized components of the relaxation of fluorescence from DPH or TMADPH[+] incorporated into DMPC vesicles around the main phase transition temperature, T_m. (Copyright 1984, *Colloid & Polymer Sci.*, Genz and Holzwarth.)

characteristic for all the kinetic measurements, not only for fluorescence but also in turbidity.

DISCUSSION

In turbidity-time measurements at temperatures around T_m we found five well-separated relaxation signals from vesicles of DMPC and DPPC. The two fastest signals showed no cooperativity, which would be

reflected in maxima of their relaxation times as well as their respective relaxation amplitudes. Ultrasonic experiments by Gamble *et al.*
(1978), Sano *et al.* (1982) and others have also shown relaxation
phenomena in the nanosecond time regime. Signal 1 has been attributed
to the formation of kinks connected with a contraction of the bilayer
(Gruenewald *et al.*, 1981; Holzwarth *et al.*, 1982). The second signal
can be explained by disordering of the headgroups, as suggested by
Eck and Holzwarth (1983), because of changes in the signal correlated
with weakening of the headgroup interaction.

Here we would like to concentrate on signals 3, 4 and 5. They all
show strong cooperativity in τ as well as in the respective amplitudes.
The very sharp decrease of the signals above T_m is typical of all
three relaxations. Its theoretical interpretation is not yet clear,
but we believe that the drastic decrease is caused by the part of the
phase transition which is of the first order type. The other part of
the phase transition, mainly below T_m, seems to be of higher order.

In Fig. 16 we have tried to reconstruct the equilibrium turbidity
curve of DPPC by the five relaxation amplitudes which we found by
normalizing them at 46°C. The figure clearly shows that all five
amplitudes are needed to achieve the equilibrium result. The almost
perfect agreement between the thermodynamic turbidity-temperature
curve and the summation of our kinetic amplitudes strongly confirms
the quality of our measurements and the representation of the whole
main phase transition by our five relaxation times.

Figure 17 summarizes the fluoresence measurements. The lack of
signal 4 around 200 μs allows us to attribute this signal to the formation of clusters (Kanehisa and Tsong, 1978) which retain their
order during the disordering of the surrounding phospholipid molecules.
Because of the tendency of probe molecules to affect their immediate
environment they are insensitive to the disordering which occurs some
molecules away. The slowest signal around a millisecond therefore is
due to the disordering of the whole bilayer by the disappearance of
more rigid clusters of phospholipids.

The only signal which is now left without interpretation is signal
3. We believe that it is connected with the formation of gauche conformers at the ends of the hydrocarbon chains which enhances the

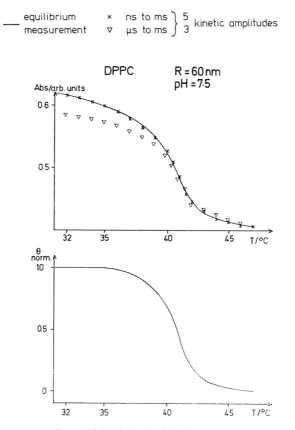

FIG. 16   Reconstructed equilibrium turbidity-temperature dependence
of DPPC vesicles using the amplitudes of 3 and of all 5 relaxation
signals at the same temperature compared with the same relation
measured in static experiments.

lateral diffusion of single phospholipid molecules by weakening the
contact between the two monolayers forming the bilayer. This is confirmed by experiments with cholesterol which is located in the hydrophobic area (Yeagle *et al.*, 1975) and accelerates the third relaxation
signal, as measurements by Eck and Holzwarth (1983) have demonstrated.
A summary of these results is given in Table 3.

CONCLUSION

We have developed a very fast and flexible iodine-laser T-jump method
for the detection of dynamic processes in biological systems. This
method not only avoids all the artifacts and unwanted side-reactions

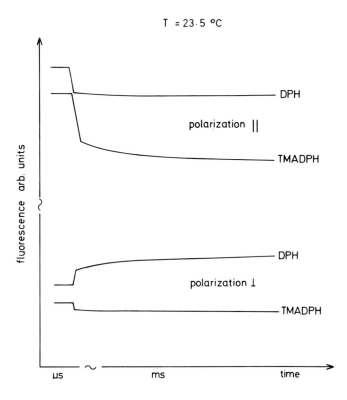

FIG. 17 Schematic summary of the relaxation signals measured as polarized components of fluorescence from the probe molecules DPH and TMADPH⁺ parallel (‖) and perpendicular (⊥) to the polarization of the exciting photons.

which are produced in complex systems by Joule-heating techniques, but it also offers for the first time a kinetic instrument for measurements over the whole time-range between seconds and 100 picoseconds, independent of the composition of the solutions. As detection parameters, conductance, light absorption, light scattering and fluorescence have already been used. The iodine-laser T-jump has been applied to investigate the main phase transition in single-bilayer vesicles of phospholipids. The whole order to disorder transition could be resolved into five well-separated relaxation processes, and a molecular interpretation for these signals has been presented. The reconstruction of the equilibrium turbidity-temperature dependence by the five relaxation times and amplitudes obtained is convincing

TABLE 3 Molecular models for the five relaxation phenomena representing the main phase transition measured in solutions of single-bilayer vesicles using the iodine laser T-jump arrangement.

Time-Range Seconds	Dynamic Processes	Molecular Model Schematic
4×10^{-9}	kink formation in hydrocarbon chains of PL molecules not cooperative decreasing amplitude with increasing temperature increasing BL - density	127Å
$10^{-7} - 10^{-6}$	lateral diffusion of single PL molecules "hopping" hexagonal headgroup order is disappearing very weak cooperative decreasing BL - density	2.54Å
$\sim 10^{-5}$	gauche formation in hydrocarbon chains of PL molecules cooperative maximum at $T=T_M$ weakening of contact between monolayers start of river - like disordered areas decreasing BL - density	

Time-Range Seconds	Dynamic Processes	Molecular Model Schematic
$\sim 10^{-4}$	formation of clusters of higher order inside the monolayers strongly cooperative disappearing at $T > T_M$ decreasing PL - density no signal from fluorescence of DPH	
$10^{-3} - 10^{-2}$	clusters are disappearing leaving the BL in a completely disordered state and only with weak contact between the monolayers strongly cooperative decreasing PL - density decreasing fluorescence	
10^{0}	completely disordered BL in fluid state	

evidence for the reliability of the whole arrangement in time-resolution as well as in the reproduction of the amplitudes of relaxations
between nanoseconds and seconds. Further experiments with modified
phospholipid molecules and proteins incorporated into the phospholipid
bilayer of vesicles are in progress and may contribute to the understanding of the dynamic processes controlling the function of natural
cell-membranes.

ACKNOWLEDGEMENTS

We should like to thank the Deutsche Forschungsgemeinschaft for a
research grant, Mrs M. Rokosch who drew the figures, and E. Lerch for
the computer programme to process the relaxation signals.

REFERENCES

Bannister, J.J., Gormally, J., Holzwarth, J.F. and King, T.A. (1984).
    Chem. Brit. 20, 227.
Batzri, S. and Korn, D. (1973). *Biochim. Biophys. Acta* 298, 1015-
    1019.
Brederlow, G., Fill, E. and Witte, K.J. (1983). In "The High-Power
    Iodine Laser". (Ed. K. Shimoda). Springer-Verlag,
    New York.
Chong, C.S. and Colbow, K. (1976). *Biochim. Biophys. Acta* 436, 260-
    282.
Dale, R.E., Chen, L.A. and Brand, L. (1977). *J. Biol. Chem.* 252,
    7500-7510.
Eck, V. and Holzwarth, J.F. (1983). *Surfactants in Solution* 3, 2059-
    2080. (Ed. K.L. Mittal), Plenum Press, New York.
Frisch, W., Schmidt, A., Holzwarth, J.F. and Volk, R. (1979). *In*
    "Techniques and Applications of Fast Reactions in Solution". (Eds
    W.J. Gettins and E. Wyn-Jones), pp. 61-70. Reidel Publ. Col,
    Dordrecht-Holland.
Gamble, R.C. and Schimmel, P.R. (1978). *Proc. Natl. Acad. Sci. USA*
    75, 3011-3014.
Genz, A. and Holzwarth, J.F. (1984). *Colloid and Polymer Sci.* 262.
    In press.
Goodall, D.M. and Greenhow, R.C. (1971). *Chem. Phys. Lett.* 9, 583-585.
Gruenewald, B., Frisch, W. and Holzwarth, J.F. (1981). *Biochim. Bio-
    phys. Acta* 641, 311-319.
Holzwarth, J.F. (1979). *In* "Techniques and Fast Reactions in Solution"
    (Eds W.J. Gettins and E. Wyn-Jones), pp. 47-69. Reidel Publ. Co.,
    Dordrecht-Holland.
Holzwarth, J.F., Frisch, W. and Gruenewald, B. (1982). *In* "Micro-
    emulsions". (Ed. I.D. Robb), pp. 185-205. Plenum Press, New York.
Holzwarth, J.F. and Rys, F. (1984). *Colloid and Polymer Sci.* 262.
    In press.
Kanehisa, M.I. and Tsong, T.Y. (1978). *J. Amer. Chem. Soc.* 100,
    424-432.
Kremer, J.M.H., Esker, M.W.T., Pathmamanoharan, C. and Wiersema, P.H.

(1977). *Biochemistry* 16, 3932-3935.
Natzle, W.C., Moore, B., Goodall, D.M., Frisch, W. and Holzwarth, J.F. (1981). *J. Phys. Chem.* 85, 2882-2884.
Prendergast, F.G., Haugland, R.P. and Callahan, P.J. (1981). *Biolochem.* 20, 7333-7338.
Sano, T., Tanaka, J., Jasunaga, T. and Toyoshima, Y. (1982). *J. Phys. Chem.* 86, 3013-3016.
Shinitzky, M. and Barenholz, Y. (1974). *J. Biol. Chem.* 249, 2652-2657.
Yaegle, P.L., Hutton, W.C., Huang, C.-H. and Martin, B. (1975). *Proc. Natl. Acad. Sci. USA* 72, 3477-3481.

Some Notes on Picosecond Spectroscopy

P.M. RENTZEPIS and J.B. HOPKINS

INTRODUCTION

Picosecond laser pulses made their entrance without the splash common
to many other fields of science. In fact, the number of true practitioners for the first few years was minute. Notwithstanding this,
the field has flourished, its practitioners have expanded to several
thousand and it has spread to every part of the scientific globe.
Today, picosecond pulses are generated in many countries of the world,
and picosecond spectroscopy is commonplace in universities (particularly American) and other scientific institutes.

Recalling the early days of picosecond lasers is similar to recalling those in the development of a child. In the midst of a new era,
everything which was learned as a result of each experiment was a
first. This endeavour certainly provided the continuous excitement of
learning something new with every laser shot, but one hardly had any
time to sit back and enjoy those early days. Everything was new.
Every pulse produced a new event - every molecule that was studied
showed new behaviour. Although the number of active investigators was
small, the number of experiments pursued was incredibly large, particularly compared to our meager understanding of these ultrafast
phenomena.

In the beginning, besides our group at Bell Laboratories with Joe
Giordmaine, Stan Shapiro and myself, there were the group at United

Aircraft with Tony Demaria, Brienza and others, John Armstrong at IBM, and a few more in other laboratories. The effort was to identify and understand how picosecond pulses are generated and to measure their characteristics, namely pulse width, bandwidth and chirping. Nd^{3+}/glass was the oscillating medium, and cyanine dyes, for example Kodak 9740 and 9860, were the experimental media for mode-locking.

Our original effort at Bell Laboratories was devoted to devising means for measuring the pulse width, since available oscilloscopes were were not fast enough to display the pulse widths. A simple, inexpensive, yet rather elegant method was invented for their display (Giordmaine et al., 1967). It was based on the two-photon virtual transition concept. Two-photon transition theory had been described as early as 1932 by Maria Goeppert-Mayer. However, because the cross section is small, i.e. 10^{-52} versus 10^{-18} for a one-photon allowed transition, the experimental application of two photon processes remained unsuccessful until the development of the laser.

METHODS

The two-photon technique used for the measurement of ultrashort pulses is shown in Fig. 1. A train of pulses was generated in a simple oscillator cavity composed of a Brewster's angle cut ND^{3+}/glass rod, two reflecting mirrors, and a mode-locking dye cell. The laser light which emanated through the 70% front mirror consisted of a train of $\sim$50, 1060 nm pulses separated by $\sim$6 ns or $\frac{2\ell}{c}$ where ℓ is the cavity's length, and c is the speed of light. The 1060 nm pulses were collimated into a cell containing a fluorescing dye, such as rhodamine 6G. This dye was selected not to absorb at the fundamental wavelength of the laser pulse, i.e. 1060 nm, but possesses a strongly absorbing level at the two-photon energy, i.e. 530 nm. The pulses, after entering the dye cell, propagated through the dye solution without inducing appreciable fluorescence. After traversing through the cell, the pulses were reflected by a mirror situated at the end of the cell and returned through the same path where they overlapped with the oncoming trailing pulse in the pulse train. At the point of the two-pulse overlap, a fluorescence spot appeared whose length was related to the pulse time, width t by $\ell = \frac{act}{n}$ where a is a shape factor, and n is the index of refraction. This first method for the measurement of picosecond pulses

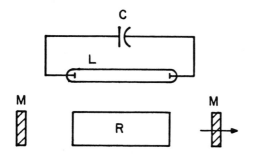

SIMPLE LASER CAVITY

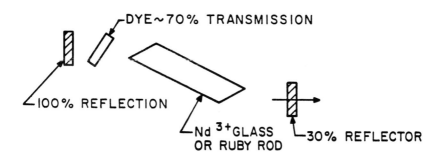

MODE LOCKED LASER

FIG. 1 a) Laser configurations; (b) pulse train emitted by a Nd^{3+}/ glass mode-locked laser; (c) the first two photon apparatus and pulses; (d) fluorescence spot of a 6 ps pulse generated by the TPF method.

 provided a rather accurate display of picosecond pulses. However, it did have disadvantages, which included intensity contrast ratio for background to pulse of 1/3. Other methods, such as three-photon fluorescence, can decrease this ratio to 1/10. Later, streak cameras were developed which could display pulses with as little as ∿2 ps duration electronically. The pulses generated at that time, i.e.

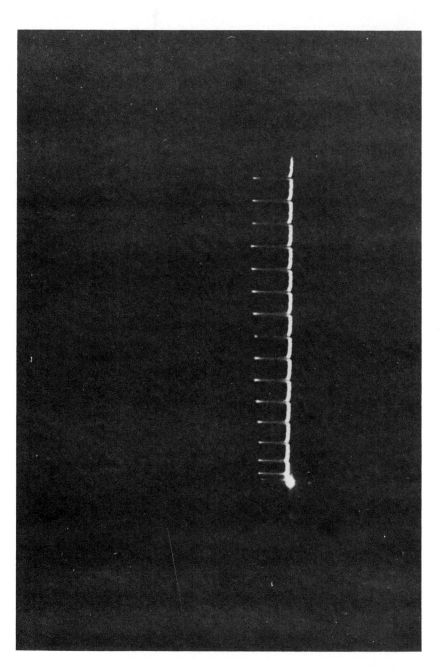

FIG. 1b

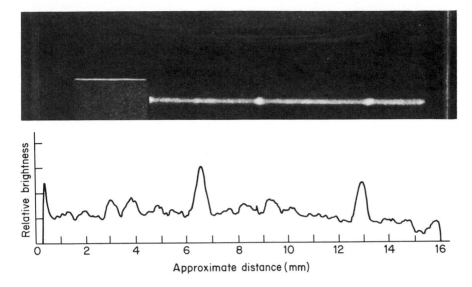

FIG. 1c

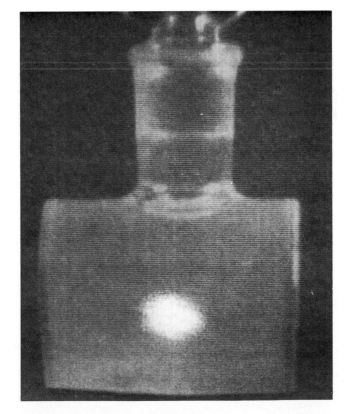

FIG. 1d

1967, exhibited the following characteristics: pulse width, t $\sim$ 6 ps
FWHM; bandwidth, $\Delta\nu$ $\sim$ 50 cm^{-1}; energy, $\sim$30 mJ; peak power $\sim$2 GW. The
wavelengths available were the fundamentals of the Nd^{3+} and ruby at 1060
nm and 694.3 nm, respectively, and their second harmonics at 530 and
347.2 nm. Even though these were rather meager by the standard of
present capabilities, several pivotal experiments were performed with
them.

The first experiment in picosecond spectroscopy (Rentzepis, 1968a,
b) was designed to measure the radiationless decay of a large molecule,
azulene, which exhibited a dense vibrational manifold in the excited
electronic state. The experimental system and energy level diagram
of azulene are shown in Figs. 2 and 3, respectively. The initial
530 nm pulse excited the azulene to an upper vibronic level of its
first excited state. No fluorescence is observed from this level
because of fast radiationless decay to the ground state. However,
when the second pulse, at 1060 nm, overlapped the 530 nm pulse, the
molecule was excited into its second electronic excited state from
which it fluoresced. If the two pulses are propagated from different
directions, then the period of their overlap and hence the fluorescence spot length - is determined by the lifetime of the vibrational
level populated by the 530 nm pulse. The fluorescence spot (Fig. 4)
was recorded by means of a polaroid camera, and its intensity measured
by means of a photodensitometer. The same experiment can be performed
by delaying one beam with respect to the other. Such pulse/probe and
delayed coincidence techniques are very often used even today to

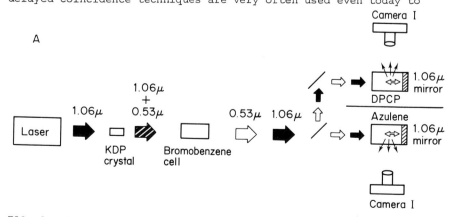

FIG. 2 Pictorial representation of the system utilized for the
measurement of vibrational relaxation of azulene.

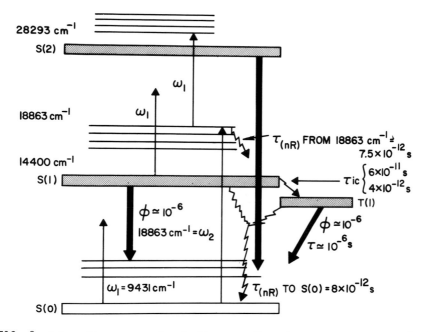

FIG. 3   Schematic energy level diagram of azulene, ω_1, corresponds to the fundamental Nd^{3+} laser wavelength, ω_2, to the second harmonic.

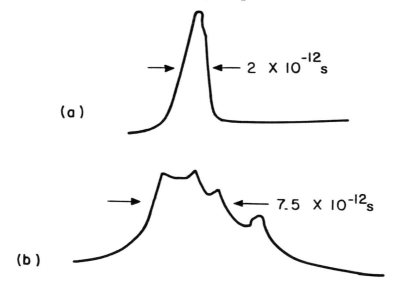

FIG. 4   Densitometer traces of (a) the picosecond pulse, (b) the azulene signal.

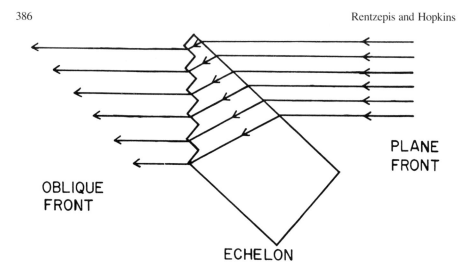

FIG. 5 a) The echelon; (b) light segment of an echelon (the distance
between segments is 8 ps); (c) picosecond continuum generated by
a ∿ 530 nm, 6 ps pulse focussed on a 10 cm H_2O cell; the wavelength
range of the continuum is 800 nm - 350 nm.

elucidate the mechanisms of vibrational relaxation, ionization and,
in general, to study intermediate states and reactive intermediates,
such as radicals, radical ions and other species.

These single-shot, single wavelength and time experiments were
extended to simultaneous measurement of preselected time intervals
extending from a few picoseconds to hundreds of picoseconds by inserting an echelon between the probe pulse and sample. The echelon is a
simple, yet accurate, picosecond clock which is fabricated from optically contacted fused silica blocks. The staircase structure, shown
in Fig. 5, allows the single picosecond pulse to be segmented as it
passes through the successive steps. The difference in the step
thickness, d, causes a delay between pulses: $\Delta t = \frac{d}{c}[(n^2-\sin^2\theta)^{1/2}-$
$-\cos\theta]$ where θ and n are the angle of incidence and refractive index,
respectively, of the echelon material. The very early applications
of the echelon in picosecond spectroscopy were performed with photo-
graphic detection, wherein 10-15 echelon pulses separated by ∿10 ps
were imaged onto a polaroid film after passing through the sample.
This was the method by which, for example, the kinetics of the first
intermediate involved in the visual transduction process, bathorhodopsin, were determined by Busch *et al.* (1972). In this experimental

FIG. 5b

FIG. 5c

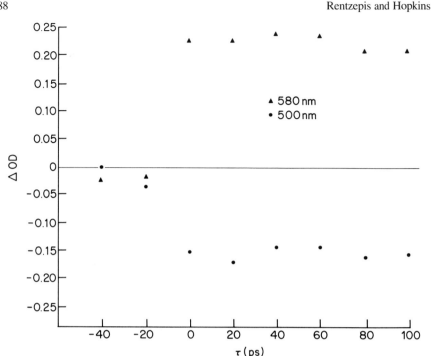

FIG. 6 Formation of bathorhodopsin (upper) and bleaching of rhodopsin (lower) at 300°K after excitation with a 532 nm, 6 ps pulse.

study, the appearance of the transient absorption of bathorhodopsin
at 561 nm was monitored immediately after excitation of rhodopsin
with a 532 nm pulse. The formation of this first intermediate, bathorhodopsin, was found to occur in less than 6 ps (Fig. 6) because even
the first echelon pulse, coincident with the excitation pulse was
absorbed by the bathorhodopsin.

    With the advent of vidicons, photographic recording was eclipsed.
It was now possible to record the optical density changes electronically in a more efficient and reliable manner. Netzel et al. (1973)
applied this device to the identification and study of the picosecond
kinetics of the bacteriochlorophyll photosynthetic reaction centres
R-26. Near infrared and visible continuum pulses were generated by
focussing the 1060 nm and 530 nm pulses into a 10 cm cell containing
a water D_2O mixture. These continuum pulses were segmented, delayed
relative to each other by an echelon, and passed through an R26 reaction centre sample which was excited by a single 530 nm pulse.

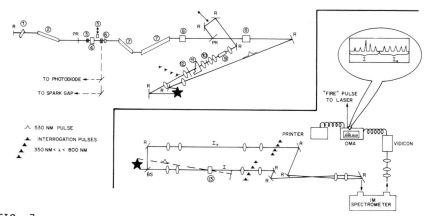

FIG. 7 Double-beam picosecond spectrometer.

Transmitted light intensity at 864 nm was isolated by means of a 1-
meter monochromator prior to imaging on the vidicon. A later modification transformed this method into a picosecond analogue of a double-
beam absorption spectrometer (Fig. 7). The segmented echelon pulses
were split into reference and probe beam continuum trains. The
reference beam traversed a cell containing only the solvent while the
probe train was partially focussed onto the sample region intersected
by the excitation pulse. The changes in intensity encountered by the
probe pulses were compared to the intensities of the reference beam,
enabling the direct computation of absorption changes in the sample
as a function of time.

Experiments based on double-beam picosecond absorption spectroscopy
form the basis for most picosecond research today. It can resolve
absorption changes as low as 0.01 as studies by Struve, Eisenthal,
Alfano, Kaufmann and Netzel have shown. The two-dimensional vidicon
has been employed for the simultaneous time- and wavelength-resolution
of optical density. Figure 8 displays the transient absorption spectrum, obtained on the double-beam instrument depicted schematically
in Fig. 9, of cytochrome c between 540 and 640 nm at uniformly stepped
time intervals of 6 ps upon excitation by a 6 ps, 530 nm single pulse.

Flexibility in pump and probe wavelengths is a strongly desired
objective. However, although the high energy content of the $Nd^{3+}/$
glass or Nd^{3+}/YAG pulses facilitate high energy second harmonic pulses

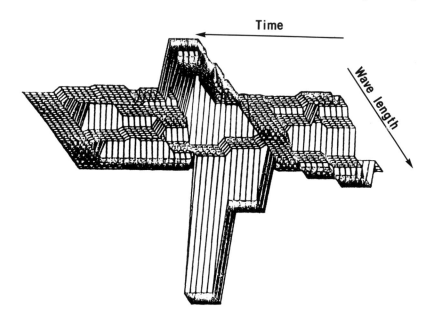

FIG. 8 Three dimensional I vs t vs λ) data output from system depicted in Fig. 9.

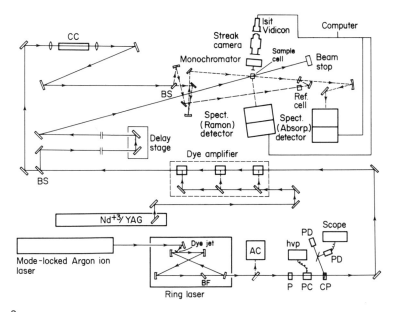

FIG. 9 Schematic diagram of a synchronously pumped dye laser system. AC: autocorrelator, P: polarizer, PC: Pockels cell, CP: crossed polarizer, hvp: high voltage pulser, PD: photodiode, SHG: second harmonic generating crystal, BS: beam splitter, CC: continuum cell.

and interrogating continua, they offer no avenues for tunability or
for a high repetition rate which could circumvent the problem of low-
energy pulses through extensive signal averaging.

A dye laser (Ruddock and Bradley, 1976), especially one which is
synchronously pumped by an acousto-optically mode-locked argon or
krypton ion laser, or a passively mode-locked dye laser pumped by a
cw gas ion laser produces 1 nJ ultrashort visible light pulses at a
repetition rate of ∿90 MHz. Passively mode-locked dye lasers yield
sub-picosecond pulses (∿0.5 ps FWHM) but with more restricted tuning
ranges (<20 Å) than are available with synchronously pumped lasers
(fundamental wavelengths from 400 to 900 nm with suitable dyes and
pulse durations of 2-10 ps FWHM). Acousto-optic cavity-dumping is
advantageous because it allows periodic pulse extraction at greatly
reduced repetition rates (10^4-10^5 Hz) which permit sample re-equilibration between pulses.

The minuscule absorption transients induced by nJ pulses from a cw
dye laser may be recovered using a lock-in amplifier for phase-sensitive detection. In a transient absorption study of conformational relaxation in excited singlet states of stilbenes, Teschke *et al.* (1971)
prepared 615 nm fundamental and 307 nm second harmonic pulses, modulated the 307 nm pulses prior to excitation of *trans*-stilbene in *n*-
hexane or lucite, and probed the sample $S_n \leftarrow S_1$ absorption with the 615
nm pulses at variable optical delays introduced by a programmable step-
ping motor. Optical density transients thus obtained in a single delay
sweep exhibit ≤15 per cent noise, and would readily be improved by
averaging. However, these studies offer little flexibility in pump
or probe wavelengths because tuning bandwidths in passively mode-
locked cw dye lasers are stringently limited by compatibility of the
lasing and mode-locking dyes, and because second harmonic generation
is the only practical nonlinear wavelength conversion technique for
such weak pulses. This limitation could be partially relaxed by using
synchronously pumped cw lasers, which trade off subpicosecond resolution for fundamental wavelength tunability throughout the visible spectrum. In turn, amplification of the dye pulses by Nd^{3+}/YAG pulses has
been demonstrated, but the repetition rate advantage is eliminated
since the Nd^{3+}/YAG laser operates at about a 10-30 Hz repetition rate.

RESULTS

Subpicosecond pulses as short as 30×10^{-15} s have been reported to have been generated in modified dye lasers. These pulses, although short in duration, possess the inherent disadvantage of a very broad, ~ 1000 cm^{-1} bandwidth which makes them inappropriate for the study of molecular changes, such as isomerization, vibrational relaxation and other dynamic processes. This is a result of the fact that for many molecular species the original and induced states are shifted by less than 1000 cm^{-1}. Nevertheless, femtosecond pulses are of considerable use for the study of other physical phenomena.

Picosecond fluorescence presents some challenging experimental problems, but the rewards for overcoming these are substantial. Since the radiative lifetimes of most fluorescing excited states range between ~ 10 ns and 30 ns, the quantum yield observed over a 10 ps interval is typically in the order of 10^{-3} to 10^{-6} requiring photomultipliers or image intensifiers in addition to gating. Electro-optical streak cameras have provided probably the most fruitful source of picosecond fluorescence data. The streak camera coupled to a vidicon, optical multichannel analyser (OMA) and a minicomputer provides a rapid and accurate measurement of detailed fluorescence decay data. The resulting information has stimulated widespread applications of picosecond streak cameras in chemical and biological studies. Intramolecular proton transfer in excited states of methyl salicylate, vibrational relaxation of s-tetrazine in n-hexane and of tetracene in solid argon, lifetime measurements of laser dyes such as 4-methyl-umbelliferone and polymethines, intramolecular heteroexcimer formation, double proton transfer in 7-azaindole dimers, exciton migration in pentacenedoped tetracene, rotational diffusion or organic dyes in solution as reflected in picosecond-resolved fluorescence depolarization and pumping-intensity dependence of chlorophyll fluorescence lifetimes in *Chlorella* are representative of topics pursued by streak camera fluorometry. A typical fluorescence histogram, for acridine in hexane, is shown in Fig. 10.

Pulses from a picosecond cw dye laser are far too weak for single-trace measurements, and the triggering jitter (≥ 2 ps) inherent in conventional streak tubes precludes use of facile signal averaging to exploit their tunability, repetition rates, and (in passively

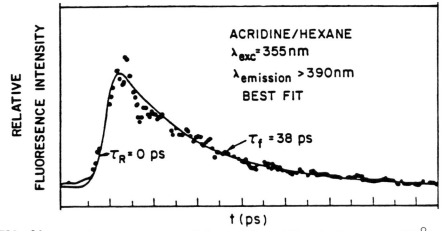

FIG. 10   Streak camera trace of decay of acridine in hexane at $300°K$.

mode-locked systems) subpicosecond durations. The synchroscan streak
camera circumvents these drawbacks by using the picosecond trigger
pulse signal to gate a 140 MHz tunnel diode oscillator whose amplified
output is coupled to the streak tube deflection plates. This syn-
chronized triggering mode creates precisely superimposed streak images
for successive fluorescence pulses, and integration of $\sim1.4 \times 10^8$
fluorescence traces per second renders image intensification super-
fluous even for fluorescences excited by a synchronously pumped cw
rhodamine 6G laser (0.6 nJ/pulse, 2 ps FWHM, tunable from 565 to 630
nm). To date, synchroscan cameras have been applied to fluorescence
studies on viscosity dependence of excited-state lifetimes of organic
molecules and the mechanism of dissociation of haloaromatics.

A typical application of picosecond spectroscopy to biology is
offered by the study of primary events in the visual transduction
process. The eye is the organ responsible more than any other, for
the information and knowledge we have about our surroundings. Yet
the primary sequence of events involved in the visual transduction
process which is initiated by photons absorbed by the chromophore, is
not well established. Rhodopsin is a substance composed of the protein, opsin, and a polyene aldehyde, 11-*cis*-retinal:

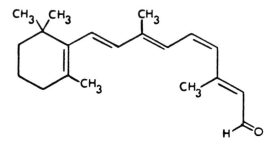

The protein and retinal are attached by means of a Schiff base linkage between the carbonyl group of retinal and an ε-amino group of a lysine residue contained in the protein. Many investigations of this photo-initiated sequence of events, a highly complex and efficient communi-cation process, have been performed, and many extensive reviews have been written on this subject.

Although much effort has been exerted to obtain an understanding of the initial process and establish the nature of the first inter-mediate by means of absorption and Raman spectroscopy, controversy still exists with regard to the nature of the first intermediate, bathorhodopsin, and the process responsible for its formation. The primary changes in the absorption spectrum of rhodopsin were first observed at 77 K and reported by Yoshizawa and Kito (1958), subsequent-ly by Grellman et al. (1962), and by Yoshizawa and Wald (1963), and were attributed to the first intermediate, bathorhodopsin. Using pico-second absorption spectroscopy, Busch et al. (1972) were able, for the first time, to measure the room temperature kinetics of formation and decay of bathorhodopsin. This work revealed that the first-formed intermediate, bathorhodopsin, was formed within a time of less than six picoseconds and decayed to the second intermediate, lumirhodopsin, with a time constant of ∿30 ns. This very fast formation of bathoro-dopsin prompted the low-temperature studies of Peters et al. (1977). The formation kinetics of bathorhodopsin were studied in optically clear glasses of the rhodopsin preparation in ethylene glycol over temperatures ranging from 300°K to 4°K (Fig. 11). Even at the lowest temperature studied, 4°K, the risetime of the appearance of bathoro-dopsin was found, at 36 ps, to be extremely fast.

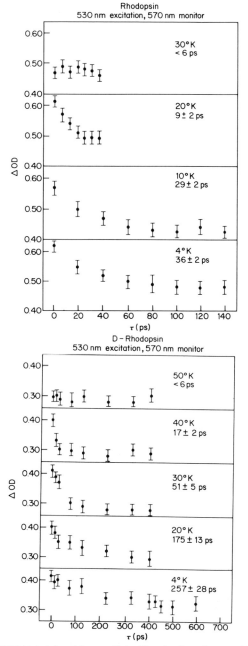

FIG. 11   (a) Foramtion kinetics of bathorhodopsin monitored at 570 nm
and at various temperatures; (b) formation kinetics of deuterium-
exchanged bathorhodopsin at 570 nm and at various temperatures. In
both cases, excitation was performed with a 5 nJ, 530 nm, 6 ps pulse;
the rise time is given for each temperature.

Classically the primary event which results in the formation of bathorhodopsin has been described as the isomerization of the 11-*cis*-
retinal moiety of rhodopsin in the all-*trans* form (Wald, 1968; Rosenfeld *et al.*, 1977; for a recent review see Ottolenghi, 1980). Prior
to the low temperature studies of Peters *et al.* (1977), the possibility of the occurrence of complete isomerization of the retinal chromophore within <6 ps had been questioned (Yoshizawa and Kito, 1858;
Grellman *et al.*, 1962; Yoshizawa and Wald, 1963). In view of the very
fast formation of bathorhodopsin at $4^{\circ}K$, Peters *et al.* (1977) proposed
the possibility of proton translocation as an alternate mechanism to
that of *cis-trans* isomerization for the primary event of visual transduction, at least at low temperature. A deuterated rhodopsin derivative (D-rhodopsin) was prepared by shaking the rhodopsin preparation
with D_2O. Under these conditions, only the readily exchanged protons
are substituted with deuterium, while other hydrogen atoms, those for
example bonded directly to the hydrocarbon skeleton of the retinal
moiety, are unaffected. A pronounced deuterium isotope effect was
observed for the rate of formation of bathorhodopsin, k_H/k_D = 7 (Fig.
11). The temperature dependence of the rate of formation of bathorhodopsin also exhibited a non-Arrhenius behaviour: at very low
temperatures the rate was practically independent of temperature.
These data thus uncovered an interesting, and previously undetected,
process, namely proton translocation, which occurs via quantum mechanical tunnelling at low temperatures. Analysis of such kinetic data
permits the calculation of the energy barrier through which the proton tunnels and, therefore, the distance required for the proton
translocation. The calculated distance of ~ 0.5 Å coupled with the
fact that the Schiff base proton was the only deuterium-exchangeable
proton within the retinal moiety provides a strong basis for suggest-
ing that translocation of the Schiff base proton is at least a prominent feature of the first event in vision.

    A criterion that has been used to support the hypothesis that *cis-*
trans isomerization is the primary event in the visual transduction
process is the commonly recurring statement that both 11-cis-rhodop-
sin and 9-*cis*-rhodopsin lead to a common bathorhodopsin. Raman
experiments (Mathies *et al.*, 1976) data presented by Rosenfeld *et al.*

(1977), Green *et al.* (1977) and Monger *et al.* (1979), and several
theoretical models (Warshel, 1976; Birge and Hubbard, 1980) all support the proposal of a common all-*trans* intermediate and appear to
provide a strong argument for *cis-trans* isomerization and not merely
a slight skeletal deformation, as was proposed by Busch *et al.* (1972),
and Peters *et al.* (1977) as the process generating bathorhodopsin.
However, in a recent picosecond absorption study, Spalink *et al.*
(1984) obtained evidence which demonstrates that bathorhodopsin
formed from photoexcitation of 9-*cis*-rhodopsin is not the same as the
bathorhodopsin generated from 11-*cis*-rhodopsin. Using a 25 ps, 532 nm
pulse for excitation of 9-*cis* and 11-*cis*-rhodopsin, they measured difference absorption spectra with an interrogating picosecond continuum.
The probing time was 85 ps after excitation of the rhodopsin sample,
which was sufficient to permit excited-state rhodopsin to decay to
bathorhodopsin (time constant of ∿6 ps) and orders of magnitude
shorter than the decay of bathorhodopsin to the next intermediate,
lumirhodopsin (time constant of ∿30 ns). Therefore, the absorption
spectra measured correspond to bathorhodopsin. The difference spectra
were calculated according to:

$$\Delta A(\lambda) = -\log_{10} \frac{S^{ex}(\lambda)}{R^{ex}(\lambda)} \Big/ \frac{S^{noex}(\lambda)}{R^{noex}(\lambda)}$$

where ΔA is the change in absorbance, $S^{ex}(\lambda)$ and $R^{ex}(\lambda)$ are the intensities of the sample probe and of the reference probe, respectively,
as a function of wavelength, when the sample is subjected to photoexcitation by the laser pulse, while $S^{noex}(\lambda)$ and $R^{noex}(\lambda)$ are the
intensities of the sample probe and of the reference probe, respectively, as a function of wavelength when the sample is not subjected
to photoexcitation by the laser pulse. Low excitation energies were
used to ensure that difference spectra were not artifacts resulting
from saturation of the electronic transition or biphotonic events.

Figure 12 shows the difference spectra obtained 85 ps after excitation of 9-*cis*-rhodopsin and 11-*cis*-rhodopsin at 290 K. Two important
observations not previously noted are quite evident upon inspection
of the two spectra. First, the absorption and bleached bands of the

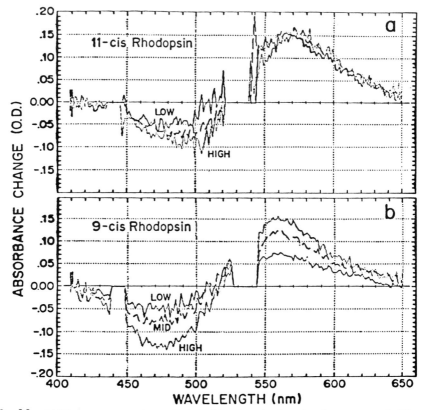

FIG. 12 Difference spectra of (a) 11-*cis*-rhodopsin (average of 58 laser shots) and (b) 9-*cis*-rhodopsin (average of 55 laser shots) obtained 85 ps after excitation with a 532 nm pulse.

difference spectrum resulting from excitation of the 9-*cis* form and their isobestic point are shifted to longer wavelengths by ∿10 nm relative to the corresponding positions in the difference spectrum of the 11-*cis* form. This shift is similar to that observed for the ground-state absorptions of 11-*cis* and 9-*cis*-rhodopsin. Second, while both 11-*cis*-rhodopsin and 9-*cis*-rhodopsin exhibit the same amount of maximum bleach (absorption decrease), the ratio of their absorption maxima is not 1:1 but 1.5:1. The absorption spectra (Fig. 13) of bathorhodopsin formed 85 ps after excitation of 11-*cis*-rhodopsin and 9-*cis*-rhodopsin can be obtained by subtracting the bleach from the difference spectrum. Scaling of the absorption bands clearly illustrates that they are not superimpossible. This spectral difference persits to a time of 8 ns after excitation of the sample. These

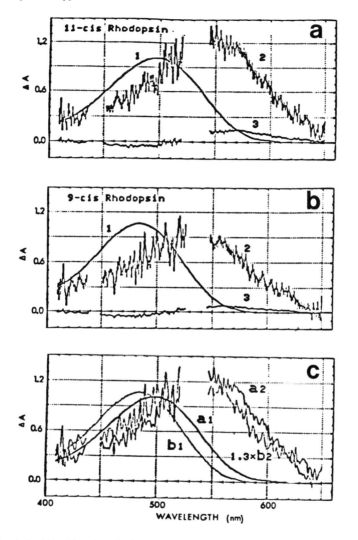

FIG. 13 (a) (1) 11-*cis*-rhodopsin spectrum; (2) calculated spectrum of bathorhodopsin formed from 11-*cis*-rhodopsin (resolved from difference spectrum (curve 3) by assuming a 15% bleach); (3) difference spectrum equivalent to Fig. 12a. (b) (1) 9-*cis*-rhodpsin spectrum; (2) calculated spectrum of bathorhodopsin formed from 9-*cis*-rhodopsin (resolved from difference spectrum (curve 3) by assuming a 13% bleach); (3) difference spectra equivalent to Fig. 12b. (c) Comparison of the two calculated bathorhodopsin spectra (a_2 and b_2). Both are scaled to the same maximum absorption to illustrate the spectral shift. a_1 and b_1 are the spective 11-*cis*- and 9-*cis*-rhodopsin spectra.

data clearly show that a common intermediate is *not* formed when 9-*cis*-rhodopsin and 11-*cis*-rhodopsin are excited with 532 nm light!

In view of the fact that photoexcited rhodopsin is transformed to the first intermediate bathorhodopsin, within 6 ps after excitation and to the second intermediate lumirhodopsin with about a 30 ns decay time, selection of 85 ps after excitation for the monitoring of the absorption spectra assures that the species observed is bathorhodopsin. The data presented therefore strongly suggest that the bathorhodopsin of the 11-*cis*- and 9-*cis*-rhodopsin are not the same species. Although these data do not record nor can examine skeletal deformations, that is, *cis-trans* isomerization, they do show that one should not rest the isomerization mechanism on the premise of a common 11-*cis*- and 9-*cis*-batho-intermediate. Raman data (Braiman and Mathies, 1982) have recently shown that there is rather little configurational change, and infrared spectra (Rothschild and Marreo, 1982) of bacteriorhodopsin, which contains essentially the same chromophore, show that indeed the proton attached to the Schiff base translocates as we originally proposed. The picosecond absorption data really had shown that a proton translocation takes place during the first 6 ps after excitation. However, such experiments neither distinguish nor identify the proton involved in the translocation. It was proposed (Mathies *et al.*, 1976) that the proton attached to the Schiff base would translocate and further that it would move towards the Schiff base nitrogen. Recent infrared studies by Sandorfy and co-workers (Favrot, Leclerq *et al.*, 1978; Favrot *et al.*, 1978) indicate that indeed the Schiff base proton translocates away from the counter ion and suggest that this proposal is correct.

Lately, picosecond studies have started to probe organic reactions and energy transfer in inorganic molecules and so continue the applications to more fields and the acquisition of more in-depth knowledge.

To venture into forecasting the future, I shall merely extrapolate from the past: I tend to the belief that the large number of practitioners of picosecond spectroscopy, now an established field, will continue to pump new species and probe new phenomena unknown to us at present and will continue to invent novel devices, not only for basic research, but for communication and photonics technology.

The narrative presented is unfortunately very brief, very biased, and represents only a small fraction of the early research performed with picosecond pulses. I have done something of an injustice to many investigators by not mentioning their work and excellent contributions even by title. For this my only refuge lies in the present time and space allowance.

REFERENCES

Birge, R.R. and Hubbard, L.M. (1980). *J. Amer. Chem. Soc.* 102, 2205.

Braiman, M. and Mathies, R. (1982). *Proc. Natl. Acad. Sci. USA* 79, 4045.

Busch, G.E., Applebury, M.L., Lamola, A.A. and Rentzepis, P.M. (1972). *Proc. Natl. Acad. Sci. USA* 69, 2802.

Favrot, J., Leclerq, J., Roberge, R., Sandorfy, C. and Vocelle, D. (1978). *Chem. Phys. Lett.* 53, 433.

Favrot, J., Sandorfy, C. and Vocelle, D. (1978). *Photochem. Photobiol.* 28, 271.

Giordmaine, J.A., Rentzepis, P.M., Shapiro, S.L. and Wecht, K. (1967). *Appl. Phys. Lett.* 11, 216.

Green, B.K., Monger, T.G., Alfano, R.R., Aton, B. and Callender, R.H. (1977). *Nature* 269, 179.

Grellman, K.H., Livingston, R. and Pratt, D. (1962). *Nature* 193, 1258.

Mathies, R., Oseroff, A.R. and Stryer, L. (1976). *Proc. Natl. Acad. Sci. USA* 73, 1.

Monger, T.G., Alfano, R.R. and Callender, R.H. (1979). *Biophys. J.* 27, 106.

Netzel, T.L., Rentzepis, P.M. and Leigh, T. (1973). *Science* 182, 238.

Ottolenghi, M. (1980). *In* "Advances in Photochemistry". Vol. 12, (Eds J.N. Pitts, G. Hammond, K. Gollnide and D. Grosjean), p. 97. Wiley, New York.

Peters, K., Applebury, M.L. and Rentzepis, P.M. (1977). *Proc. Natl. Acad. Sci. USA* 74, 3119.

Rentzepis, P.M. (1968a). *Chem. Phys. Lett.* 2, 117.

Rentzepis, P.M. (1968b). *Photochem. Photobiol.* 8, 579.

Rosenfeld, T., Honig, B., Ottolenghi, M., Hurley, J. and Ebrey, T.G. (1977). *Pure Appl. Chem.* 49, 341.

Rothschild, K.J. and Marrero, H. (1982). *Proc. Natl. Acad. Sci. USA* 79, 4045.

Ruddock, I.S. and Bradley, D.J. (1976). *Appl. Phys. Lett.* 29, 296.

Spalink, J.D., Reynolds, A.H., Rentzepis, P.M., Applebury, M.L. and Sperling, W. (1984). *Proc. Natl. Acad. Sci. USA* (in press).

Teschke, O., Ippen, E.P. and Holton, G.R. (1971). *Chem. Phys. Lett.* 52, 233.

Wald, G. (1968). *Nature* 219, 800.

Warshel, A. (1976). *Nature* 260, 679.

Yoshizawa, T. and Wald, G. (1963). *Nature* 197, 1279.

Yoshizawa, T. and Kito, Y. (1958). *Nature* 182, 1604.

INDEX